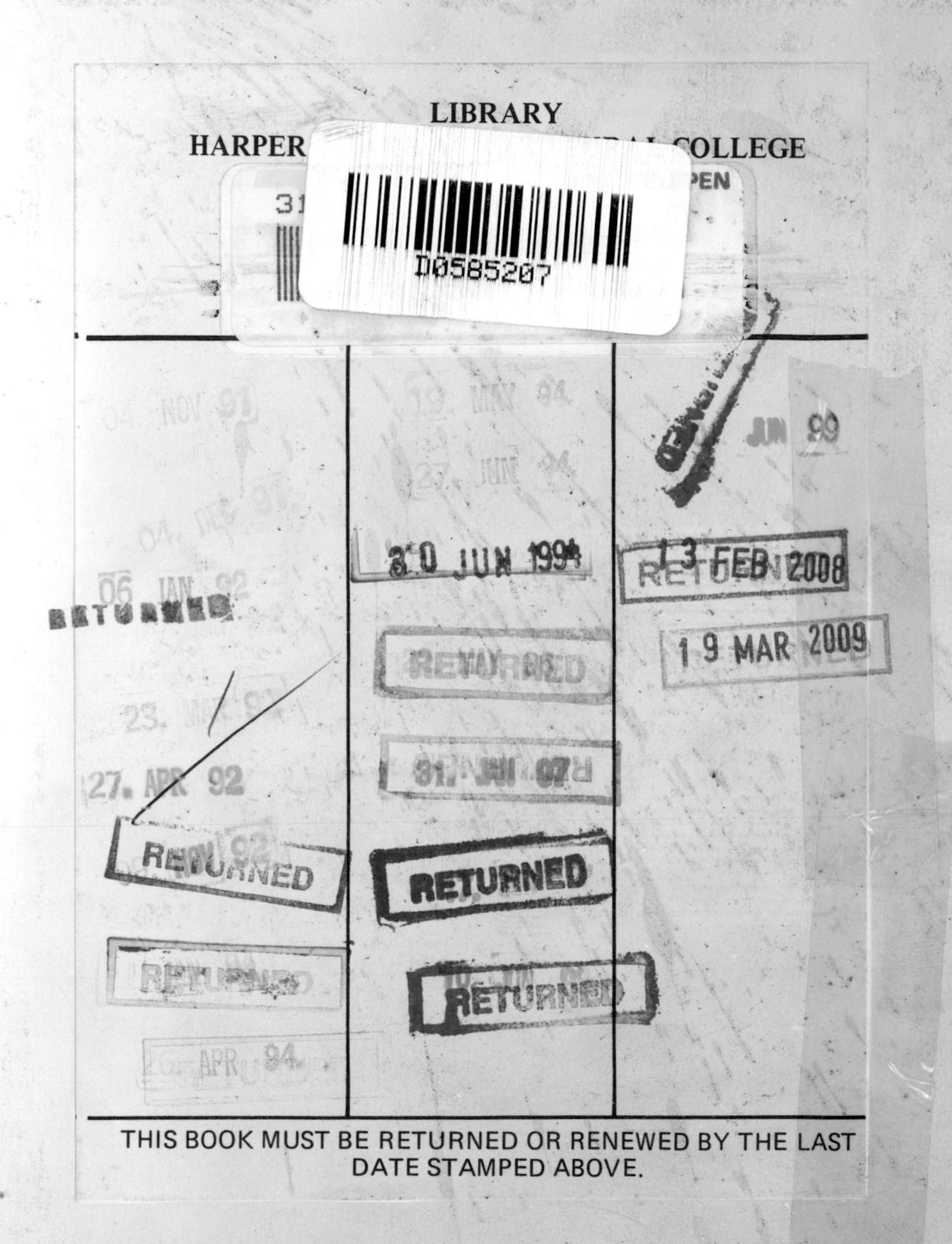

WITHDRAWN

Treatment and Use of Sewage Effluent for Irrigation

Treatment and Use of Sewage Effluent for Irrigation

Proceedings of the FAO Regional Seminar on the treatment and use of sewage effluent for irrigation
held in Nicosia, Cyprus, 7–9 October, 1985

Edited by

M. B. Pescod
Tyne and Wear Professor of Environmental Control Engineering and Head, Department of Civil Engineering, University of Newcastle upon Tyne, UK

and

A. Arar
Senior Regional Officer, Water Resources, Development and Management Service, Land and Water Development Division, FAO, Rome, Italy

Published by arrangement with the
Food and Agriculture Organization of the United Nations
by

Butterworths
London Boston Singapore Sydney Toronto Wellington

First published 1988

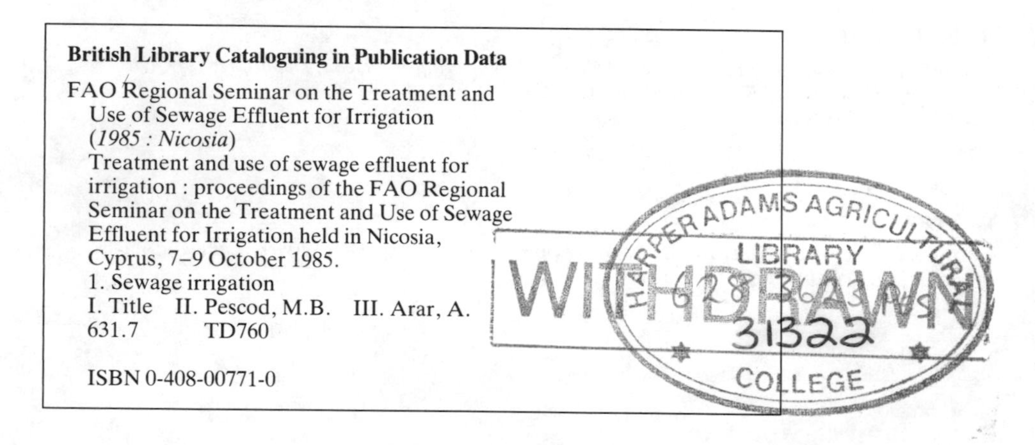

British Library Cataloguing in Publication Data

FAO Regional Seminar on the Treatment and Use of Sewage Effluent for Irrigation (*1985 : Nicosia*)
Treatment and use of sewage effluent for irrigation : proceedings of the FAO Regional Seminar on the Treatment and Use of Sewage Effluent for Irrigation held in Nicosia, Cyprus, 7–9 October 1985.
1. Sewage irrigation
I. Title II. Pescod, M.B. III. Arar, A.
631.7 TD760

ISBN 0-408-00771-0

Library of Congress Cataloging-in-Publication Data

FAO Regional Seminar on the Treatment and Use of Sewage Effluent for Irrigation (1985 : Nicosia, Cyprus)
Treatment and use of sewage effluent for irrigation : proceedings of the FAO Regional Seminar on the Treatment and Use of Sewage Effluent for Irrigation, held in Nicosia, Cyprus, 7–9 October 1985/ edited by M.B. Pescod, A. Arar.
p. cm.
Bibliography: P.
Includes index.
ISBN 0-408-00771-0
1. Sewage irrigation–Congresses. 2. Sewage–Purification–Congresses. I. Pescod, M.B. II. Arar, Abdullah. III. Food and Agriculture Organization of the United Nations. IV. Title.
TD760.F23 1985 87-32568
628.3′623–dc19 CIP

Photoset by Butterworths Litho Preparation Department
Printed and bound in Great Britain by Anchor Brendon Ltd., Tiptree, Essex

Foreword

This book comprises the Proceedings of the FAO Regional Seminar on The Treatment and Use of Sewage Effluent for Irrigation, held in Nicosia, Cyprus, 7–9 October, 1985.

The main purpose of the Seminar was to review available information and experience on the treatment and reuse of sewage effluent for increased agricultural production. Papers were presented by international experts on health and agricultural guidelines for effluent quality and on the short-term and long-term effects of effluent reuse on public health, soil fertility and crop productivity. Appropriate sewage treatment systems were considered and sewage sludge treatment and agricultural utilization were discussed. Case studies of sewage effluent treatment and reuse in irrigation in the Near East Region and elsewhere were presented.

The papers and discussions were edited by Professor M. B. Pescod, Head, Department of Civil Engineering, University of Newcastle upon Tyne, UK. The Seminar was organized by Dr A. Arar, Senior Regional Officer, Water Resources, Development and Management Service, Land and Water Development Division, FAO, Rome, Italy.

Contents

Chapter 1

Introduction

Introduction to the Seminar

Purpose of the Seminar

In acknowledgement of the limitations in supply of good quality water in the Near East Region and the great importance of the recycling of wastewater for irrigation, the Seventh Session of FAO's Regional Commission for Land and Water Use in the Near East recommended that a Regional Seminar on the subject should be organized in 1984–85. The Seminar was designed to draw together international and regional experts on the treatment and reuse of sewage effluents for increased agricultural production. Through exchange of information and experience it was expected that appropriate systems for effluent reuse in irrigation would be reviewed and evaluated in respect of their suitability for application in the Near East. Effluent quality guidelines related to public health, soil fertility and agricultural productivity were to be discussed and sewage treatment technology considered in the context of agricultural reuse of effluents and sludges.

The findings of the Seminar were presented to the Eighth Session of the Regional Commission on Land and Water Use in the Near East in October 1985.

Attendance at the Seminar

The Seminar was attended by about 85 participants and observers from 16 countries of the Near East Region (namely Algeria, Bahrain, Cyprus, Djibouti, Egypt, Jordan, Kuwait, Morocco, Pakistan, People's Democratic Republic of Yemen, Palestine Liberation Organization, Qatar, Saudi Arabia, Syria, Tunisia and Yemen Arabic Republic), eight countries outside the FAO Near East Region and Regional and International Organizations.

Opening session

The Seminar was opened by His Excellency the Minister of Agriculture and Natural Resources, Dr A. Papasolomontos, on behalf of the Government of Cyprus. In his opening speech, His Excellency The Minister welcomed the participants and thanked FAO for the timely organization of the Seminar. He expressed his desire for FAO to continue its efforts to help the countries of the Region, including his own country, in the utilization of sewage effluents for irrigation, through appropriate treatment and management methods.

Dr H. M. Horning, Director, Land and Water Development Division, welcomed the participants on behalf of the Director-General of FAO, Mr Edouard Saouma, and thanked the Government of Cyprus for hosting the Seminar. He extended his appreciation to His Excellency the Minister of Agriculture and his staff for the excellent arrangements and facilities they had provided to ensure the success of the Seminar. In his address, Dr Horning briefly discussed the important role that the use of treated wastewater plays in agricultural production, particularly in the Near East Region where water is scarce. He also indicated that FAO organized this Seminar in compliance with the recommendations of the FAO Regional Commission for Land and Water Use in the Near East, which represents the Member Governments of the Commission.

Summary and conclusions

Guidelines for water reuse in agriculture

The value of wastewater for crop production has been recognized in many countries, including India and China, and more recently in the Middle East. However, its use must be approached with caution. Current concern about environmental quality requires a more responsible and integrated approach to effluent reuse so that both health risks and costs are minimized. Health considerations are centred around survival rates of pathogenic organisms, particularly in relation to the irrigation of health-sensitive crops (including fruits and vegetables eaten uncooked). Associated environmental hazards are: contamination of groundwater and accumulation of heavy metals and toxic organics in surface soils and water. In agriculture, the major concerns are salinity, reduction in soil permeability and specific ion toxicity.

Several effluent quality guidelines and standards have been established for irrigation reuse but most of these are too stringent for realistic application in most developing countries. Standards should be established on a site-specific basis. In establishing quality criteria for use of sewage effluent, certain physical, chemical and biological characteristics have to be defined. Comprehensive chemical criteria for irrigation water have been established by FAO, in which salinity, infiltration, toxicity and trace element limits are covered. WHO has suggested health criteria and treatment processes for wastewater reuse and these are at present being reviewed. By combining the guidelines of FAO and WHO, a potential user of effluent should be able, as a first step, to assess the quality limitations of his wastewater source. Another important aspect of using wastewater for agriculture is the necessity of selecting appropriate crops, soil types and irrigation methods to suit local conditions.

In 1973, WHO published *Guidelines for Reuse of Water for Irrigation*, where health criteria and treatment requirements were indicated for irrigated crops grown for various purposes. In the light of recent experience it has been necessary to reassess these standards and include the consideration of hazards caused by helminthic pathogens, specifically intestinal nematodes and protozoa. The new standards, according to the 'Engelberg' guidelines, classify effluent for restricted and unrestricted use based on nematode egg counts and faecal coliform populations. These guidelines are more likely to be adopted since they are not so restrictive as previous standards. Depending on the local situation, limiting concentrations of certain toxic heavy metals will have to be defined among the effluent quality criteria. The dangers of heavy metal contamination of municipal and industrial effluents and sludges should be kept in mind where these are to be used for agricultural purposes. Japan's experience in this regard provides information on the severity of the problem and possible solutions. Judging from its serious hazard to human health and its easy incorporation into the food chain, cadmium is of major concern and great attention has been paid to cadmium contamination of rice grown on polluted paddy fields in Japan. Under existing environmental quality standards in Japan it is recommended that cadmium in irrigation water for rice cultivation should not exceed 10 ppb Cd.

In conclusion, it is necessary to have guidelines when using effluent for agriculture in order to minimize public health hazards, environmental degradation and decline in crop productivity. Such guidelines should not be so restrictive that they will not be used, nor should they be so broad that they will not be effective.

Treatment of sewage effluents

The principal objective of sewage treatment is to allow municipal and industrial effluents to be disposed of without danger to public health or to the environment. When wastewaters are treated with a view to producing effluents for irrigation, the most appropriate method will be that which meets the recommended quality criteria specified for the intended use at low cost and with minimal operational and maintenance requirements. Conventional treatment processes, involving primary, secondary and tertiary treatments, although they produce good quality effluents, may not be economical or practical in many local situations. Hence alternative treatment methods which are cheap but efficient and reliable are needed in treating effluents for irrigation purposes.

In meeting biological standards, particularly in terms of the removal of helminths, protozoa, bacteria and viruses, waste stabilization ponds are known to be very efficient. These ponds are cheap, easy to maintain and require very little energy. Design, operation and maintenance of anaerobic, facultative and maturation ponds have been studied and evaluated so that acceptable quality effluents for irrigation can now be produced. Recommendations have been made for the irrigation of different crops depending on biochemical oxygen demand and faecal coliform levels in the effluent from waste stabilization ponds. However, further research is still needed to improve the design and management of waste stabilization ponds, particularly in hot climates. Currently, waste stabilization ponds are used in treating effluent for irrigation in Kenya, South America and Egypt.

The soil–aquifer treatment (SAT) process converts sewage effluent into renovated water which can be collected from the aquifer by wells or drains. Methods to prevent the contamination of native groundwater by the percolating effluent have been developed through the use of pumping wells and drains. In Phoenix, Arizona, various aspects of the design, operation and maintenance of the SAT system have been tested and evaluated. Under Arizona conditions, suspended solids, nitrogen, phosphate, fluoride, heavy metals (Zn, Cu, Cd and Pb), faecal coliforms, viruses and total organic carbon are reduced to acceptable limits by SAT. The cost of the system is low and it does not require highly trained operators. Thus, where land availability and hydrogeological conditions are favourable for groundwater recharge by surface spreading, SAT can play an important role in the reuse and recycling of municipal wastewater and release natural water resources for potable use.

In the United Kingdom the effects of artificial recharge of sewage on groundwater quality have been studied in chalk and sandstone formations. In general, removal of organic matter, bacteria and viruses has been excellent. Recharge of crude or settled sewage to the soil profile in chalk results in satisfactory BOD and partial nitrogen removals. However, in the Triassic sandstone, nitrogen is not removed during the recharge of settled sewage. Artificial recharge of sewage effluent has potential as a means of storing water intended for irrigation. The presence of large fissures or karstic conditions reduces the effectiveness of an aquifer for purifying sewage and might result in dangerous microbial contamination of groundwater. Even though the first 1 m of a soil profile is the most active removal zone, at least 10 m is needed for safe removal of all contaminants.

Use of the soil–aquifer treatment process to obtain effluents of acceptable quality was illustrated in two pilot projects in France, one in the sand dunes on the Mediterranean coast and the other in the chalky basin south-west of Paris. For optimum results it may be necessary to pretreat the sewage in order to remove suspended solids, which decrease basin infiltration rates. Clogging causes problems but can be overcome by removal of dried crusts. Results obtained in the chalk formation indicate that, to achieve a high level of purification, the thickness of the sand layer placed at the bottom of the infiltration basins should be at least 1.50 m and two infiltration basins may be required in order to alternate their use.

Studies carried out in Beltsville, USA, have shown that sewage sludge can be used safely in agriculture after composting. However, the success of composting will depend on the process control and characteristics of the inputs, mainly the sludge and bulking material. The sludge characteristics that can influence the composting process are its digestibility, pH, carbon:nitrogen ratio and contents of soluble salts, heavy metals and microbial population. Some bulking materials used in sewage sludge composting are wood chips, leaves, groundnut hulls and municipal refuse. The aerated pile method of composting developed in Beltsville has been found to be superior to other composting methods as it is effective in pathogen destruction and odour control, with a minimum of equipment and facilities. Typical uses of sewage sludge compost include soil conditioning, reclamation of contaminated land and, where permitted, farming and gardening.

Use of sewage effluent and sludge in agriculture – country experiences

West Germany

Because of a shortage of fertilizers up to the middle of this century in Germany, wastewater was used on farmland not only to water crops but also to increase the productivity of light soils. During the early days, raw sewage was utilized but, in the mid-1960s, pretreatments were introduced. In a sewage irrigation farm at Braunschweig, effluents were mechanically treated and a variety of crops were irrigated, using the mechanized sprinkler system. During winter, effluent was discharged to infiltration basins but this created operational difficulties. Sludge disposal, accumulation of heavy metals and high concentration of nitrates in groundwater are now causing problems in this system. The Wastewater Utilization Association is, therefore, designing a pretreatment plant incorporating nitrogen removal by denitrification to overcome these difficulties. It should be kept in mind that the experience acquired under the climatic conditions of northern Germany might not be transferable to arid climatic regions without modification.

Asian and Pacific Region
Sewage treatment in the Asian and Pacific Region is insignificant, due to the high cost of conventional treatment methods. Land application of wastewater appears to be a viable alternative as it costs little, conserves nutrients and is a source of irrigation water. In India, land disposal of wastewater has been practised since the second quarter of this century, and the associated problems have been related to the level of sewage treatment, dilution, hydraulic and nutrient loadings, and the crops. A recent survey conducted by the National Environmental Engineering Research Institute (NEERI) in Nagpur, India, showed that many of the environmental and public health problems could be attributed to application rates far in excess of the capacity of the soil system to accommodate such wastewaters. Experimental results show that primary-treated and secondary-treated sewage effluents are superior to untreated sewage in terms of crop yields and nutrient utilization efficiency. An experiment on a multiple-reuse system, including fish culture and crop production, has yielded useful information. Similar results are available on the use of partially treated wastewater for forest irrigation, particularly of eucalyptus. NEERI is conducting research on the effects of wastewater irrigation on soil productivity, heavy metal pollution and health hazards. A tentative guideline for selection of crops to suit Asian conditions is available. Further research and development in this field is necessary to attain efficient utilization of waste effluents in the region.

Egypt
In Egypt, acute shortage of water and the low organic content of the soils necessitate the utilization of sewage for crop production when sufficient quantities are available. Since the first decade of this century, primary sewage effluent of Cairo has been used for irrigation at El Gabal El Asfar citrus farm. Although several changes in soils were observed during the 50 years of sewage irrigation at this farm, only a few are considered unfavourable. Favourable effects include increase in organic matter, clay content and cation exchange capacity. Increases in Cr, Cu, Pb and B have been noticed but they have not reached levels that affect yields. At Abou Rawash farms near Giza, 3 years of sewage irrigation have lowered the total solids and pH to more favourable levels. There has been a significant increase in soil organic matter and water-holding capacity and the yield of corn-maize has doubled during the 3 year period. Encouraging results on the use of sewage for growing fruit crops have been obtained in Alexandria. In all cases sewage irrigation has produced high yields, with a significant increase in the case of olives. Experiments conducted on the use of sludge in crop production indicated high yields of wheat, beans and alfalfa and a higher accumulation of heavy metals in the leaves of the plant species studied. In Alexandria, sewage effluent is considered as a good source of irrigation water to increase agricultural production under controlled management. The quality criteria adopted in China appear to be less stringent than those of California and are likely to be adopted for the Alexandria sewerage scheme.

Cyprus
In Cyprus, the need to use treated sewage effluent for irrigation is viewed as an integral part of the planning and development of water resources for the island and an efficient means of recycling scarce water resources on economic, social and environmental grounds. The utilization of treated effluent for irrigation and

groundwater recharge constitutes an important aspect of agricultural development in a semi-arid country like Cyprus. Since 1979 studies have been conducted under FAO/UNDP technical assistance to evaluate water, soil and crop suitability for sewage irrigation. These studies and national research programmes have yielded valuable information on treatment requirements, crop production potentials and environmental effects. Currently, effluent is being reused from the Akrotiri sewage treatment plant, which has received secondary treatment and disinfection, and from the Nicosia sewerage system, after being treated in stabilization ponds and chlorinated. The Zenon-Kamares sewage treatment plant effluent is used to irrigate alfalfa after receiving tertiary treatment and chlorination. Preliminary results of research conducted at the Agricultural Research Institute of Cyprus indicate the superiority of treated effluent over freshwater for cotton production.

Cyprus has undertaken a multi-faceted research programme on the use of effluent for irrigation. It covers the use of secondary-treated effluent for irrigation and fertilization, and evaluation of soil and groundwater contamination as a result of sewage irrigation. Preliminary results indicate that hazards associated with total soluble salts and toxic elements in the effluent are negligible when compared with those resulting from the use of well water. Results also show the superiority of treated effluent over well water regarding nutrient supply and crop yields. By using trickle irrigation public health hazards are minimized. Up to now, the results are very promising and research is in progress to ascertain the effects of sewage irrigation on a variety of crops under varying soil conditions.

Mexico

As a consequence of rapid development of agriculture and the urban population, farmers in Mexico have been utilizing sewage and stormwater directly from discharge drains for agricultural production in the Mezquital Valley. Some of the most serious problems in this area are related to pathogens, heavy metal contamination and salinity build-up. Consumption of raw vegetables and fruit produced in this valley is the greatest health hazard to consumers, despite the fact that restrictions have been imposed against the consumption of these products in the raw state. It has not been possible to eliminate the occurrence of intestinal infections and other endemic diseases in spite of regular warnings. On the other hand, the continued use of this effluent has increased agricultural production, employment and incomes. The current effluent and drainage disposal system does not permit improvement of effluent quality through standard treatment processes. It appears that efforts should be made to improve crop and water management practices and increase the awareness of the public to the health hazards.

Jordan

In Jordan, the use of treated wastewater is important for agricultural development as well as for the general economic welfare of the country. Effluent reuse can make a significant saving in the diversion of potable water for irrigation, as well as providing a safe method of effluent disposal. According to estimates, by using effluents, about 38% of the 1983 import of barley and an equal quantity of alfalfa can be produced in the country. Currently, in Amman, treated sewage effluent is mixed with stormwater in the wadis and stored in the King Talal Reservoir before being used for irrigation. In some other cities, sewage effluent that percolates from septic tanks is mixed with natural recharge water; this is pumped through boreholes from wadi beds for irrigation purposes. Regulatory standards have been established for treatment plants to meet quality standards of secondary treatment level when discharging effluents. In addition, measures to minimize negative impacts on water and environmental quality have been proposed for implementation.

Saudi Arabia

In Saudi Arabia, the policy is to utilize all available treated municipal wastewater in the most beneficial manner for several purposes, among which the agricultural sector is given top priority. Realizing the importance of providing an adequate degree of wastewater treatment to ensure public health protection, wastewater regulations are being framed to enforce a minimum of tertiary treatment, producing a quality level required for unrestricted irrigation. In Riyadh, a total effluent discharge of 210 000 m^3/day is used by industry and agriculture. The agricultural volume is used to irrigate 4000 hectares of fruit, cereals, vegetables and fodder crops. Recently, treated and dried sewage sludge has also been used by farmers as fertilizer and soil conditioner. Several projects are being planned to expand the use of sewage effluent in agriculture.

Kuwait

Kuwait presents a good example of an efficient way of utilizing sewage through multiple use and efficient management. For a long time, untreated sewage effluent has been used to irrigate forestry projects which are far from inhabited areas. Since 1956 secondary-treated effluent from the Ardiya treatment plant has been used to irrigate some plantations and government-supervised production farms. Several studies and expert consultations have been carried out in Kuwait to evaluate quality standards, treatment methods and irrigation practices best suited to effluent irrigation under local conditions. New sewage treatment plants are being designed and constructed, incorporating new ideas, such as waste stabilization ponds, for irrigation purposes. Special management, operation and maintenance facilities are being established for sewage effluent irrigation, which will also include testing, monitoring and coordination of works. The government strategy in effluent use is to give high priority to intensive cultivation in enclosed farm complexes and environmental forestry. An estimate indicates that the cost of producing treated effluent for irrigation is about US \$0.45/m^3, which is about half the cost of desalinated water. The case study from Kuwait on the use of treated sewage effluent for irrigation brought out the rational approach adopted by the country. A masterplan for effluent reuse has been prepared, taking into account effluent quality, treatment and distribution methods, crop selection, land and soil suitability and economic feasibility.

Syria

In Syria, municipal and industrial wastewaters have been used for irrigation in the Damascus plain for a long time. Serious problems are being encountered with the present form of utilization, due to high microbial population, salinity and high concentrations of chlorides, heavy metals and phytotoxins. This leads to the conclusion that a proper sewage treatment and distribution system should be established before utilizing these wastes for irrigation. Plans are being drawn up to construct sewage treatment plants which would enable the eventual safe use of effluent for irrigation.

Portugal
In order to overcome the water shortage problem for crop production in Portugal, a research project has been undertaken by LNEC in collaboration with selected national and international universities to evaluate the possible use of sewage effluent in agriculture. The main objective of the project is to determine the productivity of crops and possible public health risks. Experiments are in progress and preliminary findings are encouraging.

France
In France, studies have been conducted to analyse the microbial population of soils that were irrigated with effluents from tertiary lagoons and secondary treatment plants. Results indicate that effluents from tertiary lagoons do not pose health risks due to microbial infections from soils. However, effluent from secondary treatment plants results in microbial build-up in the soil but the population count varies with the type of irrigation carried out. Further studies are required to obtain more reliable results.

Conclusions

In view of the fact that the Near East Region is importing 50% of its agricultural needs, it is imperative that food production be increased through irrigated agriculture. The value of using treated sewage effluent for irrigation to increase agricultural production in the Region is well acknowledged and it is gratifying to note from several papers that serious efforts have been made to augment water resources from non-conventional sources, including treated sewage effluent. With careful planning, various agricultural and industrial needs may be met by reclaimed wastewater, thereby making freshwater available for domestic use. The use of treated effluent will minimize the need for the addition of fertilizers in crop production.

One of the major objectives of the Seminar was to consider guidelines and criteria for the use of sewage effluent in agriculture. It was generally felt that the guidelines should not be restrictive but should be suitable for use as the basis for national and site-specific standards. Accordingly, the general opinion was that a combination of the WHO health guidelines proposed in the 'Engelberg Report' and the agricultural guidelines proposed by FAO in *Irrigation and Drainage Paper No. 29 (Revision 1)* should be taken as the basis for establishing national and project-level effluent quality criteria.

The opinion of the Seminar was that appropriate treatment has to be applied to enable sewage effluent use in crop production and precautions are necessary to safeguard public health and protect the environment. Ideally, the method of treatment should produce an effluent which meets the recommended microbial and chemical guidelines for unrestricted irrigation with the lowest investment costs and minimum operational and maintenance requirements. However, it was recognized that if the irrigation system could be designed to apply low-grade effluent without health risk and if crop selection could be controlled, a minimum degree of sewage treatment would lower investment costs and be more manageable in most developing countries.

The Seminar discussed several treatment methods, comparing them with conventional primary, secondary and tertiary treatments (pretreatments). Among these were waste stabilization ponds, aerated lagoons and soil–aquifer treatment.

The merits of these methods include their low cost and easy operation and maintenance, when land is not limiting and geological conditions are favourable.

One important conclusion of the Seminar was that greater attention should be given to multipurpose uses of sewage effluent. The example was given of utilizing effluent for agricultural crops, productive forestry (timber), protective forestry (shelter belts), ornamentals, landscaping, etc. In this regard it was recognized that national policy and planning should be formulated to meet the requirements of multipurpose utilization of treated sewage. International guidelines and national legislation would help to enforce these objectives, and measures to enlighten public opinion would assist in achieving their acceptance. It was felt equally important to create an awareness among decision-makers of the subject and findings of the Seminar.

The paucity of research and operational data on effluent treatment methods and reuse necessitates that further research be undertaken to provide a sound base for future programmes. In the Near East Region, a coordinated programme could be carried out in existing national institutions. Complementary to research should be a monitoring and surveillance programme.

Recommendations

The following general recommendations were agreed:

- Guidelines for effluent reuse in irrigation should be established at national and project levels, taking account of international guidelines.
- National policies should be drawn up on the basis of recognized concepts for multipurpose use of effluents.
- Attention should be given to adapting or developing low-cost technology for effluent treatment and reuse.
- Research and monitoring programmes should be established in connection with effluent reuse programmes.
- FAO should take appropriate action to disseminate information on this subject.

Chapter 2

Background to treatment and use of sewage effluent

Abdullah Arar*

General background

The potential of irrigation water in raising both food production and the living standards of the rural poor has long been recognized. Irrigated agriculture represents only 13% of the world's total arable land, but the value of crop production from irrigated land is 34% of the world total. This potential is more pronounced in semi-arid and arid areas like the Near East Region. In this Region the provision of irrigation water is one of the most important factors for increasing agricultural production. The present irrigated area in this Region is only 30% of the cultivated area but its production amounts to some 75% of the total agricultural production. In large parts of the Region no crops can be grown without irrigation water.

The irrigation systems in this Region belong to the oldest in the world. Irrigation has led to considerable and sometimes dramatic increases in agricultural production. It can bring independence from erratic rainfall and thus reliability of production and stable incomes.

On the other hand, past and existing trends in food and agricultural production in the Region have led to a situation which, despite noticeable achievements, is fundamentally unsatisfactory. At present the Region is importing more than 50% of its food requirements, and the rate of increase in demand for food exceeds the rate of increase in agricultural production. Because aridity is the major constraint for increased agricultural production, most of the countries in the Near East Region consider irrigation development a prime way of raising agricultural production.

Countries' action programmes in the field of irrigation

Until the end of the Second World War the expansion rate of land and water development in the Region was very low. Rapid development started in the 1950s and gained full momentum during the 1960s. Many countries introduced national development plans in which the agricultural sector, and particularly the development of irrigation, was allocated top priority. Consequently, the easily

* Senior Regional Officer, Water Resources, Development and Management Service, Land and Water Development Division, FAO, Rome, Italy

accessible water resources, such as river flows and shallow groundwater of good quality, have now been almost entirely committed.

At present, great efforts are being made in the Region to make additional water available. In all large river basins, major surface storage reservoirs have been built or are under construction (Indus, Euphrates, Tigris and Nile). In other parts of the Region (Iran, Afghanistan, Syria, Jordan, Saudi Arabia and North Africa) a number of smaller dams are in different stages of planning or execution. Saudi Arabia, Yemen Arab Republic and the People's Democratic Republic of Yemen are planning to convert the traditional spate irrigation to perennial irrigation by better control of the floodwater from seasonal wadis and use of the groundwater reservoirs in the alluvial plains of these wadis. The large groundwater basins known so far (Egypt, Sudan, Libya, Saudi Arabia and the Arab Gulf States) are being developed.

This process of rapid agricultural development under irrigation has been accompanied by the process of desertification, as marked by increasing microaridity and declining productivity. In many countries of the Region, desertification manifestations of waterlogging and salinity are major problems in irrigated areas due to poor management of irrigation water in the conveyance system as well as in the field. Increasing salinity of underground water and falling water tables due to overpumping are also serious problems. In Saudi Arabia and the Gulf States, for example, the artesian flow of springs and wells is decreasing, water quality is deteriorating and the water level is falling, due to increased extraction and perhaps decreased recharge, thus causing salt water intrusion. On the other hand, the scarcity of water supplies, which are badly needed to meet the needs of population growth and rapid development in agriculture as well as industry, has given cause for concern in formulating national development plans in these countries. It is gratifying to report that decision-makers are being increasingly involved in devising ways of optimizing the use of available supplies as well as augmenting available water resources by non-conventional means. The latter includes two programmes: one is for increasing domestic water supply through desalination of saline water; the other is for the treatment of sewage effluent and its use for irrigation purposes. This latter programme is covered in detail by four other papers presented to this Seminar, three national papers (Jordan, Kuwait and Saudi Arabia) and one Regional paper (ECWA); hence it will not be discussed in this chapter.

From the above it becomes apparent that in arid areas, as is the case with the Near East Region, recycling of water is likely to have a greater impact on future usable water supply than any of the other technologies aimed at increasing water supply, such as water harvesting, desalination of sea water, weather modification (artificial rain), etc. Treated sewage effluent can be used for irrigation, industry, recharge of groundwater and, in special cases, properly treated wastewater could be used for municipal supply. With careful planning, various industrial and agricultural demands could be met by treated wastewaters, thereby freeing freshwater sources for municipal use.

It is because of the great importance of the recycling of wastewater for irrigation that the Regional Commission for Land and Water Use in the Near East recommended at its Seventh Session, held in Rome in 1983, to organize in 1984–85 a regional seminar on this subject. Hence, this Seminar is the implementation of this recommendation, which is, in fact, a recommendation by the governments of the Region.

Topics of the Seminar

The purpose of the Seminar is to exchange worldwide information and experience on the treatment and reuse of sewage effluent for increased agricultural production in view of the limitations in supply of good quality water. It is hoped that participants in the Seminar will make full use of the knowledge acquired and impart to others any new methods and techniques learned.

The topics to be covered by the Seminar could be organized under the following major headings:

- Irrigation water standards (chemical and biological) and the required degree of sewage treatment for irrigation.
- The short-term effects of the use of sewage effluent for irrigation on soil productivity and public health. This will include improved soil fertility and increased productivity, salinity and sodicity hazards, toxicity by heavy metals and synthetic organic chemicals and danger of infection by pathogens. The latter include viruses, bacteria, protozoa, nematodes, trematodes and cestodes.
- Appropriate sewage treatment in connection with irrigation, such as the use of waste stabilization ponds, aerated lagoons, deep (more than 3 m) facultative and maturation ponds, soil profile and the recharge of groundwater aquifers.
- Sludge treatment and disposal in connection with its agricultural utilization.
- Case studies from several countries inside and ouside the Near East Region.

Major issues

Several questions arise when contemplating the treatment and use of sewage effluent for irrigation, for which it is hoped that the proceedings of the Seminar will give satisfactory answers. Some of the most important issues in this regard could be listed under the following headings.

Appropriate sewage treatment methods with a view to using the effluent for irrigation

It is important to remember that the principal object of sewage treatment is to allow human and industrial effluents to be disposed of without danger to human health or unacceptable damage to the natural environment. Hence, the most appropriate wastewater treatment to be used for irrigation is that which will produce an effluent that meets the recommended microbiologial and chemical quality guidelines for unrestricted irrigation, both at low cost and with minimal operational and maintenance requirements. Undoubtedly, the Seminar will give special attention to this most important question and will update the information which is briefly outlined in the following paragraphs.

Extensive information is available on the occurrence and survival in the environment of most excreted pathogens and on their removal by various sewage treatment technologies. Many of these pathogens may be present in sewage in high numbers, may be only moderately reduced in passing through sewage treatment plants, and may survive for several days or more on irrigated soils and crops. These facts have been taken to indicate a potential risk from wastewater reuse and have led some authorities to set stringent quality standards for effluents to be used for irrigation. An epidemiological rather than an environmental approach to the problem is advocated here, in which decisions are based on actual risks rather than potential risks. The data on actual risks are very scarce at the present time and do not permit the health consequences of a particular reuse project at a particular site to be predicted. Hence, at this stage we do not have adequate methods for measuring possible deleterious health impacts against more easily demonstrable advantages.

In considering the diseases that may be associated with wastewater reuse, distinction should be made between communicable and non-communicable diseases. The former are related to the microbiological quality of the wastewater, while the latter are related to its chemical constituents. Communicable diseases are caused by viruses, bacteria, protozoa, nematodes, trematodes and cestodes. Conventional sewage treatment plants incorporating primary and secondary treatment stages are notoriously inefficient systems for the removal of excreted pathogens. Excreted viruses and bacteria are present in significant numbers in both the effluent and the sludge and the heavier protozoal cysts and helminth eggs tend to be concentrated in the sludge. On the other hand, selection of high technology tertiary processes, including effluent chlorination, can virtually eliminate pathogens from the effluent and produce a microbiological quality approaching that of drinking water. However, wastewater treatment plants are sophisticated in construction and expensive to finance; at the same time, secondary treatment is relatively inefficient for removing nutrients such as nitrogen and phosphorus. Although tertiary treatment is much more efficient for the removal of pollutants, it also requires considerable funds for capital investment and maintenance.

On the other hand, sewage treatment using waste stabilization ponds, if well designed and operated, can eliminate protozoal cysts and helminth eggs and reduce the concentrations of excreted bacteria and viruses to low levels. This type of treatment is capable of producing an effluent which meets the recommended quality guidelines for unrestricted irrigation, both at low cost and with minimal operational and maintenance requirements. However, it has extensive land requirements, which is not a serious handicap in arid and semi-arid areas where land is normally available and cheap. In locations where land is limited and/or expensive there is an urgent need to evaluate the efficiency of land-saving systems such as deep (>3 m) facultative and maturation ponds and aerated lagoons of various designs which can be used whenever conventional pond systems cannot be adopted. There is also a need to develop and evaluate wastewater-treatment technologies which could be used as a second stage of treatment after aerated lagoons, such as filtration, microstraining, chemical coagulation and ovicidal disinfection.

The Seminar proceedings include three chapters on recharging partially treated and untreated sewage effluent to groundwater, which contain the experience of three countries (France in Chapter 10, United Kingdom in Chapter 11, and United States in Chapter 10) in this field. This type of treatment of sewage effluent would appear to have a greater role to play in the Near East Region in the future.

In conclusion, it could be stated that the advantages of utilizing a land (plant–soil) system for wastewater treatment in the Region seem to be several. First, it will maximize the efficient use of the limited water supply for the production of food and fibre. Second, it will improve public health and improve water quality at low cost. In so far that such a system is an agriculture-based operation, it is most suitable for less-developed countries. Developing countries normally do not have sufficient financial resources available for the secondary treatment of most of their wastewaters, let alone tertiary treatment. If a city cannot afford to build a sewage treatment plant of sufficient capacity, the use of an additional lagoon–land treatment system would be of practical significance (*Figure 2.1*). Aerobic-lagoon treatment followed by land application is becoming a popular method of treating wastewater in which the self-purifying capacity of the water–soil–plant system is utilized. The principal advantages of this treatment process lie in its low cost of construction and simplicity of operation and maintenance. Plants, such as reeds, algae, etc. can be grown in lagoons or reservoirs, and even harvested for their food value to livestock. In addition, duck and fish can also be raised in the lagoons. It should also be mentioned that recharge of the untreated and partially treated sewage effluent to the groundwater might, under certain conditions, prove to be the most appropriate method of treating sewage effluent.

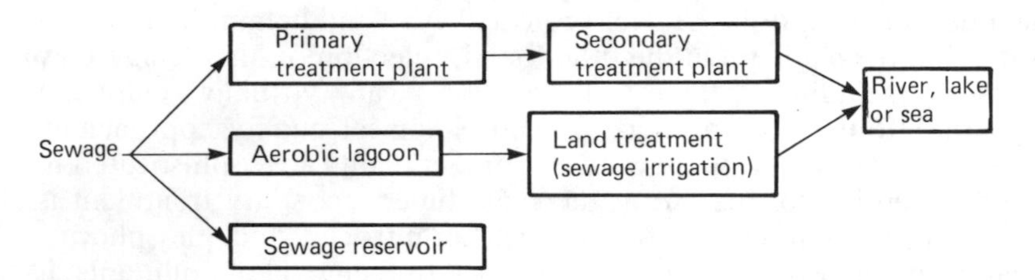

Figure 2.1 Schematic arrangement of the lagoon–sewage irrigation system

Sewage irrigation

Sewage irrigation as used in wastewater treatment can save energy by comparison with conventional treatment systems. In addition, the resources present in the wastewater can be recovered with significant benefits when it is used on land and the discharge of pollutants to water courses is eliminated. Over the past few years much effort has been spent on developing land treatment technology and on improving methods of control. In China, various types of land treatment systems have been accepted as viable wastewater-treatment techniques and the number of land-treatment systems has steadily grown. Sewage *irrigation* developed rapidly since 1958 and now over 1.33 million hectares are irrigated by this method. There, sewage is mainly utilized to irrigate crops, but not for turf, silviculture, etc., and the flooding technique from a surface border strip is generally used. China's experiences and pollution problems associated with sewage irrigation could be summarized by saying that, generally, sewage irrigation is to be encouraged from an agricultural standpoint but many pollution problems that arise have to be faced. Hence, it should be pointed out that the development of sewage irrigation can be justified from the viewpoint of agricultural and economic necessity but care should be taken to minimize its environmental impact. Many pollutants at low levels are beneficial, or even essential, but can be toxic at high levels. A pollution problem can occur when certain toxic constitutents in the sewage are taken up by the crops, thus leading to reduced yields and/or toxic effects on animal and human health. Some of the factors related to the pollution hazard of sewage irrigation are briefly discussed, as follows.

The soil

Many soil properties are important in considering the use of sewage effluent on land. The most important are pH, oxidation potential (Eh), cation exchange capacity and organic matter. Of these, pH is perhaps the most important since most heavy metals are more available under acid conditions than in neutral to alkaline conditions. Studies on sewage treatment carried out by Pescod and Alka (1984) have shown that this level of pH is easily reached by the use of lime. Thus, the use of lime provides an additional mechanism for pathogen reduction, as well as for the precipitation of most heavy metals in solution. Heavy metals are also converted to their insoluble sulphides under reducing conditions (low Eh) by anaerobic bacteria. In general, the higher the exchange capacity of the soil the more heavy metals a soil can accept without potential hazard. A soil–plant system has a good self-purification potential to degrade and utilize certain pollutants but attention should be paid to the accumulation of non-biodegradable materials in soils treated with sewage. Heavy metals accumulate in the upper soil because of strong adsorption and precipitation. If a soil is highly permeable (sandy soils) or if the groundwater table is close to the surface, then care must be taken to prevent the pollution of groundwater. Lastly, it must be pointed out that in any investigation of pollution of soil and crops by sewage irrigation, it is of primary importance to obtain information on the natural background level of such substances in the soil.

Crops

Many pollutants can accumulate in sufficient concentrations in crops to pose a serious hazard to the health of animals and humans, but without any visible effects on the growth of crops. The tolerance of plants to the levels of these pollutants in soils varies widely with plant species as well as with soil properties. Uptake is strongly a function of plant species and of plant organs. For example, the order of adsorption and accumulation of mercury in the edible parts of crops is as follows: rice > cabbage > turnip > maize > sorghum > wheat, but in the case of cadmium it is: turnip > cabbage > wheat > rice. Fortunately, most fruits, seeds and grains have a lower content of heavy metals than leaves and roots. Hence, pollution problems could be alleviated by selecting crops with a low rate of uptake or by selecting non-edible crops, such as cotton.

Inorganic pollutants

Of all the possible pollutants in sewage, heavy metals are non-degradable and, when present in sufficient quantities, pose the most serious long-term environmental threat. Cadmium absorbed by certain crops can easily render them unfit for human consumption. Mercury contamination can arise from the application of sludge and sewage on land. Uptake of lead by plants is less efficient than that of cadmium. Soils are able to bind lead more efficiently and consequently to reduce the resultant toxicity. Crops can grow well even in a soil with a concentration of 2100 ppm of lead. Boron, arsenic, chromium and nickel are other heavy and toxic metals usually found in sewage. Boron at low levels is essential, but it is toxic at high levels (greater than 1 mg/l in sewage).

Organic pollutants

Persistent synthetic organic compounds and some organochlorine insecticides are potential hazards both to the environment and public health. Knowledge of the health effects of these chemicals and the technology for their monitoring and removal from municipal water supply will always lag behind the development of new chemicals. They are introduced into commerce and industry at the rate of some 1000 annually and ultimately find their way into the water courses that drain urban and industrial areas. In China it is reported that toxicity to crops often occurred and yield was markedly reduced when trichloroacetaldehyde was present in sewage, affecting about 1.5% of the total sewage-irrigated area. Hence, these compounds should be kept within acceptable limits by reducing the discharge of contaminants into water courses and removing the contaminants by treatment at source. Since the most dramatic and severe impact of pollution generally occurs from the wastewater discharged by industry (heavy metals and synthetic organic compounds), source treatment of pollutants should be carried out and made the responsibility of the concerned industry by law.

Agricultural control measures
The following precautions need to be taken to control pollution, especially that from the uptake and accumulation of heavy metals by crops grown on land treated with wastewater:

- Maintain soil pH at or above 7.0, thus lowering the availability of most of the toxic heavy metals.
- If possible, maintain low oxidation potential, for example by keeping paddy fields waterlogged for as long a period as possible during the growing season.
- Grow crops which exclude particular heavy metals from the whole plant, or from the reproductive tissues; for example rice in a cadmium-contaminated field.
- Grow non-edible crops, such as cotton.
- Use low annual application rates; if the pollutant content of the sewage exceeds specified levels, decrease the pollutant input by mixing sewage with clear water for irrigation.
- Control sewage irrigation, avoiding excessive application to protect groundwater from becoming contaminated with nitrates; this concern has prompted a demand for reliable guidelines which can be used to determine the safe application rates for sewage.
- Maintain the organic matter status of the soil.

Monitoring
Since the use of sewage for irrigation carries the greatest risk from substances with irreversible toxic effects on humans and the environment, monitoring of sewage irrigation must be well-regulated. In sewage-irrigated areas, the toxic constituents of the sewage used, and their concentration in soil vegetation and groundwater, must be monitored and public health hazards assessed. Monitoring pollutants in soils and crops is useful for estimating pollution grades and for determining the effectiveness of control programmes.

Conclusions

Sewage irrigation ensures the reuse of resources and achieves the best treatment of municipal wastewater. Efforts should be focused on maximizing the benefits and on minimizing any detrimental effects to the environment and to man himself. Sewage irrigation involves complex interactions and it is difficult to assess its long-term impact. Appropriate management is of primary importance and requires experience for the realization of its full benefits; routine monitoring is also essential. Further research should be carried out by a cooperative team composed of specialists from different scientific disciplines. Nothing short of a long-term research programme can provide the answers to questions about the effects of sewage irrigation on the environment and on agricultural productivity.

Lastly, it must be emphasized that the purpose of this brief summary of the major issues of the subject matter of this Seminar was to invite participants to comment critically on the points raised and to bring to the attention of the meeting their up-to-date knowledge and experience.

Bibliography

ARAR, A. (1981) Techniques for water management and conservation under semi-arid and arid conditions. *Regional Meeting of Management, Conservation and Development of Agricultural Resources.* ECWA, Damascus, 9–15 May

ARAR, A. (1983) Water harvesting and other techniques for the optimization of water use in arid areas. *FAO/Finland Training Course on Watershed Management for Africa.* Nairobi, Kenya, 10–29 January

ARAR, A. (1984) The role of supplementary irrigation in rainfed agriculture. *Report on a Consultancy to ICARDA, January 1984.* ICARDA, Aleppo, Syria

COWAN, J. P. and JOHNSON, P. R. (1984) Re-use of effluent for agriculture – The Middle East. *International Symposium on Re-use of Sewage Effluent.* The Institution of Civil Engineers, London, 30–31 October

DIELEMAN, P. J. and ARAR, A. (1981) *Irrigation in the Near East Region.* Entwicklung and landlicher Raum, March 1981, 10–14

FEACHEM, R. G. and BLUM, D. (1984) Health aspects of agricultural and aquacultural reuse of wastewater in developing countries. *International Symposium on Re-Use of Sewage Effluent.* The Institution of Civil Engineers, London, 30–31 October

INTERNATIONAL REFERENCE CENTRE FOR WASTES DISPOSAL (IRCWD), DUBENDORF, SWITZERLAND (1985) Health aspects of wastewater and excreta use in agriculture and aquaculture. *Report of a Review Meeting of Environmental Specialists and Epidemiologists*, held at Engelberg, Switzerland, 1–4 July, sponsored by The World Bank and WHO

ISAAC, P. C. G. (1984) Conclusions and future considerations. *International Symposium on Re-use of Sewage Effluent.* The Institution of Civil Engineers, London, 30–31 October

OKUN, D. (1984) Keynote Address and Overview of Direct Reuse. *International Symposium on Re-Use of Sewage Effluent.* The Institution of Civil Engineers, London, 30–31 October

PESCOD, M. B. and ALKA, A. (1984) Urban effluent reuse for agriculture in arid and semi-arid zones. *International Symposium on Re-use of Sewage Effluent.* The Institution of Civil Engineers, London, 30–31 October

WANG, H. K. (1984) Sewage irrigation in China. *International Journal for Development Technology,* **2,** 291–301

Section I

Guidelines for water reuse in agriculture

Chapter 3

Guidelines for wastewater reuse in agriculture

M. B. Pescod and U. Alka*

Introduction

High population growth rates and high rural-to-urban migration are common factors giving rise to rapidly expanding urban centres in most countries in arid and semi-arid regions. The pressure of expanding urban populations and the desire for an improved standard of living have focused attention on problems connected with water economy in recent years. Although domestic and industrial consumption of water is considerable in urban areas, it is in relation to agricultural production that demand for water becomes critical. The correct combination of water and land, in space and time, will set the upper limit to the population-carrying capacity of any region.

In arid and semi-arid regions groundwater is often the only source of water and its occurrence at reasonable depth and with acceptable quality has, for a long time, been a prerequisite for the existence of human settlements and support of livestock. Modern technologies for groundwater prospecting, well construction and deep drilling have facilitated the development of this resource but such operations have their limitations. In many inland locations fossil groundwater is being mined, with very little recharge taking place in the short term, and certainly at a rate much lower than that of extraction. In coastal areas excessive pumping of groundwater leads to saline water intrusion into aquifers with resultant deterioration in water quality. Consequently, in arid and semi-arid regions it is more imperative than anywhere else to plan for rational development of water resources. A number of measures currently being considered include conservation by restricting usage to vital and essential needs, substitution of low-quality for high-quality water whenever possible and the reuse of wastewater from sewered urban centres. The latter is a source of water which cannot be ignored and it should be recognized that it is too valuable to be discharged into the ground or to a surface channel. Where desalination is necessary to provide potable water for urban communities it is particularly irresponsible not to collect and reuse the effluent.

Controlled wastewater reuse in agriculture has been practised in Europe, North America and Australia since the beginning of this century, and many sewage farms are still operating today. The value of wastewater for crop irrigation has also been recognized in India, China and, latterly, the Middle East. Even in water-plentiful

* Professor and Research Student, Department of Civil Engineering, University of Newcastle upon Tyne, UK

areas reuse of effluent is increasing in importance as a beneficial water conservation measure. The nitrogen and phosphorus contained in municipal and some industrial wastewaters are valuable plant nutrients, which can be of significant economic benefit in irrigation. Scientific literature over the past few years has reported the results of many studies indicating the beneficial effects of wastewater nutrients on plant growth (Quinn, 1979; Kipnis *et al.*, 1979; Overman, 1978, 1979a, 1979b).

Characteristics of wastewater

Sewage is the spent water of a community, its liquid portion (almost entirely water) becoming a vehicle for the transport of wastes in sewers. This sewage comprises a complex mixture of mineral and organic matter in many forms, including large and small particles of solid matter, floating and in suspension; colloidal and pseudocolloidal dispersions; and true solutions. Among the organic substances present in sewage are carbohydrates, lignin, fats, soaps, synthetic detergents, proteins and their decomposition products, as well as various synthetic and natural organic chemicals from the process industries. Ammonia and ammonium salts are always present, being produced by the decomposition of complex nitrogenous organic matter. Also present are sulphur-containing and phosphorus-containing compounds, the decomposition of which leads to the objectionable odour often associated with sewage. Wastewater accumulates a variety of inorganic substances as a result of different domestic and industrial uses. These might include a number of potentially toxic elements such as arsenic, cadmium, chromium, copper, lead, mercury, etc., with the possibility that each might exist in a number of different chemical states differing in solubility, reactivity and toxicity.

Sewage also contains macro-organisms and micro-organisms and is an excellent medium for their dissemination. It is known that human excrement is a principal vehicle for the transmission and spread of a wide range of communicable diseases and sewage can thus serve as an efficient vector of human pathogens. It has now been established (Feachem *et al.*, 1983) that raw wastewater from a community usually carries a wide spectrum of pathogenic bacteria, viruses, protozoa and helminths which are excreted by those suffering from, or by carriers of, the particular diseases endemic in the community.

Therefore, although the use of reclaimed water offers the potential for exploiting a 'new' resource which can be substituted for existing sources, it must be approached with caution. In the past, ignorance about the nature of disease in some societies, coupled with the absence of substantial epidemiological evidence in general, have engendered an empirical approach to wastewater reuse in many countries. Current concern about environmental quality requires a more responsible approach to effluent reuse so that both health risks and costs are minimized.

Problems of wastewater reuse in agriculture

Because of the nature of sewage, fears have been expressed about the possible hazards associated with effluent reuse. In assessing these hazards various pathways for the dissemination of undesirable pollutants have been examined. Two aspects of wastewater reuse in agriculture have become subjects of paramount importance:

the possible risks to health and the potential environmental damages. Health considerations are centred around the pathogenic organisms that are, or could be, present in the effluent and the build-up of toxic materials within the soil, and subsequently within plant and animal tissues which might eventually reach the human food chain. The leaching of materials such as nitrates and toxic soluble chemicals into the groundwater is also a matter for concern. Environmental risks involve the effects of the use of wastewater containing dissolved substances which have deleterious effects on the soil as well as inhibitory effects on the growth and development of plants.

Health aspects of wastewater reuse in agriculture

Although studies have indicated that bacteria, viruses, protozoa and helminths do not penetrate healthy undamaged surfaces of vegetables and crops and die away rapidly on crop surfaces exposed to sunlight, pathogens can survive for extended periods inside leafy vegetation or in protected cracks or stems. It is quite clear from the many studies to date, reviewed by Feachem *et al.* (1983) and Gerba, Wallis and Melnick (1975), that under favourable conditions enteric pathogens can survive for extremely long periods of time on crops, in water or in the soil. Factors that affect survival include the number and type of organisms, soil organic matter content, temperature, humidity, pH, amount of rainfall, amount of sunlight, protection provided by foliage, and competitive microbial flora. The survival of pathogens in soil has been reported to vary from a few hours to several months. A summary of pathogen survival rates is given in *Table 3.1* from which it is clear that whereas organisms such as *Salmonella typhosa* have relatively short survival times, other pathogens, including some bacterial species, *Ascaris* ova and enteric viruses, appear to be highly resistant to environmental stress. Survival times among different types of bacteria and viruses vary greatly and are difficult to assess without studying each type individually. In most cases it appears that 2–3 months is

Table 3.1 Survival of excreted pathogens (at 20–30°C)

Type of pathogen	*Survival times in days*			
	In faeces, nightsoil and sludge	*In fresh water and sewage*	*In the soil*	*On crops*
Viruses				
Enteroviruses	<100 (<20)	<120 (<50)	<100 (<20)	<60 (<15)*
Bacteria				
Faecal coliforms	<90 (<50)	<60 (<30)	<70 (<20)	<30 (<15)
Salmonella spp.	<60 (<30)	<60 (<30)	<70 (<20)	<30 (<15)
Shigella spp.	<30 (<10)	<30 (<10)	–	<10 (<5)
Vibrio cholerae	<30 (<5)	<30 (<10)	<20 (<10)	<5 (<2)
Protozoa	<30 (<15)	<30 (<15)	<20 (<10)	<10 (<2)
Entamoeba histolytica cysts	<30 (<15)	<30 (<15)	<20 (<10)	<10 (<2)
Helminths				
Ascaris lumbricoides eggs	Many months	Many months	Many months	<60 (<30)

*Figures in brackets show the usual survival time.

Source: Feachem *et al.* (1983)

sufficient for the reduction of pathogenic bacteria to negligible numbers once they have been applied to the soil, with the period for reduction of viruses being slightly higher. However, survival times as long as 5 years have also been reported. It is thus evident that irrigation of health-sensitive crops (including fruits and vegetables eaten uncooked) with raw or partially treated wastewater can present real health risks.

Despite the extensive worldwide practice of nightsoil and sludge fertilization and wastewater irrigation dating back many years, there are few epidemiological studies that have established definitive adverse health impacts of the consumption

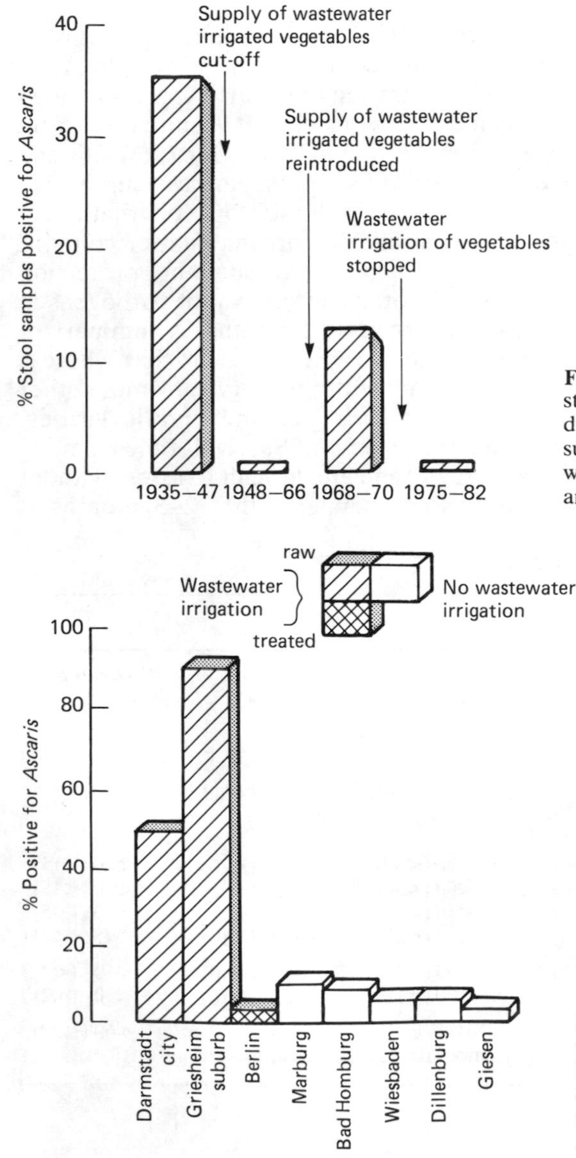

Figure 3.1 Prevalence of *Ascaris*-positive stool samples in West Jerusalem population during various periods, with and without supply of vegetables and salad crops irrigated with raw wastewater (Gunnerson, Shuval and Arlosoroff, 1984)

Figure 3.2 Wastewater irrigation of vegetables and *Ascaris* prevalence in Darmstadt and Berlin, compared with other cities in Germany not practising wastewater irrigation (Gunnerson, Shuval and Arlosoroff, 1984)

of food grown in this way. Shuval and Fattal (1984) have reported one of the earliest evidences connecting agricultural wastewater reuse with the occurrence of disease (*Figure 3.1*). In areas of the world where helminthic diseases caused by *Ascaris* and *Trichuris* spp. are endemic in the population, and where raw untreated wastewater is used to irrigate salad crops and/or other vegetables which are generally eaten raw, transmission of these infections has been found to occur through the reuse channel. A study in West Germany (reported by Gunnerson, Shuval and Arlosoroff, 1984) provides additional evidence (*Figure 3.2*) to support this hypothesis. Further evidence exists (as reported by Shuval and Fattal, 1984; Gunnerson, Shuval and Arlosoroff, 1984) showing that cholera could be transmitted through the same channel. There is also limited epidemiological evidence indicating that beef tapeworm (*Taenia saginata*) has been transmitted to the population consuming the meat of cattle grazing on wastewater-irrigated fields, or fed crops from such fields. Reports from Melbourne, Australia, and from Denmark (reported by Gunnerson, Shuval and Arlosoroff, 1984) strongly confirmed this. Although the reported incidence of diseases among workers on sewage farms has been inconclusive, there is always a potential risk associated with direct contact of wastewater with hands, which might then contaminate food. Another envisaged problem is that of possible inhalation of aerosolized sewage containing pathogens from spray irrigation. Shuval (1977) estimated that between 0.1 and 1% of the sewage sprayed into the air forms aerosols which are capable of being carried considerable distances by wind.

The risk to the health of animals grazing on sewage-irrigated pasture has been studied by many workers. Tuberculosis and beef tapeworm are among the diseases which are known to be transmitted to animals in this way and which can subsequently infect humans.

The possibility of long-term exposure of the population to low levels of toxic chemicals through the consumption of groundwater into which these materials have leached is very apparent. Although studies have indicated that only negligible amounts of such toxic chemicals move beyond 30 cm within the soil, it is possible that long-term reuse and eventual accumulation of toxic materials in the soil might lead ultimately to their mobilization and result in an increasing concentration being present in groundwater. Numerous studies have indicated that the content of certain toxic metals in plant tissues is directly proportional to the concentration of such metals within the soil root zone. Thus, long-term application of wastewater in irrigation poses the risk of plants having high levels of toxic materials in their tissues, and the FAO paper by Ayers and Westcot (1985) recommended some maximum concentrations for phytotoxic elements in irrigation water.

Environmental implications of wastewater reuse in agriculture

Although irrigation has been practised throughout the world for several millenia, it is only in this century that the importance of the quality of irrigation water has been recognized. Irrigation water quality is of particular importance in arid zones where extremes of temperature and low relative humidity result in high rates of evaporation, with consequent salt deposition from the applied water which tends to accumulate in the soil profile. The physical and mechanical properties of the soil, such as dispersion of particles, stability of aggregates, soil structure and permeability, are very sensitive to the type of exchangeable ions present in irrigation water. Thus, when effluent reuse is being planned, several factors related to soil properties must be taken into consideration.

It has been established that the productivity of irrigated land is fundamentally dependent on its internal drainage, which is a function of the soil profile morphology, pore size distribution and stability of pore structure. The first two factors are of paramount importance in relation to effluent reuse in irrigation. No irrigation scheme can succeed unless the soil profile remains permeable, and this depends both on the proportion of exchangeable cations, other than sodium, held by the soil (termed the exchangeable sodium percentage or ESP) and on the total concentration of soluble salts in the percolating water. Considerable laboratory evidence exists to indicate that pore structural stability is very important in determining the hydraulic properties of soils. MacNeal and Coleman (1966) showed that the hydraulic conductivity of a soil is a function of the ESP and is related to a solution parameter termed the sodium adsorption ratio (SAR). Quirk and Schoefield (1955) proved that the higher the SAR and the lower the electrolyte concentration of the percolating solution, the larger the hydraulic conductivity reduction (*Figure 3.3*). A further factor to consider in respect of wastewater irrigation is the high content of nutrients which might promote microbial growth with consequent reduction in soil permeability and hydraulic conductivity.

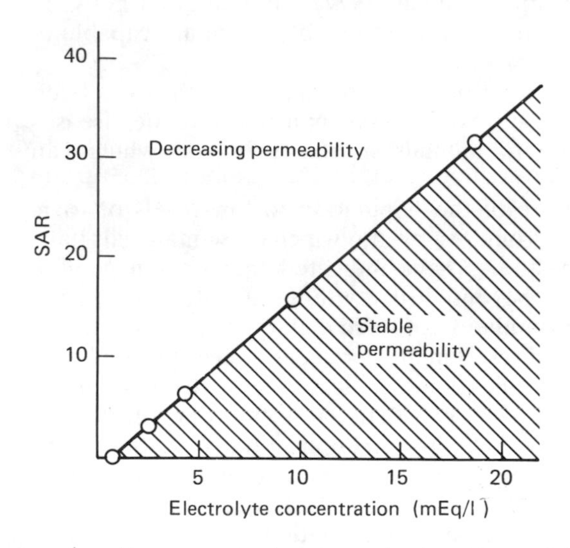

Figure 3.3 Threshold concentrations for irrigation water in relation to sodium adsorption ratio (SAR) (Quirk, 1971)

Another aspect of concern is the effect of dissolved solids in the irrigation water on the growth of plants. Dissolved salts increase the osmotic potential of soil water and an increase in osmotic pressure of the soil solution increases the amount of energy which plants must expend to take up water from soil. As a result, respiration is increased and the growth and yield of most plants decline progressively as osmotic pressure increases. *Figure 3.4* shows the effects of salinity on the yield of some selected crops. Although most plants respond to salinity as a function of the total osmotic potential of soil water, some plants are susceptible to specific ion toxicity. Many of the ions which are harmless or even useful at relatively low concentrations may become toxic to plants at high concentration, either through direct interference with metabolic processes or through indirect effects on other nutrients which might be rendered inaccessible.

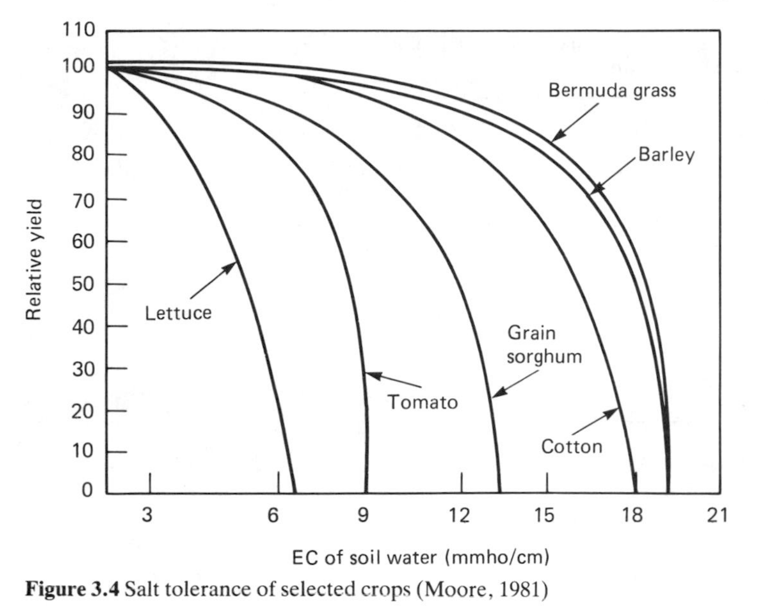

Figure 3.4 Salt tolerance of selected crops (Moore, 1981)

Quality criteria for reuse

The fears reviewed above have led to efforts to set up conditions and criteria that would allow for safe reuse of wastewater in irrigation. Developments of standards and water quality criteria for effluent reuse in irrigation have mainly evolved from an examination of health risks. In the United States, state health departments or agencies responsible for reuse activities formulate policy or decide on specific projects primarily on the basis of concern about infectious agents, accepting that most other constituents in reclaimed water would pose no immediate substantial harm in the rare case of accidental ingestion. For example, the State of California has established standards (Camp, Dresser and McKee, 1980) which require that the reclaimed water for irrigated food crops at all times must be adequately disinfected and filtered, with median coliform count no more than 2.2/100 ml. A World Health Organization (WHO) Committee of Experts on the subject (WHO, 1973) recommended that crops eaten raw should be irrigated only with biologically treated effluent that has been disinfected to achieve a coliform level of not more than 100/100 ml in 80% of the samples. *Table 3.2* shows recommended treatment processes to meet given health criteria.

These WHO standards have generally been accepted as a reasonable basis for design of wastewater treatment systems for reuse. However, in those Middle East countries which have recently developed systems for wastewater reuse, including Bahrain, Kuwait, Qatar, Saudi Arabia and the United Arab Emirates, the tendency has been to adopt more stringent health criteria. This arises out of the desire to protect an already high standard of public health by preventing at any expense the introduction of pathogens into the human food chain. A 30 000 m^3/d reclamation plant in Jeddah, Saudi Arabia, commissioned during the first half of 1985, has been designed to produce water to WHO drinking water standards from the sewage treatment plant effluent, using the reverse osmosis system for final polishing.

Table 3.2 Suggested treatment processes to meet the given health criteria for wastewater reuse in agriculture

Unit treatment process	*Type of agricultural reuse*		
	Crops not for direct human consumption	*Crops eaten cooked*	*Crops eaten raw*
Primary treatment	+++	+++	+++
Secondary treatment		+++	+++
Sand filtration		+	+
Disinfection		+	+++
Health criteria	A + F	D + F	

Key: +++ = Essential
+ = May sometimes be required
A = Freedom from gross solids; significant removal of parasite eggs
D = Not more than 100 coliforms per 100 ml in 80% of samples
F = No chemicals that lead to undesirable residues in crops

Source: WHO (1973)

The concern which forms the basis of quality criteria and standards is mainly with respect to crop contamination and focuses particularly on surface contamination and the persistence of pathogens until consumed by man or animals. Because of this preoccupation with the risk to public health, attention has been directed towards the type and nature of crops. Thus, timber crops can be grown with practically untreated wastewater. Some authors have suggested, as a reuse standard, that effluent irrigation should be entirely confined to non-edible crops, such as cotton or timber, or to crops which must be cooked before eating. It is evident, however, that heavy metals cannot be removed by cooking and it is necessary, therefore, to examine not only this limited effect in setting up standards for reuse but to consider other aspects as well.

It has already been pointed out that in arid and semi-arid regions there is a preponderance of the use of groundwater. The main problem with groundwater is that its pollution is not easily observable because the effects, particularly of low doses of conservative pollutants, are not manifested until it is too late to rectify them. Historically, groundwater has been considered a safe source of water since it is protected from surface contamination by the soil. Because groundwater is often used for human and animal consumption without any treatment, it is important to understand the sources and potential routes by which pollutants might gain entry into the groundwater, and these must be taken into account in deriving appropriate standards and criteria for effluent reuse.

The potential for groundwater contamination by micro-organisms depends upon the rate of removal of pathogens by the soil, the depth to the groundwater and the survival of pathogens in both soil and groundwater. Their rate of removal depends upon the type of soil, the rate and duration of pathogen leaching as well as the rate of water percolation within the soil. Migration of wastewater pathogens in soil involves transport by percolation through the soil profile to the groundwater. The main factors limiting transport of bacteria, ova of intestinal worms and cysts of protozoa through the soil are straining, sedimentation and, to a small extent, adsorption. Unlike bacteria and other larger pathogenic micro-organisms, virus removal from percolating wastewater is almost totally dependent on adsorption to various soil components. It has been observed that the removal of micro-organisms from percolating water is inversely proportional to the soil particle size. The infiltration rate of water in soil influences the removal efficiency, with low flow rate favouring greater retention. However, the flow rate is dependent on soil texture, structure, type of clays present and nutrients supplied by the wastewater to the indigenous soil microfauna (flora). Several studies have shown that as much as 90–95% of the faecal organisms are concentrated in the surface layer during the passage of wastewater through the soil and the remainder are concentrated in subsurface layers. Information on the movement of bacteria and viruses through soil in relation to wastewater application has been compiled by Gerba, Wallis and Melnick (1975) and is given in *Tables 3.3* and *3.4*.

It must be appreciated that bacteria and viruses are living things and are subject to death and decay. Prolonged storage, both within the environment and after abstraction, could serve as safeguards against these hazards. Because trace

Table 3.3 Movement of bacteria through soils

Nature of fluid	*Type of organism*	*Soil type*	*Maximum distance of travel* (m)
Tertiary treated wastewater	Coliforms	Fine to medium sand	6.1
Secondary effluent on percolation beds	Faecal coliforms	Fine to loamy sand to gravel	9.1
Primary sewage in infiltration beds	Faecal streptococci	Silty sand and gravel	183
Inoculated water and sewage injected subsurface	*Bacillus stearothemophilis*	Crystalline bedrock	28.7
Sewage in buried latrine intersecting groundwater	*Bacillus coli*	Sand and sandy clay	10.7
Canal water in infiltration basins	*Escherichia coli*	Sand dunes	3.1

Source: Frankenberger (1984)

Table 3.4 Movement of viruses through soils

Nature of fluid	*Type of organism*	*Soil type*	*Maximum distance of travel* (m)
Distilled water with added salts	T_1, T_2, f_2	Nine types of soils	0.45–0.50
Distilled water, 10^{-5} N Ca and Mg salts	Poliovirus 1	Dune sand	0.20
Distilled water	Poliovirus 2	Low humic latersols	0.0375–0.15
Secondary effluent	Poliovirus 2	Sandy gravel	60
Spring water	Coxsackie	Garden soils	0.9
Distilled water	T_4	Low humic latersols	0.0375–0.15
Secondary effluent	T_7	Sandy forest	0.195
Secondary effluent	Indigenous enteric viruses	Loamy sand	3–9

Source: Frankenberger (1984)

inorganic pollutants persist in the environment and do not suffer loss of toxicity (unlike bacteria and viruses, which do), long-term use of wastewater for irrigation will result, ultimately, in their gradual accumulation in and transport through the soil. It is necessary, therefore, to develop techniques for assessing the build-up and travel times of toxic materials within the soil to ensure the safety of groundwater. The passage of wastewater through the complex soil matrix induces a variety of physical and chemical reactions that influence the capacity of soil to remove wastewater pollutants. The mechanisms of removal depend on the characteristics of

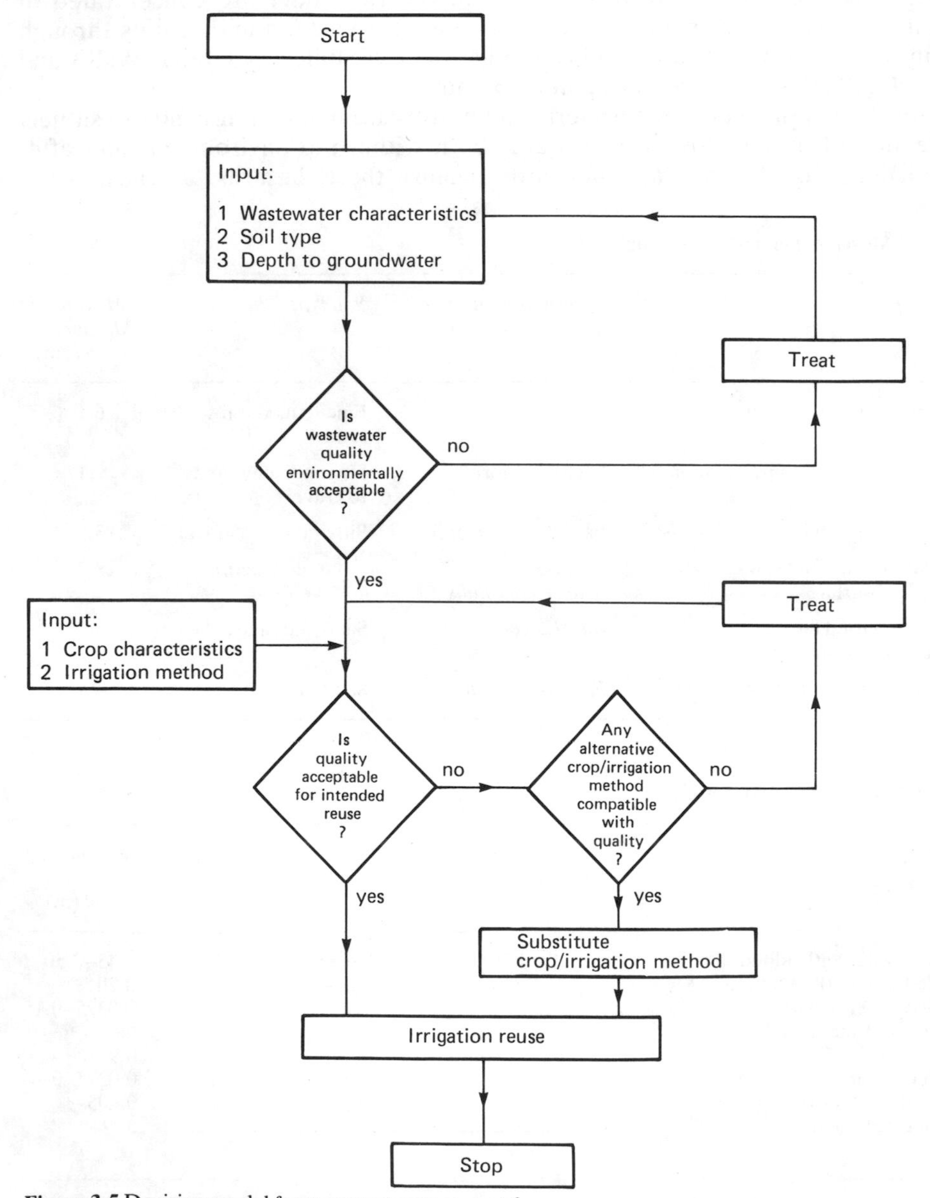

Figure 3.5 Decision model for wastewater treatment for reuse

the elements. Trace elements present in the wastewater in suspended form are removed primarily through filtration. For all practical purposes suspended solids are expected to be deposited near the surface of the soil profile during irrigation with wastewater. However, for trace elements present in wastewater in the dissolved state, filtering has no effect on their retention by the soil. Instead, chemical reactions such as ion exchange, precipitation, surface adsorption and organic complexation are important. Unless the equilibrium constants of each reaction are known, and the reaction kinetics in the soil are clearly defined, any standards for reuse are likely to remain arbitrary and their consequences unpredictable.

Thus, in assessing effluent quality criteria it is imperative that the nature of the soil as well as depth to the groundwater be taken into account. *Figure 3.5* shows a proposed decision model for setting effluent quality standards. It must be realized that it is not possible to cover all local situations when preparing water quality criteria and the approach should be to present guidelines that stress the management needed to use water of a certain quality successfully. The exact choice in practice must be made at the planning stage, taking account of the specific local conditions. Guidelines for evaluating irrigation water quality applicable to conditions encountered in California are given in *Table 3.5*.

Table 3.5 Guidelines for interpretation of water quality for irrigation (applicable in California)

Potential irrigation problem		*Units*	*Degree of restriction on use*		
			None	*Slight to moderate*	*Moderate*
Salinity (affects crop water availability)	a. EC_w	ds/m or mmho/cm	<0.7	0.7–3.0	>3.0
	b. TDS	mg/l	<450	450–2000	>2000
Permeability (affects infiltration rate of water into the soil)	SAR = 0–3		>0.7	0.7–0.2	<0.2
	3–6	EC_w	>1.2	1.2–0.3	<0.3
	6–12	and ds/m or	>1.9	1.9–0.5	<0.5
	12–20	mmho/cm	>2.9	2.9–1.3	<1.3
Evaluate using EC_w and SAR together	20–40		>5.0	5.0–2.9	<2.9
Specific ion toxicity					
Sodium					
a_1 surface irrigation		SAR	<3	3–9	>9
a_2 sprinkler irrigation		mg/l	<70	>70	–
Chloride					
b_1 surface irrigation		mg/l	<140	140–350	>350
b_2 sprinkler irrigation		mg/l	<100	>100	–
c boron		mg/l	<0.7	0.7–3.0	>3.0
Miscellaneous effects					
1. Nitrogen (total–N)		mg/l	<5	5–30	>30
2. Bicarbonate (sprinkler only)		mg/l	<90	90–500	>500
3. pH				Normal range 6.5–8.4	
4. Residual chlorine (sprinkler only)		mg/l	<1.0	1.0–5.0	>5.0

Source: Westcot and Ayers (1984)

Wastewater treatment technology for reuse

The desired quality of effluent for reuse varies with anticipated usage. Reuse application consequently requires emphasis on the removal of selected physico-chemical and biological constituents to user-specific non-adverse levels. The degree of risk of contamination in any reuse project depends on many factors, including the efficiency of wastewater treatment processes in removing or inactivating the various toxic materials and pathogens. An overriding consideration is the reliability of present technology and processes to provide reclaimed water that continuously meets the established guidelines. There are many treatment processes which can be combined to produce an end product from urban wastewater that would be acceptable for uses ranging from grass irrigation to human consumption. The choice of treatment technology is very important because of the economic consequences of the decision. Unnecessarily costly treatment will divert scarce resources away from other development uses. Because of the importance of minimizing treatment requirements in achieving acceptable effluent quality, the removal of pathogenic and toxic metals in various treatment processes will be examined.

Heavy metals removal

Primary settling removes a proportion of metals which are either insoluble or absorbed onto particulate matter (Barth *et al.*, 1965; Oliver and Cosgrave, 1974). Further metal removal occurs in the secondary biological stage of wastewater treatment, usually through adsorption of dissolved metals or fine particulate metals on to sludge flocs, as reported by Brown *et al.* (1973) and Oliver and Cosgrave (1974). Brown *et al.* (1973) found that for some metals (chromium, copper and lead) removal efficiency was greater in secondary treatment than in a primary process, while for zinc the average removal percentage was similar at both stages. Lester (1983) has indicated, however, that metal removal at these stages shows great variability (*Table 3.6*). The removal efficiency of advanced wastewater treatment processes can also be highly variable, with respect to both the process and the metal it removes (*see Table 3.6*). In recent years there have been discussions on the utilization of phytoplanktonic algae, in algal ponds, for the

Table 3.6 Removal of selected trace elements in wastewater treatment processes

Metal	*Percentage removal*		
	*Primary**	*Secondary**	*Tertiary†*
Cadmium	72	Variable	71
Chromium	51	Variable	94
Copper	71	Variable	78
Lead	73	Variable	86
Nickel	23	Variable	–
Zinc	74	Variable	21

Source: * Lester (1983); † US Board on Toxicology and Environmental Health Hazards (1982)

removal of residual metals from wastewater. Several authors (Becker, 1983; Filip *et al.*, 1979; Oswald 1972) have concluded that this technique is an economic method for removing heavy metals from wastewater, resulting in high quality effluent and valuable algal biomass which could be used for different purposes, one being the production of biogas.

Removal of pathogens

Kruse (1962) reviewed the performance of primary sedimentation and suggested that from 5 to 40% of the bacterial population in raw municipal wastewater should be removed in this stage of treatment. In considering specific reports it would appear that primary sedimentation exhibits a variable efficiency in removing pathogens, depending in part on the type of pathogen studied (*Table 3.7*). This would indicate that bacterial pathogens can be removed from wastewater, although not entirely, by primary treatment. Biological treatment, on the other hand, appears to be efficient in removing certain wastewater pathogens but does not produce an effluent which is pathogen-free. Disinfection is usually the final defence against any remaining microbial contaminants that may have survived conventional sewage treatment processes.

Table 3.7 Micro-organism removal in wastewater removal

Type of micro-organism	*Percentage removal*	
	Primary	*Biological**
Salmonella	15	96–99.999
Mycobacterium	48–57	Slight: −99.9
Amoebic cyst	Limited removal	0–99.9
Helminth ova	72–98	0–76
Viruses	3–extensive	0–84

* Biological includes trickling filter, activated sludge and waste stabilization ponds.

Source: Feachem *et al.* (1983)

Chlorine, although it has many shortcomings, is the most widely used disinfectant and has played a major role in the prevention of waterborne diseases. Other disinfectants include ozone, chlorine dioxide, iodine, ultraviolet-radiation and gamma-radiation. However, the use of chlorine as a disinfectant for controlling micro-organisms in the treatment of sewage for reuse has some potentially serious drawbacks. When it is used as a disinfectant, chloramines, simple and complex chlorinated organics and free chlorine residuals need to be removed prior to reuse, as many of these products of chlorination may be toxic, mutagenic or carcinogenic. Other potentially harmful long-term effects of these byproducts may still be unknown. Even chloride ions are toxic to some plants. Consequently, potentially harmful byproducts of chlorination treatment need to be removed by further treatment. Thus, although chlorination is quite effective as a means of disinfection, beneficial uses of the effluent for reuse may be limited without some sophisticated and costly techniques for further purification.

Because the harmful side-effects of chlorination are of considerable concern the use of other disinfectants, particularly ozone and chlorine dioxide, has stimulated interest. However, apart from the fact that these alternatives are more costly than chlorine, ozone produces a number of unstable oxonides and hydroperoxides which decompose to form epoxides and aldehydes, while chlorine dioxide also forms predominantly oxygenated compounds which are more soluble in water and are therefore more difficult to extract. Another method currently being investigated is treatment with lime. Most bacteria and viruses die rapidly at pH levels above 11.0. This pH is easily reached using lime and the treatment not only kills but also physically removes bacteria that are attached to solids and this provides an additional mechanism for pathogenic reduction. When used in conjunction with open storage (in ponds for example), the high pH is gradually reduced due to recarbonation by CO_2 from the air.

It has long been accepted that tropical and subtropical climates provide an ideal environment for the natural treatment of sewage in stabilization ponds. By detaining raw sewage in a series of shallow ponds for 2–3 weeks, significant reductions of both BOD and pathogens can be achieved. The natural action of storage and sunlight promote the rapid growth of micro-organisms which remove BOD both aerobically and anaerobically. The die-off of pathogens in waste stabilization ponds depends on the environmental and climatological parameters already mentioned, as well as on the retention time within the pond system. Since the ratio of sodium to calcium in any irrigation water is of paramount importance, it is anticipated that integrated operation of lime-treatment with a stabilization pond system might, in some situations, provide optimal wastewater treatment for effluent reuse in agriculture.

Irrigation technology for reuse

Various irrigation methods have been used for crop production, including flooding, furrow, sprinkler and trickle irrigation, and their application to wastewater reuse has been covered in a US Environmental Protection Agency (1977) Design Manual. Flooding and furrow irrigation with treated effluent do not differ essentially from irrigating with ground or surface water. However, land levelling should be carried out carefully to avoid surface ponding of stagnant effluent. Sprinkler irrigation is commonly used in many countries but, when irrigating with treated effluent, plugging of sprinkler nozzles is a problem. Secondary treatment, including final sedimentation, produces an effluent containing little or no solids that settle, and resultant problems with sprinkler plugging have been found to be negligible. Further precautionary measures can be taken, including the installation of granular filters or strainers and enlargement of the diameter of the nozzles, preferably to not less than 5 mm. However, the health risk through airborne droplets is apparent with sprinkler systems.

In water-short arid and semi-arid areas the use of flooding or sprinkler systems is not advisable, primarily because they encourage water wastage. Moreover, in effluent irrigation, these systems promote effluent wetting of most plant parts, thus encouraging contamination. Trickle irrigation seems to be the most promising technique from both the health and water-conservation standpoints and is now being used in a number of countries, extensively in Israel. It has proved to be efficient in arid areas because it reduces water consumption and increases the yield of crops. This form of irrigation is based on applying the water to the land surface or subsurface continuously by means of trickles from small diameter pipes having ports or leaking seams at regular intervals, releasing very small flows of about 2–15 l/h.

When compared with other methods the main advantages of trickle irrigation may be summarized as:

- Increased crop growth and yield, achieved by optimizing the water, nutrients and air regimes in the root zone.
- High irrigation efficiency; no canopy interception, wind drift or conveyance losses and minimal drainage losses.
- Minimal contact between worker and effluent.
- Low energy requirements; trickle system requires water pressure of only 100–300 kPa (1–3 bar).
- Low labour requirements; trickle system could easily be automated and there is usually no shifting of pipes.

The main advantage of trickle irrigation with effluent, however, is that contact between the effluent and the crops is less than with other methods and this reduces possible contamination by pathogens.

The main disadvantages may be summarized as:

- Relatively high investment costs; however, this depends very much on design and choice of equipment and materials – recent developments in plastic pipes are reducing the costs of this system.
- Risk of clogging; without conscientious management laterals can become clogged and unserviceable.
- Risk of damage; systems can be susceptible to damage by fire, pests, etc.

Conclusions

Clearly, wastewater reuse operations impose a greater risk of public or worker exposure to pathogens or toxic substances than would the use of unpolluted waters of non-sewage origin. The objective, therefore, is to minimize the exposure and reduce the potential health and environmental hazards to acceptable levels. In general, the health concern is in proportion to the degree of human contact with the effluent, the quality of the effluent, and the reliability of the treatment processes used to treat the wastewater before reuse. Currently, there is a scarcity of information concerning the performance and reliability of key treatment processes in respect of pollutants of concern. A lot of research is needed to evaluate and improve the capability and reliability of treatment processes and equipment to consistently produce a uniform quality of effluent. However, an approach based on reusing the *lowest* grade of effluent feasible under the site conditions will usually be sensible in developing countries, where experience in operating wastewater treatment systems is very limited and resources are scarce.

The benefits of wastewater irrigation can only be realized when the total system is designed and managed properly, so there are several considerations which need to be taken into account when planning such systems. Since irrigation requires a controlled discharge of effluent, the application method selected should depend upon the soil, the type of crop, the climate and the topography. It should also be understood that effluent quality requirements can be satisfied in different ways. If only a single crop, or class of crops, is being grown, the reclaimed effluent will need to meet only one level of water quality. Should several types of crop be envisaged, some method of devising ways of satisfying variable quality requirements must be found. In many cases it will be necessary to restrict the choice for effluent irrigation in the interests of the economy of wastewater treatment.

Thus, because of the site-specific characteristics of effluent reuse in agriculture, design features and quality guidelines should not be considered as absolute but should be incorporated into an integrated process of evaluation and design. Wastewater treatment, effluent storage and distribution and crop selection must be planned at the same time if an economic and conservative integrated system is to be achieved. In practice, this will require the close cooperation of the wastewater engineer and the agricultural engineer.

References

AYERS, R. S. and WESTCOT, D. W. (1985) Water quality for agriculture. *Irrigation and Drainage Paper 29,* Rev.1, FAO, Rome

BARTH, E. F., ETTINGER, M. B., SALOTTO, B. V. and DERMOTT, G. N. (1965) Summary report on the effects of heavy metals on the biological treatment processes. *Journal Water Pollution Control Federation,* **37,** 86–96

BECKER, E. W. (1983) Limitations of heavy-metals removal from wastewater by means of algae. *Water Research,* **17,** 459–466

BROWN, H. G., HENSLEY, C. P., McKINNEY, G. L. and ROBINSON, J. L. (1973) Efficiency of heavy-metals removal in municipal sewage treatment plants. *Environmental Letters,* **5,** 103–114

CAMP, DRESSER and McKEE Inc. (1980) *Guidelines for Water Reuse.* USEPA Contract 1980, No. 68-03-2380

FEACHEM, R. G., BRADLEY, D. J., GARELICK, H. and MARA, D. D. (1983) *Sanitation and Disease: Health Aspects of Excreta and Wastewater Management.* Chichester: John Wiley

FILIP, S. D., PETERS, T., ADAMS, V. D. and MIDDLEBROOKS, E. J. (1979) Residual heavy-metal removal by an algae-intermittent sand filtration system. *Water Research,* **13,** 305–313

FRANKENBERGER, W. T. (1984) *Fate of Wastewater Constituents in Soil and Groundwater: Pathogens in Irrigation with Reclaimed Municipal Wastewater. A Guidance Manual,* edited by G. S. Pettygrove and T. Asano. California State Water Resources Control Board, 14-1–14-25

GERBA, C. P., WALLIS, C. and MELNICK, J. L. (1975) Fate of wastewater bacteria and viruses in soil. *Journal, American Society of Civil Engineers, Irrigation and Drainage Division,* **101,** 157–174

GUNNERSON, C. G., SHUVAL, H. I. and ARLOSOROFF, S. (1984) Health effects of wastewater irrigation and their control in developing countries. *Proceedings of the Water Reuse Symposium III,* 21–31 August, San Diego

KIPNIS, T., FEIGIN, A., DOVRAF, A. and LEVANON, D. (1979) Ecological and agricultural aspects of ntirogen balance in perennial pasture irrigated with municipal effluents. *Progress in Water Technology,* **11(4/5),** 127–138

KRUSE, H. (1962) Some present-day sewage treatment in use for small communities in the Federal Republic of Germany. *WHO Bulletin 26.*

LESTER, J. N. (1983) Significance and behaviour of heavy-metals in wastewater treatment processes. I. Sewage treatment and effluent discharge. *Science of Total Environment,* **30,** 1–44

MacNEAL, B. L. and COLEMAN, N. T. (1966) Effect of solution composition on soil hydraulic conductivity. *Soil Science Society of America Proceedings,* **30,** 308–312

MOORE, C. V. (1981) Economic evaluation of irrigation with saline water within the framework of a farm. Methodology and empirical findings: a case study of Imperial Valley California. In: *Salinity in Irrigation and Water Resources,* edited by D. Yaron. New York: Marcel Dekker Inc., pp. 159–172

OLIVER, B. G. and COSGRAVE, E. G. (1974) The efficiency of heavy-metal removal by a conventional activated sludge treatment plant. *Water Research,* **8,** 869–874

OSWALD, W. J. (1972) Complete waste treatment in ponds. *Proceedings of the Sixth International Water Pollution Research Conference.* Oxford: Pergamon Press

OVERMAN, A. R. (1978) Effluent irrigation of sorghum X Sudan grass and kenaf. *Journal, American Society of Civil Engineers, Environmental Engineering Division,* **104,** 1061–1066

OVERMAN, A. R. (1979a) Effluent irrigation at different frequencies. *Journal, American Society of Civil Engineers, Environmental Engineering Division,* **105,** 535–545

OVERMAN, A. R. (1979b) Effluent irrigation of coastal Bermuda grass. *Journal, American Society of Civil Engineers, Environmental Engineering Division,* **105,** 55–60

QUINN, B. F. (1979) Surface irrigation with sewage effluent in New Zealand – a case study. *Progress in Water Technology,* **11(4/5),** 103–126

QUIRK, J. P. and SCHOEFIELD, R. V. (1955) The effect of electrolyte concentration on soil permeability. *Journal of Soil Science,* **6,** 163–178

QUIRK, J. P. (1971) Chemistry of saline soils and their physical properties. In: *Salinity and Water Use,* edited by T. Talsma and J. R. Philip, pp. 79–91. London: Macmillan Press

SHUVAL, H. I. (1977) Health considerations in water renovation and reuse. In: *Water Renovation and Reuse,* edited by H. I. Shuval, pp. 33–72. London: Academic Press

SHUVAL, H. I. and FATTAL, B. (1984) Epidemiological evidence for helminth and cholera transmission by vegetables irrigated with wastewater. Jerusalem – a case study. *Proceedings of the IAWPRC Conference,* Amsterdam, September 1984

US BOARD ON TOXICOLOGY AND ENVIRONMENTAL HEALTH HAZARDS (1982) *Quality Criteria for Water Reuse.* Washington DC: National Academy Press

US ENVIRONMENTAL PROTECTION AGENCY (1977) *Process Design Manual for Land Treatment of Municipal Wastewater.* Environmental Research Information Centre Technology Transfer

WESTCOT, D.W. and AYERS, R. S. (1984) *Irrigation Water Quality Criteria. Irrigation with Reclaimed Municipal Wastewater: A Guidance Manual,* edited by G. S. Pettygrove and T. Asano. California State Water Resources Control Board, pp. 3.1–3.37

Chapter 4

Quality criteria in using sewage effluent for crop production

A. Kandiah*

Introduction

When considering the use of effluents for irrigation their microbial and biochemical properties have to be evaluated. These values are then compared with the public health standards, taking into consideration the crop, soil and irrigation system and consumption of the produce, and only when the effluent meets these standards is it evaluated in terms of chemical criteria such as dissolved salts, relative sodium content and specific toxic ions.

In quantitative terms, the volume of wastewater available for reuse by irrigated agriculture is negligible when compared with the overall volume of water used for irrigation, except in certain arid or semi-arid areas. Yet water reclamation and reuse may be the most practicable solution to water shortages in many countries in the Near East Region and they are likely to be forced on governments who have no alternatives in certain areas with increasing urgency for water. They present no insuperable technical problems, although more knowledge will lead to economies and greater reliability. Reclamation is a practical solution to water scarcity under most conditions, provided that adequate precautions are taken in the design and operation of systems to protect the health of the individual and of the community.

In many parts of the world water still remains one of the limiting factors in crop production. Available soil moisture derived from rain (or from underground waters) is not sufficient for the requirements of plant growth and production, at least through part of the growing season. Such a deficiency should hence be made up by irrigation. Irrigated agriculture is dependent on an adequate water supply of usable quality. However, with the world population growing so rapidly, the water resources of the world are becoming limited, which calls for skilled planning and careful management in their development and utilization. The problem is often complicated by the fact that water is unequally distributed about the earth and its availability varies greatly with time.

Under conditions of limited supply and increasing demand, water allocation will have to be based on some form of priority established among the various demand sectors. In allocating water on this basis, quality requirements and tolerances should be considered. Needless to say, high priority is normally given to the domestic water use sector, which meets human consumption and sanitation

---

* Land and Water Development Division, FAO, Rome

requirements. Although the quantity required for the domestic sector is not so high when compared with that for the agricultural sector, the quality requirements are stringent and hence good quality waters are allocated on a priority basis to this demand sector. Between the next two competing demands, namely agriculture and industry, priority is dependent on the economic base of the country, but usually in developing nations agriculture receives priority over industry. Although agriculture is the major user of water, it can accept lower quality water than domestic and industrial users. It is therefore inevitable that there will be a growing tendency to look towards irrigated agriculture as a potential user of sewage effluent to meet deficiencies in conventional water sources and also for solutions to the overall effluent disposal problem. However, this does not mean that agriculture can readily accept sewage effluents without quality considerations, because sewage effluents contain salts of both inorganic and organic origin and biologically active materials which could result in undesirable immediate or long-term effects on soils, plants and the farm environment, as well as on users of the agricultural products.

The objective of this chapter is to present some of the important quality criteria adopted in evaluating sewage effluents for use in crop production. In this chapter discussion is limited to municipal wastewater which is defined as the spent water of a community consisting of water-carried wastes from residences, commercial buildings and industrial plants and surface or groundwater that enters the sewerage system (WHO, 1973).

Characteristics of municipal wastewater

Municipal wastewater consists mainly of a mixture of water and wastes which generally include dissolved and suspended materials made up of human and animal wastes, soaps, oils, greases, vegetable and animal residues, household chemicals, soil, bacteria and viruses. *Table 4.1* illustrates the composition and characteristics of a typical municipal sewage effluent. The medium strength wastewater is typical

Table 4.1 Physical and chemical characteristics of domestic wastewater

Major constituents	*Concentration* (in mg/l)		
	Strong	*Medium*	*Weak*
Total solids	1200	700	350
Dissolved solids	850	500	250
Suspended solids	350	200	100
Nitrogen (as N)	85	40	20
Phosphorus (as P)	20	10	6
Chlorides*	100	50	30
Alkalinity (as $CaCo_3$)	200	100	50
Grease	150	100	50
BOD_5†	300	200	100

Source: These data are adapted from Metcalf and Eddy Inc. (1972), p. 231.

* This amount should be increased by the concentration of these constituents in the carriage water; the table shows major constituents only.

† BOD_5 is the 5-day biochemical oxygen demand at 20°C. It is a measure of the biodegradable organic content of wastewater.

of the type of wastewater found in developed countries such as the United States or Canada where an abundant supply of water is available. In areas where water is used more sparingly, owing to scarcity or cost, a stronger wastewater can be anticipated. Weak wastewater generally occurs where the sewers also collect significant amounts of storm and/or groundwater due to leakage or inlets. The key consideration in the successful reuse of wastewater is to remove from it economically those substances which will be detrimental to the use proposed for the reclaimed wastewater. The limitation on the reuse of effluent is that the cost of reliably removing these potentially harmful substances often exceeds the cost of obtaining water from another source.

Agricultural use of sewage

Reuse of wastewater in agriculture is mainly for irrigation of crops or landscape, although it has been used to a lesser degree for aquaculture, forestry and livestock watering. One advantage in using reclaimed municipal wastewater for irrigation is that the nutrients (generally phosphorus and nitrogen) in the wastewater are a plant stimulant. Because of its nutrient content wastewater irrigation can lower fertilizer requirements and increase crop yields. For example, in India non-edible crops have been cultivated with a 30–40% higher yield using municipal wastewater for irrigation rather than conventional surface water.

Irrigation with wastewater has been practised for some time in a number of countries on a variety of crops. The cities of Melbourne, Australia, and Johannesburg, South Africa, have been using treated municipal wastewater for irrigation purposes since 1892 and 1914, respectively, on pasture and silage crops. In India, municipal wastewater has been used in some areas for irrigation of crops since the 19th century. In Mexico, crops including alfalfa, maize, wheat, tomatoes and chillies have been grown successfully. Favourable experiences have also been recorded with the use of treated wastewater to irrigate sugar cane on the islands of Puerto Rico and Hawaii, as well as to enhance the yield of cotton in the southwestern region of the United States.

Although irrigation with treated wastewater can provide benefits, there are also potential problems associated with its use. Some of the major concerns are: (a) health hazards; (b) salinity build-up; and (c) toxicity hazards.

A major consideration in the utilization of wastewater is the positive acceptance by both the farmer and the consumer of the agricultural products produced using reclaimed wastewater for irrigation purposes.

Quality criteria based on chemical constituents

Guidelines for evaluation of water quality for agriculture

Conceptually, water quality refers to the characteristics of a water supply that will influence its suitability for a specific use, that is, how well the quality meets the needs of the user. Quality is defined by certain physical, chemical and biological characteristics. There have been a number of different water quality guidelines related to irrigated agriculture. Each has been useful but none is entirely satisfactory because of the wide variability in field conditions.

A most recent and comprehensive guideline for evaluation of water quality for agriculture was published by FAO (1985) in which four problem categories – salinity, infiltration, toxicity and miscellaneous – are used for evaluation of conventional sources of irrigation water. Water may be classifed into one of three categories – no restriction, slight to moderate restriction, and severe restriction for use. *Tables 4.2* and *4.3* present the FAO guidelines for interpreting water quality for irrigation.

The guidelines presented in *Tables 4.2* and *4.3* are equally applicable to evaluate wastewater for irrigation purposes in terms of the chemical constituents such as dissolved salts, relative sodium content and toxic ions. On the other hand, municipal wastewater, which may be reused for irrigation, requires guidelines to protect against public health hazards. The degree of risk associated with such efluents is related to the microbial characteristics.

Table 4.2 Guidelines for interpretation of water quality for irrigation

Potential irrigation problem	*Units*	*Degree of restriction on use*		
		None	*Slight to moderate*	*Severe*
Salinity (affects crop water availability)*				
EC_w	dS/m	<0.7	0.7–3.0	>3.0
or				
TDS	mg/l	<450	450–2000	>2000
Infiltration (affects infiltration rate of water into the soil. Evaluation using EC_w and SAR together)†				
SAR = 0–3 and EC_w =		>0.7	0.7–0.2	<0.2
= 3–6 =		>1.2	1.2–0.3	<0.3
= 6–12 =		>1.9	1.9–0.5	<0.5
= 12–20 =		>2.9	2.9–1.3	<1.3
= 20–40 =		>5.0	5.0–2.9	<2.9
Specific ion toxicity (affects sensitive crops)				
Sodium (Na)†				
surface irrigation	SAR	<3	3–9	>9
sprinkler irrigation	me/l	<3	<3	
Chloride (Cl)‡				
surface irrigation	me/l	<4	4–10	>10
sprinkler irrigation	me/l	<3	>3	
Boron (B)	mg/l	<0.7	0.7–3.0	>3.0
Trace elements (see Table 4.3)				
Miscellaneous effects (affects susceptible crops)				
Nitrogen ($NO_3 - N$)§	mg/l	>5	5–30	>30
Bicarbonate (HCO_3) (overhead sprinkling only)	me/l	<1.5	1.5–8.5	>8.5
pH		Normal range 6.5–8.4		

Source: Adapted from FAO (1985)

* EC_w means electrical conductivity, a measure of the water salinity, reported in deciSiemens per metre at 25°C (dSm) or in units millimhos per centimetre (mmho/cm). Both are equivalent. TDS means total dissolved solids, reported in milligrams per litre (mg/l).

† SAR means sodium adsorption ratio, and is sometimes reported by the symbol RNa. At a given SAR, infiltration rate increases as water salinity increases. Evaluate the potential infiltration problem by SAR as modified by EC_w.

‡ For surface irrigation, most tree crops and woody plants are sensitive to sodium and chloride; use the values shown. Most annual crops are not sensitive. With overhead sprinkler irrigation and low humidity (<30%), sodium and chloride may be absorbed through the leaves of sensitive crops.

§ NO_3-N, nitrate nitrogen, reported in terms of elemental nitrogen (NH_4-N and organic−N should be included when wastewater is being tested).

Table 4.3 Recommended maximum concentrations of trace elements in irrigation water

Element	*Recommended maximum concentration* (mg/l)	*Remarks*
Aλ (aluminium)	5.0	Can cause non-productivity in acid soils (pH <5.5), but more alkaline soils at pH <7.0 will precipitate the ion and eliminate any toxicity
As (arsenic)	0.10	Toxicity to plants varies widely, ranging from 12 mg/l for Sudan grass to less than 0.05 mg/l for rice
Be (beryllium)	0.10	Toxicity to plants varies widely, ranging from 5 mg/l for kale to 0.5 mg/l for bush beans
Cd (cadmium)	0.01	Toxic to beans, beets and turnips at concentrations as low as 0.1 mg/l in nutrient solutions; conservative limits recommended due to its potential for accumulation in plants and soils to concentrations that may be harmful to humans
Co (cobalt)	0.05	Toxic to tomato plants at 0.1 mg/l in nutrient solution, tends to be inactivated by neutral and alkaline soils
Cr (chromium)	0.10	Not generally recognized as an essential growth element; conservative limits recommended due to lack of knowledge on its toxicity to plants
Cu (copper)	0.20	Toxic to a number of plants at 0.1–1.0 mg/l in nutrient solutions
F (fluoride)	1.0	Inactivated by neutral and alkaline soils
Fe (iron)	5.0	Not toxic to plants in aerated soils, but can contribute to soil acidification and loss of availability of essential phosphorus and molybdenum; overhead sprinkling may result in unsightly deposits on plants, equipment and buildings
Li (lithium)	2.5	Tolerated by most crops up to 5 mg/l; mobile in soil; toxic to citrus at low concentrations (<0.075 mg/l); acts similarly to boron
Mn (manganese)	0.20	Toxic to a number of crops at a few tenths to a few mg/l, but usually only in acid soils
Mo (molybdenum)	0.01	Not toxic to plants at normal concentrations in soil and water; can be toxic to livestock if forage is grown in soils with high concentrations of available molybdenum
Ni (nickel)	0.20	Toxic to a number of plants at 0.5–1.0 mg/l; reduced toxicity at neutral or alkaline pH
Pb (lead)	5.0	Can inhibit plant cell growth at very high concentrations
Se (selenium)	0.02	Toxic to plants at concentrations as low as 0.025 mg/l and toxic to livestock if forage is grown in soils with relatively high levels of added selenium; an essential element to animals but in very low concentrations
Sn (tin) Ti (titanium) W (tungsten)	–	Effectively excluded by plants; specific tolerance unknown
V (vanadium)	0.10	Toxic to many plants at relatively low concentrations
Zn (zinc)	2.0	Toxic to many plants at widely varying concentrations; reduced toxicity at pH >6.0 and in fine textured or organic soils

Source: Adapted from FAO (1985)

Assumptions in the guidelines

The water quality guidelines in *Table 4.2* are intended to cover the wide range of conditions encountered in irrigated agriculture. Several basic assumptions have been used to define their range of usability. If the water is used under greatly different conditions, the guidelines may need to be adjusted. Wide deviations from the assumptions might result in wrong judgements on the usability of a particular water supply, especially if it is a borderline case. Where sufficient experience, field trials, research or observations are available, the guidelines may be modified to fit local conditions more closely.

The basic assumptions in the guidelines are the following.

Yield potential

Full production capability of all crops, without the use of special practices, is assumed when the guidelines indicate no restrictions on use. A 'restriction on use' indicates that there may be a limitation in choice of crop, or special management may be needed to maintain full production capability. A 'restriction on use' does *not* indicate that the water is unsuitable for use.

Site conditions

Soil texture ranges from sandy-loam to clay-loam with good internal drainage. The climate is semi-arid to arid and rainfall is low. Rainfall does not play a significant role in meeting crop water demand or leaching requirement. (In a monsoon climate or areas where precipitation is high for part or all of the year, the guideline restrictions are too severe. Under the higher rainfall situations infiltrated water from rainfall is effective in meeting all or part of the leaching requirement.) Drainage is assumed to be good, with no uncontrolled shallow water table present within 2 m of the surface.

Methods and timing of irrigations

Normal surface or sprinkler irrigation methods are used. Water is applied infrequently, as needed, and the crop utilizes a considerable portion of the available stored soil-water (50% or more) before the next irrigation. At least 15% of the applied water percolates below the root zone (leaching fraction [LF] 15%). The guidelines are too restrictive for specialized irrigation methods, such as localized drip irrigation, which results in near daily or frequent irrigations, but are applicable for subsurface irrigation if surface applied leaching satisfies the leaching requirements.

Water uptake by crops

Different crops have different water uptake patterns, but all take water from wherever it is most readily available within the rooting depth. On average about 40% is assumed to be taken from the upper quarter of the rooting depth, 30% from the second quarter, 20% from the third quarter, and 10% from the lowest quarter. Each irrigation leaches the upper root zone and maintains it at a relatively low salinity. Salinity increases with depth and is greatest in the lower part of the root zone. The average salinity of the soil-water is three times that of the applied water and is representative of the average root zone salinity to which the crop responds. These conditions result from a leaching fraction of 15–20% and irrigations that are timed to keep the crop adequately watered at all times.

Salts leached from the upper root zone accumulate to some extent in the lower part but a balance is achieved as salts are moved below the root zone by sufficient leaching. The higher salinity in the lower root zone becomes less important if adequate moisture is maintained in the upper, 'more active', part of the root zone and long-term leaching is accomplished.

Restriction on use
The 'restriction on use' shown in *Table 4.2* is divided into three degrees of severity: none, slight to moderate, and severe. The divisions are somewhat arbitrary since change occurs gradually and there is no clearcut breaking point. A change of 10–20% above or below a guideline value has little significance if considered in proper perspective with other factors affecting yield. Field studies, research trials and observations have led to these divisions, but management skill of the water user can alter them. Values shown are applicable under normal field conditions prevailing in most irrigated areas in the arid and semi-arid regions of the world.

Biological criteria

In assessing the implications of wastewater irrigation to public health it is useful to draw upon the experiences of regulatory agencies in some selected countries. In many parts of North America regulatory agencies have established bacteriological guidelines for water for agricultural users. These guidelines generally allow water having 1000–10 000 MPN total coliforms/100 ml and 200–2000 MPN faecal coliforms/100 ml.

In California, the MPN coliform level should be less than 2.2/100 ml for surface spray irrigation for produce eaten raw and less than 23/100 ml for spray irrigation of produce eaten cooked.

Raw or untreated sewage cannot be used for irrigation purposes in Oregon; an adequately disinfected secondary effluent (MPN coliform level 100/100 ml) may be used for surface or spray irrigation of pasture or for row irrigation of orchards. In most European countries, surface application of settled sewage may be used for the irrigation of industrial and fodder crops, orchards and vegetables which are eaten in a cooked state.

A meeting of experts convened by WHO (1973) concluded that primary treatment would be sufficient to permit reuse through irrigation of crops not for direct consumption. Secondary treatment, and most probably disinfection and filtration, were considered necessary if the effluent were to be used for irrigation of produce for direct human consumption.

The maximum concentration is based on a water application rate which is consistent with good irrigation practices (10 000 m^3/ha year). If the water application rate greatly exceeds this, the maximum concentrations should be adjusted downward accordingly. No adjustment should be made for application rates less than 10 000 m^3/ha year. The values given are for water used on a continuous basis at one site.

Table 4.4 presents WHO-suggested health criteria and treatment processes for wastewater reuse. By combining *Tables 4.2, 4.3* and *4.4*, a potential user of municipal wastewater should be able to assess as a first step the quality limitations of his irrigation wastewater source. However, this alone is not sufficient since the limitations imposed by quality factors can be minimized through adoption of

Table 4.4 Suggested treatment processes to meet the given health criteria for wastewater reuse

	Irrigation			*Recreation*		*Industrial reuse*	*Municipal reuse*	
	Crops not for direct human consumption	*Crops eaten cooked; fish culture*	*Crops eaten raw*	*No contact*	*Contact*		*Non-potable*	*Potable*
Health criteria (see below for explanation of symbols)	A + F	B + F or D + F	D + F	B	D + G	C or D	C	E
Primary treatment	●●●	●●●	●●●	●●●	●●●	●●●	●●●	●●●
Secondary treatment		●●●	●●●	●●●	●●●	●●●	●●●	●●●
Sand filtration or equivalent polishing methods		●	●		●●●	●	●●●	●●
Nitrification						●		●●●
Denitrification								●●
Chemical clarification						●		●●
Carbon adsorption								●●
Ion exchange or other means of removing ions						●		●●
Disinfection		●	●●●	●	●●●	●	●●●	●●●*

Health criteria:

A Freedom from gross solids; significant removal of parasite eggs.
B As A, plus a significant removal of bacteria.
C As A, plus more effective removal of bacteria, plus some removal of viruses.
D Not more than 100 coliform organisms/100 ml in 80% of samples.
E No faecal coliform organisms in 100 ml, plus no virus particles in 1000 ml, plus no toxic effects on man, and other drinking-water criteria.
F No chemicals that lead to undesirable residues in crops or fish.
G No chemicals that lead to irritation of mucous membranes and skin.

In order to meet the given health criteria, processes marked ●●● will be essential. In addition, one or more processes marked ●● will also be essential, and further processes marked ● may sometimes be required.

* Free chlorine after 1 h.

Source: WHO (1973)

appropriate management practices. In selecting management alternatives for using a given acceptable, but marginal, quality municipal wastewater, the objective should be to minimize risk to public health, reduce adverse long and short-term impacts on the environment such as the soil, vegetation, surface and groundwater bodies and obtain crop produce at economically viable yield levels and acceptable quality. Selection of appropriate crops, soil types and irrigation methods is one of the most commonly adopted management alternatives when using a given source of wastewater.

Crop selection

From the point of view of total dissolved salts or salinity, the relative salt tolerance of most agricultural crops is known well enough to give general salt tolerance guidelines. *Table 4.5* presents crops classified as tolerant, moderately tolerant, moderately sensitive and sensitive groups to salinity. The yield response of these four classes of salt-tolerant groups to soil salinity and irrigation water salinity is presented in *Figure 4.1*. Similar crop tolerance tables have been developed for exchangeable sodium and boron (FAO, 1985).

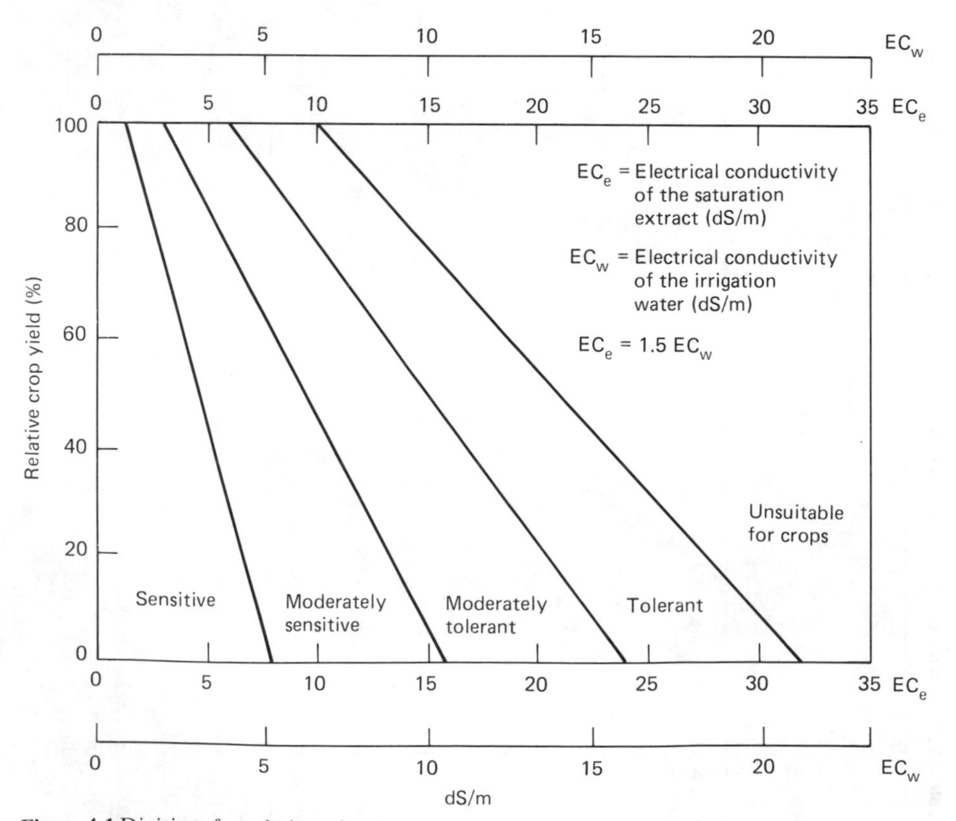

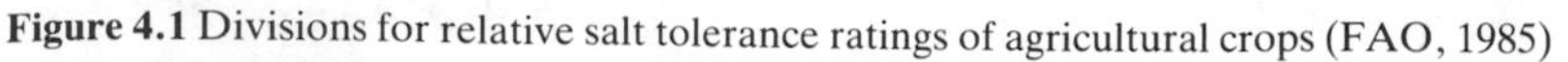

Figure 4.1 Divisions for relative salt tolerance ratings of agricultural crops (FAO, 1985)

In using sewage effluents, crop selection should also be considered from the point of view of public health hazards. Given below is some useful information on crop selection based on public health criteria:

- Forest crops probably offer greater opportunities than any other agricultural ventures for effluent irrigation with the fewest associated risks to public health.
- The irrigation of industrial crops not intended for consumption by humans or animals offers an alternative to forest irrigation, with a similar lower degree of risk to public health.
- Forage crops are more suited to sewage irrigation than vegetables which are likely to be eaten raw by humans. It should be noted that when forages are processed (dried) and stored before being fed to livestock (such as hay or pellets) they carry a lower associated health risk because of the time lag between irrigation and consumption.

Some established guidelines

Some countries have developed standards for the use of effluents in terms of crops, treatment required, type of irrigation system most suited and crop management. *Table 4.6* presents a summary of existing standards governing the use of renovated water in agriculture.

Perhaps Hawaii is one of the leading states in the United States which has developed a comprehensive set of principles and guidelines for irrigation with sewage effluent. A summary of the checklist as developed in Hawaii is presented below (Lau, 1979).

- Effluent quality requirement
  - Secondary treatment and chlorination
  - Domestic and municipal origin
  - Minimal toxic chemicals
  - Low concentration of total dissolved solids, boron, sodium, suspended solids and grease
  - Reasonably consistent quality over time
- Soils and crops
  - Soils suitable for crop growth
  - Soils with high sorptive capacity and high iron oxide preferred
  - Crops with high tolerance to nitrogen (as cane variety H59-3775) and/or salinity
  - Grass with thickly matted root system (as bermuda grass)
  - Exclude vegetable crops that are generally eaten without cooking
- Irrigation and fertilization
  - Maintain a no-water stress condition: for furrow irrigated sugar cane, 1 mg d supply for 150–200 acres at 4.2 inches per round every 2 weeks for an annual rain of about 40 inches; for sprinkler irrigated grassland, 1 mg d supply for 100 acres
  - Apply no excess of irrigation water for pollution control, assuming the effluent is not too saline to require leaching
  - Provide a storage or bypass facility for non-irrigation period
  - Apply commercial fertilizers to give cane a fast growth start

Table 4.5 Relative salt tolerance of agricultural crops*

Common name	Botanical name
Tolerant†	
Fibre, seed and sugar crops	
Barley	*Hordeum vulgare*
Cotton	*Gossypium hirsutum*
Jojoba	*Simmondsia chinensis*
Sugarbeet	*Beta vulgaris*
Grasses and forage crops	
Alkali grass, Nuttall	*Puccinellia airoides*
Alkali sacaton	*Sporobulus airoides*
Bermuda grass	*Cynodon dactylon*
Kallae grass	*Diplachne fusca*
Saltgrass, desert	*Distichlis stricta*
Wheatgrass, fairway crested	*Agropyron cristatum*
Wheatgrass, tall	*Agropyron elongatum*
Wildrye, Altai	*Elymus angustus*
Wildrye, Russian	*Elymus junceus*
Vegetable crops	
Asparagus	*Asparagus officinalis*
Fruit and nut crops	
Date palm	*Phoenix dactylifera*
Moderately tolerant	
Fibre, seed and sugar crops	
Cowpea	*Vigna unguiculata*
Oats	*Avena sativa*
Rye	*Secale cereale*
Safflower	*Carthamus tinctorius*
Sorghum	*Sorghum bicolor*
Soybean	*Glycine max*
Triticale	*X triticosecale*
Wheat	*Triticum aestivum*
Wheat, durum	*Triticum turgidum*
Grasses and forage crops	
Barley (forage)	*Hordeum vulgare*
Brome, mountain	*Bromus marginatus*
Canary grass, reed	*Phalaris arundinacea*
Clover, Hubam	*Mililotus alba*
Clover, sweet	*Melilotus*
Fescue, meadow	*Festuca pratensis*
Fescue, tall	*Festuca elatior*
Harding grass	*Phalaris tuberosa*
Panic grass, blue	*Panicum antidotale*
Rape	*Brassica napus*
Rescue grass	*Bromus unioloides*
Rhodes grass	*Chloris gayana*
Ryegrass, Italian	*Lolium italicum multiflorum*
Ryegrass, perennial	*Lolium perenne*
Sudan grass	*Sorghum sudanense*
Trefoil, narrowleaf birdsfoot	*Lotus corniculatus tenuifolium*
Trefoil, broadleaf birdsfoot	*Lotus corniculatus arvenis*
Wheat (forage)	*Triticum aestivum*
Wheatgrass, standard crested	*Agropyron sibiricum*
Wheatgrass, intermediate	*Agropyron intermedium*
Wheatgrass, slender	*Agropyron trachycaulum*
Wheatgrass, western	*Agropyron smithii*
Wildrye, beardless	*Elymus triticoides*
Wildrye, Canadian	*Elymus canadensis*
Vegetable crops	
Artichoke	*Helianthus tuberosus*
Beet, red	*Beta vulgaris*
Squash, zucchini	*Cucurbita pepo melopepo*
Fruit and nut crops	
Fig	*Ficus carica*
Jujube	*Ziziphus jujuba*
Olive	*Olea europaea*
Papaya	*Carica papaya*
Pineapple	*Ananas comosus*
Pomegranate	*Punica granatum*
Moderately sensitive	
Fibre, seed and sugar crops	
Broadbean	*Vicia faba*
Castorbean	*Ricinus communis*
Maize	*Zea mays*
Flax	*Linum usitatissimum*
Millet, foxtail	*Setaria italica*
Groundnut/peanut	*Arachis hypogaea*
Rice, paddy	*Oryza sativa*
Sugarcane	*Saccharum officinarum*
Sunflower	*Helianthus annuus*
Grasses and forage crops	
Alfalfa	*Medicago sativa*
Bentgrass	*Agrostis stolonifera palustris*
Bluestem, Angleton	*Dichanthium aristatum*
Brome, smooth	*Bromus inermis*
Buffelgrass	*Cenchrus ciliaris*
Burnet	*Poterium sanguisorba*

Clover, alsike	*Trifolium hybridium*
Clover, Berseem	*Trifolium alexandrinum*
Clover, ladino	*Trifolium repens*
Clover, red	*Trifolium pratense*
Clover, strawberry	*Trifolium fragiferum*
Clover, white Dutch	*Trifolium repens*
Corn (forage) (maize)	*Zea mays*
Cowpea (forage)	*Vigna unguiculata*
Dallis grass	*Paspalum dilatatum*
Foxtail, meadow	*Alopecurus pratensis*
Grama, blue	*Bouteloua gracilis*
Lovegrass	*Eragrostis sp.*
Milkvetch, Cicer	*Astragalus cicer*
Oatgrass, tall	*Arrhenatherum, Danthonia*
Oats (forage)	*Avena sativa*
Orchard grass	*Dactylis glomerata*
Rye (forage)	*Secale cereale*
Sesbania	*Sesbania exaltata*
Siratro	*Macroptilium atropurpureum*
Sphaerophysa	*Sphaerophysa salsula*
Timothy	*Phleum pratense*
Trefoil, big	*Lotus uliginosus*
Vetch, common	*Vivia angustifolia*
Vegetable crops	
Broccoli	*Brassica oleracea botrytis*
Brussels sprouts	*B. oleracea gemmifera*
Cabbage	*B. oleracea capitata*
Cauliflower	*B. oleracea botrytis*
Celery	*Apium graveolens*
Corn, sweet	*Zea mays*
Cucumber	*Cucumis sativus*
Eggplant	*Solanum melongena esculentum*
Kale	*Brassica oleracea acephala*
Kohlrabi	*Brassica oleracea gongylode*
Lettuce	*Latuca sativa*
Muskmelon	*Cucumis melo*
Pepper	*Capsicum annum*
Potato	*Salonum tuberosum*
Pumpkin	*Cucurbita pepo pepo*
Radish	*Raphanus sativus*
Spinach	*Spinacia oleracea*
Squash, scallop	*Cucurbita pepo melopepo*
Sweet potato	*Ipomoea batatas*
Tomato	*Lycopersicon lycopersicum*
Turnip	*Brassica rapa*
Watermelon	*Citrullus lanatus*
Fruit and nut crops	
Grape	*Vitis sp.*
Sensitive	
Fibre, seed and sugar crops	
Bean	*Phaseolus vulgaris*
Guayule	*Parthenium argentatum*
Sesame	*Sesamum inareum*
Vegetable crops	
Bean	*Phaseolus vulgaris*
Carrot	*Daucus carota*
Okra	*Abelmoscrus esculentus*
Onion	*Allium cepa*
Parsnip	*Pastinaca sativa*
Fruit and nut crops	
Almond	*Prunus dulcis*
Apple	*Malus sylvestris*
Apricot	*Prunus armeniaca*
Avocado	*Persea americana*
Blackberry	*Rubis sp.*
Boysenberry	*Rubus ursinus*
Cherimoya	*Annona cherimola*
Cherry, sweet	*Prunus avium*
Cherry, sand	*Prunus besseyi*
Currant	*Ribes sp.*
Gooseberry	*Ribes sp.*
Grapefruit	*Citrus paradisi*
lemon	*Citrus limon*
Lime	*Citrus aurantiifolia*
Loquat	*Eriobotrya japonica*
Mango	*Mangifera indica*
Orange	*Citrus sinensis*
Passion fruit	*Passiflora edulis*
Peach	*Prunus persica*
Pear	*Pyrus communis*
Persimmon	*Diospyros virginiana*
Plum: Prune	*Prunus domestica*
Pummelo	*Citrus maxima*
Raspberry	*Rubus idaeus*
Rose apple	*Syzygium jambos*
Sapote, white	*Casimiroa edulis*
Strawberry	*Fragaria sp.*
Tangerine	*Citrus reticulata*

Source: Data taken from FAO (1985)

* These data serve only as a guide to the relative tolerances among crops; absolute tolerances vary with climate, soil conditions and cultural practices.

† The relative tolerance ratings are defined by the boundaries in *Figure 4.1*.

Table 4.6 Existing standards governing the use of renovated water in agriculture

	California	*Israel*	*South Africa*	*Federal Republic of Germany*
Orchards and vineyards	Primary effluent; no spray irrigation; no use of dropped fruit	Secondary effluent	Tertiary effluent, heavily chlorinated where possible; no spray irrigation	No spray irrigation in the vicinity
Fodder, fibre crops, and seed crops	Primary effluent; surface or spray irrigation	Secondary effluent, but irrigation of seed crops for producing edible vegetables not permitted	Tertiary effluent	Pretreatment with screening and setting tanks; for spray irrigation, biological treatment and chlorination
Crops for human consumption that will be processed to kill pathogens	For surface irrigation, primary effluent. For spray irrigation, disinfected secondary effluent (no more than 23 coliform organisms/ 100 ml)	Vegetables for human consumption not to be irrigated with renovated wastewater unless it has been properly disinfected (<1000 coliform organisms/100 ml in 80% of samples)	Tertiary effluent	Irrigation up to 4 weeks before harvesting only
Crops for human consumption in a raw state	For surface irrigation, no more than 2.2 coliform organisms/100 ml. For spray irrigation, disinfected, filtered wastewater with turbidity of 10 units permitted, providing it has been treated by coagulation	Not to be irrigated with renovated wastewater unless they consist of fruits that are peeled before eating		Potatoes and cereals – irrigation through flowering stage only

Source: California State Department of Public Health (1968); Indian Standards Institution (1965); Israel, Ministry of Agriculture, Water Commission (1969); Müller (1969); Peru, Ministry of Health, Department of Environmental Sanitation (1970); Shuval (1976).

- Geohydrologic considerations of application site
  - Conduct a geohydrological survey to ascertain the probable pathway of deep percolation and to determine ground water occurrence, circulation, quality, recharge and discharge
  - Select areas of minimum soil thickness of 1.5 m with high absorptive capacity
  - Determine minimum allowable depth to water table on a case-by-case basis of geology and potable groundwater quality
- Monitoring methodology
  - Monitor chemical, microbiological and viral water quality to include STP effluent, leachate at bottom of root zone and groundwater
  - Monitor soil in terms of chemical properties and viruses
  - Monitor cane growth and sugar yield
  - Monitor microbiological air quality in case of sprinkler irrigation
- Disinfection of sewage effluent: public health hazards
  - Disinfect effluent before application
  - Precautionary measures for field workers to minimize the risk of contracting viral infection.

References

FAO (1985) *Water Quality for Agriculture 8* (Ayers, R. S. and Westcot, D. W.), FAO Irrigation and Drainage Paper No. 29, Rev. 1. Rome, FAO

LAU, S. L. (1979) Water re-use from sewage effluent by irrigation. A perspective for Hawaii. *Water Resources Bulletin Minnesota,* 15/3

METCALF and EDDY INC. (1972) *Wastewater Engineering: Treatment, Disposal, Re-use.* New York: McGraw-Hill

WATER RESOURCES SERVICE (1975) *Health Aspects of Sewage Effluent Irrigation.* Pollution Control Branch, Department of Lands, Forests and Water Resources, British Columbia

WHO (1973) *Re-use of Effluents: Methods of Wastewater Treatment and Health Safeguards. Report of a WHO Meeting of Experts.* Geneva, WHO Technical Report Series No. 517

Chapter 5

Health aspects of reuse of treated wastewater for irrigation

P. J. Hillman*

Introduction

In a great many countries there is an increasing need, and in some an urgent need, to conserve and protect resources. This is particularly so of water which is a vital resource but a severely limited one for countries in the Near East Region. The reuse of wastewater is, therefore, an obvious consequence of the need to conserve water and should be encouraged.

Many communities have practised excreta reuse and effluent reuse for hundreds of years and it is part of their culture. The quality of river water used in some irrigation projects is such that reuse of human and animal waste is continually taking place, albeit in an uncontrolled, haphazard fashion. Rapid increases in population and industrial growth have led to more treatment of wastewater in order to reduce pollution and protect receiving waters. It is, then, a natural progression to seek direct reuse of this treated effluent for lower grade purposes such as irrigation.

There are, unfortunately, hazards to the health of the public inherent in such reuse schemes and the bulk of this paper is devoted to a discussion of them. However, the potential benefits to the community may be substantial, provided that the health risks are minimized. These include:

- Substitution of wastewater to meet irrigation demand and thus conservation of better quality water sources for other uses, particularly potable supplies.
- Extending irrigated areas to produce cash crops aimed at export markets.
- Reduction in the costs associated with producing or importing and applying fertilizers since wastewater effluents contain major plant nutrients and many trace elements essential for plant growth.
- Reduction in the costs of water supply for irrigation, particularly where this is pumped from aquifers.

The principal benefit must be, of course, the greater availability of cheaper and perhaps better crops to improve the nutritional status and well-being of the population.

* WHO Consultant Representing the Regional Office for the Eastern Mediterranean, Alexandria, Egypt

Hazards to health

The use of wastewater for irrigation poses a health hazard, directly to the field worker and in some instances to people living close to the irrigated area, and indirectly to the field worker's family and the general public. Since the aim of irrigation is to distribute as widely as possible a limited water resource at a rate sufficient to sustain plant growth, then this threat to public health is also distributed widely.

The term 'wastewater', as used here, refers to all liquid wastes, including domestic sewage, sullage, trade effluents and industrial wastes. Since domestic sewage is generally a substantial component of this wastewater, all the micro-organisms and parasites commonly found in excreta are present and many infectious diseases may be transmitted through these agents. It may also contain chemical substances potentially hazardous to health, originating in discharges of trade effluents and industrial wastes.

Viruses

All viruses are potentially pathogenic, although not all to humans. The intestinal tract may become infected by many different types (enteric viruses). They are passed in the faeces by infected persons, some of whom have no symptoms of any diseases (carriers). In contrast to other micro-organisms the ingestion of a single virus may be sufficient to induce infection in man. Viruses do not multiply outside a suitable host cell but are very persistent in the environment and may survive in wastewater and soil for several months (WHO, 1979).

Counts of 10^5 viral units/litre have been reported in raw sewage (Shuval, 1977) and, while a reduction in number as a result of conventional wastewater treatment processes can be expected, viruses will remain in the effluent. For drinking waters the WHO (1984) guidelines suggest an operational procedure for the control of viruses; namely that a water with a turbidity of 1 NTU (nephelometric turbidity units) or less should be disinfected with at least 0.5 mg/l of free residual chlorine after 30 min contact period with a pH less than 8. The much higher turbidity, ammonia concentration and organic matter present in wastewater effluents will greatly hinder the process of disinfection using chlorine and thus, even a well treated, chlorinated wastewater free from bacteria may contain active viruses.

Many factors influence the survival of viruses in soil or on crops (WHO-EURO, 1981); pH level, moisture content, temperature, exposure to sunlight and the presence of organic matter have all been quoted. Viruses are adsorbed on to the surface of soil particles and this appears to prolong their survival. Survival on crops is usually of shorter duration because of exposure to sunlight and lack of moisture, although figures of 170 days survival of enteric viruses in soil and 23 days on crops have been recorded (WHO, 1979).

Bacteria

Many species of bacteria are normal inhabitants of the intestine of healthy persons and are excreted with the faeces. *Escherichia coli*, the most widely used indicator of faecal pollution, is only one of them. In addition, however, the intestinal tract may also harbour a number of bacterial pathogens (enteric bacteria) which, when excreted and subsequently ingested or inhaled by another individual, may cause disease. An infected person might also be a carrier who exhibits no symptoms of the disease (or only very minor ones) but will still be excreting these bacteria.

Very large concentrations (10^6/ml) of coliform indicator bacteria can occur in raw sewage. Highly efficient removal processes achieving 99% reduction will still leave large numbers of bacteria in the effluent. Chlorine is a good bactericide, particularly if it is available in its uncombined form. However, the problems of turbidity, organic content and ammonia in wastewater, referred to earlier, greatly inhibit the efficiency of chlorination. The variability of organic load in wastewater effluents adds further complication and there is a need for careful control if chlorination of effluents is to meet set standards reliably.

The survival of faecal bacteria in wastewater depends largely on environmental conditions, particularly temperature and organic content, but generally there is a substantial reduction in their numbers after one week (Pickford, 1978). However, in moist soil with adequate organic content, faecal bacteria can survive for months. Generally, on the surface of crops the bactericidal effect of sunlight reduces their number rapidly, although leafy crops that may retain moisture over long periods may sustain bacteria in significant numbers.

Protozoa

These unicellular microscopic animals are usually foreign to the human organism but several species can infect the intestinal tract of man and animals and may cause diarrhoea or dysentery. They are excreted in the faeces as cysts, a more resistant stage, which are capable of being transmitted to a new host by ingestion. Their survival time in wastewater is thought to be around 20 days (Schaefer, 1985). It is possible that in areas where protozoal infections are endemic, several hundred cysts/litre may be found in raw sewage (Shuval, 1981).

Heavy dose chlorination and contact times of 1 hour, it is claimed, can inactivate protozoal cysts (WHO, 1973). However, conventional treatment processes of sedimentation, followed by activated sludge or biological filtration, while reducing the concentration of cysts are unlikely to produce an adequate effluent. Further polishing processes, such as sand filtration or lagooning, will be required. The much longer retention times and natural processes of degradation and competition within waste stabilization ponds provide far greater opportunity for the removal of protozoa.

Helminths

Many species of parasitic worms (helminths) may infect the intestine of man which, with time and repeated infection, can lead to damage to the intestine and other organs, in many cases leading to chronic debilitating diseases. The eggs (ova), or sometimes larvae, are passed out in the excreta, with the exception of the eggs of *Schistosoma haematobium*, which are passed out in the urine, and guineaworm, which releases its eggs through the ruptured skin of the infected person. The developmental stages which the helminths have to undergo before reaching man again can be complex.

The survival of helminth ova in wastewater and the environment varies very much from species to species. The most persistent of all are *Ascaris* ova, which may survive for a year or more in moist, organic-rich situations. The close association between the spread of certain helminthic diseases, notably schistosomiasis, and irrigation development is well documented (Bradley, 1978; WHO-EURO, 1983). The long survival of some helminths in the environment, and the ability of certain helminth larvae to penetrate skin, makes control particularly difficult. Sedimentation processes to remove eggs have been shown to be only partially successful unless prolonged retention is given. Some helminth eggs hatch into free swimming larvae and so other tertiary treatments may be required, such as postchlorination or protected storage before use.

Vectors of disease

Irrigation schemes provide many sites for the breeding of mosquitoes and flies. Apart from the nuisance and problems with hygiene that these insects create, they can also become the vectors of serious disease. The close association between dramatic increases in the incidence of such diseases as malaria, filariasis and onchocerciasis and water resource developments is well known.

Table 5.1 Organic constituents of health significance

Constituent	*Unit*	*Guideline value*	*Remarks*
Aldrin and dieldrin	μg/l	0.03	
Benzene	μg/l	10*	
Benzo-a-pyrene	μg/l	0.01*	
Carbon tetrachloride	μg/l	3*	Tentative guideline value†
Chlordane	μg/l	0.3	
Chlorobenzenes	μg/l	No health-related guideline value set	Odour threshold concentration between 0.1 and 3 μg/l
Chlorophenols	μg/l	No health-related guideline value set	Odour threshold concentration 0.1μg/l
Chloroform	μg/l	30*	Disinfection efficiency must not be compromised when controlling chloroform content
2,4,D	μg/l	100‡	
DDT	μg/l	1	
1,2 Dichloroethane	μg/l	10*	
1,1 Dichlorethylene	μg/l	0.3*	
Heptachlor and heptachlor epoxide	μg/l	0.1	
Hexachlorobenzene	μg/l	0.01*	
Lindane	μg/l	3	
Methoxychlor	μg/l	30	
Pentachlorophenol	μg/l	10	
Tetrachlorethylene	μg/l	10*	Tentative guideline value†
Trichloroethylene	μg/l	30*	Tentative guideline value†
2,4,6 Trichlorophenol	μg/l	10‡	Odour threshold concentration in 0.1 μg/l
Trihalomethanes		No guideline value set	See chloroform

* These guideline values were computed from a conservative hypothetical mathematical model which cannot be experimentally verified and values should therefore be interpreted differently. Uncertainties involved may amount to two orders of magnitude (that is, from 0.1 to ten times the number).

† May be detectable by taste and odour at lower concentrations.

‡ When available carcinogenicity data could not support a guideline value, but the compounds were judged to be of importance in drinking water quality and guidance was considered essential, a tentative guideline value was set on available health-related data.

Source: WHO (1984)

This problem is common to all irrigation schemes in the tropics, not just those asssociated with wastewater reuse, and so will not be dealt with further in this chapter.

Chemicals

These substances enter the wastewater treatment plant largely as a result of industrial and trade effluents discharged to the sewer. They exist in wastewater usually at very low concentrations and ingestion over prolonged periods is necessary in order to produce detrimental health effects. High concentrations are of course possible but these would need to be controlled in order to protect the investment in the sewerage system and treatment plant, maintain a viable treatment process and protect the environment. They should not appear in wastewater effluents used for irrigation.

The WHO (1984) *Guidelines for Drinking Water* give values for nine inorganic and 15 organic substances (*Tables 5.1* and *5.2*), plus tentative values for three other organics. The values given are those which, if present in a water supply, create no significant risk to the consumer, assumed to be of average weight and having a daily intake of 2 litres of water over a 70 year lifespan. Clearly, provided that cross-connections with potable supplies can be prevented and adequate instruction is given to irrigation workers and their families, then this level of intake is extremely unlikely with water intended for irrigation.

The principal health hazards associated with chemicals in wastewater used for irrigation, therefore, arise from contamination of crops. Particular concern is attached to the cumulative poisons, principally associated with heavy metals, and carcinogens, chiefly associated with organic chemicals.

Table 5.2 Inorganic constituents of health significance

Constituent	*Unit*	*Guideline value*	*Remarks*
Arsenic	mg/l	0.05	
Asbestos	–	No guideline value set	
Barium	–	No guideline value set	
Beryllium	–	No guideline set	
Cadmium (total)	mg/l	0.005	
Chromium (total)	mg/l	0.05	
Cyanide	mg/l	0.1	
Fluoride	mg/l	1.5	Natural or deliberately added; local or climatic conditions may necessitate adaptation
Hardness	–	No health-related guideline value set	
Lead (total)	mg/l	0.05	
Mercury (total)	mg/l	0.001	
Nickel	–	No guideline value set	
Nitrate	mg/l (N)	10	
Nitrite	–	No guideline value set	
Selenium	mg/l	0.01	
Silver	–	No guideline value set	
Sodium	–	No guideline value set	

Source: WHO (1984)

It is known that some of the heavy metals, notably cadmium and selenium, may be taken up by and accumulated in crops (WHO, 1972). This is particularly so when the soil is acidic or is made so by the addition of acidic wastewater effluent.

Copper and zinc are also sometimes quoted as they are useful in indexing the toxicity of the soil to plant growth. However, they are less toxic to man. Many of the metals form oxides or hydroxides in wastewater and are precipitated out during treatment and thus form part of the sludge. The application of this sludge is, therefore, much more likely to contribute to the problems outlined above than the use of treated wastewater effluent: *Table 5.3* gives the FAO (1979) recommended maximum concentrations of elements in irrigation water.

Table 5.3 FAO recommended maximum concentrations of elements in irrigation water

Constituent	*Allowable concentration* (mg/l)	
	For water used continuously on all soils	*For use up to 20 years on fine-textured soils of pH 6.0 to 8.5*
Aluminium (Al)	5.0	20.0
Arsenic (As)	0.10	2.0
Beryllium (Be)	0.10	0.5
Boron (B)	0.75	2.0
Cadmium (Cd)	0.01	0.05
Chromium (Cr)	0.10	1.0
Cobalt (Co)	0.05	5.0
Copper (Cu)	0.20	5.0
Fluoride (F)	1.0	15.0
Iron (Fe)	5.0	20.0
Lead (Pb)	5.0	0.075*
Lithium (Li)	0.075*	0.075*
Manganese (Mn)	0.20	10.0
Molybdenum (Mo)	0.01	0.05
Nickel (Ni)	0.20	2.0
Selenium (Se)	0.02	0.02
Vanadium (V)	0.1	1.0
Zinc (Zn)	2.0	10.0

* Recommended maximum concentration for irrigating citrus

Many of the organic chemicals found in wastewater are stable and persistent in the environment. Most of the compounds in question are either recognized carcinogens or carcinogenic promoters. A number of them have been demonstrated to be mutagenic or otherwise affect health. They include chlorinated alkanes, chlorinated phenols, pesticides and others. Pesticides and herbicide residues on crops are of concern and have received considerable attention (WHO, 1977). Both have been detected in wastewaters. However, the health hazard associated with these particular compounds is far more likely to arise from the uncontrolled use of such materials in irrigated areas than from the reuse of wastewaters containing them.

Similarly, the complex organic chemicals referred to, if present in wastewater, may well, due to their persistence, be present in other water sources used for irrigation. Careful monitoring of crops and the identification of the sources of such contamination is called for. When considering the reuse of wastewater for irrigation, studies of the industries contributing to the wastewater, their potential development and the degree of trade effluent control that exists, will rapidly identify the scale of this problem.

Exposure to infection

Clearly it is the role of responsible authorities to minimize the health hazards associated with reuse of wastewater and subsequent sections will deal with this in more detail. However, it is equally important for the individual at risk to take what precautions he or she can against infection. In order to do this they must be aware that the risk exists, and understand the importance of following guidance they are given. Health education programmes and careful instruction of field workers should therefore form part of wastewater reuse projects.

Various techniques are employed in the transfer of water from the bulk source to crops in the field. When wastewater is used for irrigation these techniques can have a direct effect on transmission of some diseases. Field workers are directly at risk and they may transfer infection to their families. The use of wastewater for the irrigation of amenities such as parks and lawns obviously puts the general public at risk.

Direct skin contact with unsafe wastewater creates a health hazard in areas where certain helminthic diseases are common. The free swimming cercariae of schistosomiasis can penetrate healthy skin exposed to irrigation water, while the larvae of hookworm developed on warm moist soils can penetrate the skin, usually through the foot. Irrigation procedures that minimize the direct contact with water for field workers and the provision of footwear will obviously reduce the incidence of these diseases. However, there are very real practical difficulties associated with this approach, and the provision of treated irrigation water free of such larvae must accompany these measures.

The accidental ingestion of unsafe irigation water cannot be excluded, although to date it does not seem to have played a major role in disease transmission. However, if standards of personal hygiene are low (for example, hands are not washed before eating) infections may occur.

This mode of transmission of pathogens may be more of a risk when the management of wastewater used for the irrigation of amenity areas, such as parks, lawns and roadside trees and grass verges, is poor. Piped supplies must be clearly labelled as unsafe and careful precautions taken to prevent cross-contamination with potable supply pipelines.

When spray or sprinkler irrigation is used, small droplets (aerosols) are formed and these could contain enteric bacteria and viruses if the wastewater used is poorly treated. Many of these pathogens are more infective when inhaled than when ingested and so spray irrigation has a great potential to disseminate enteric bacteria and viruses. Harmful chemical substances, if present, will also be inhaled with the droplets. Field workers are clearly at risk but the effect of wind in carrying aerosols considerable distances may also put people in nearby housing at risk. Aerosol transmission of enteric micro-organisms over distances of 1.2 km and more have been quoted (WHO, 1979).

Having considered the health risks to those who work on or live near irrigation schemes, it is necessary to consider the hazards to people who handle, prepare, or eat the crop. Pathogen survival on crops was briefly discussed earlier. Generally, exposure to sunshine and desiccation will reduce these survival times considerably. However, they remain long enough to permit the transport of pathogens via crops into markets, food processing factories and the home. Evidence given by Cohen and Davies (1971) relating to the cholera outbreak in Jerusalem in 1970 graphically makes this point.

Whether or not the pathogens become attached to the surface of crops depends upon the method of irrigation and type of crop. Crops grown on or near the ground are almost certain to become contaminated, as are crops grown under spray irrigation. Subsurface irrigation, drip or trickle irrigation and furrow irrigation can all be used to limit the contamination of crops. Alternatively, wastewater may only be applied prior to planting, or application may be discontinued 2 weeks before harvesting in the hope that all pathogens will die before the harvest. None of these techniques are likely to be effective if untreated wastewater is applied to fields.

In addition to the direct transmission of pathogens to man from crops, there are a few diseases that develop in animals and may be transferred to man when poorly cooked or raw meat is consumed. The use of wastewater to irrigate pastures or fodder crops may contribute to an increase in the incidence of these, unless adequate treatment of the wastewater is provided.

Chemical contamination of crops as a direct result of irrigation with treated wastewater effluent is likely to be principally associated with the possible uptake of trace elements, notably heavy metals.

As outlined earlier, the WHO (1974) guidelines for drinking water quality have suggested guideline values based on acceptable daily intakes (ADI) of toxic substances. Certain toxic elements, such as cadmium, may accumulate in crops. It is therefore important to study carefully the quality of the wastewater effluent and of the crops produced to ensure there is no substantial increase in the intake of toxic materials. This may be particularly important where the crop is a major component of the staple diet of the community.

Enhanced routine surveillance of the quality of crops grown for human consumption on land irrigated with wastewater effluent would seem worthwhile.

Setting guidelines for reuse of wastewater in irrigation

In 1973, WHO published a technical report on the reuse of effluents and methods of wastewater treatment and health safeguards. The document was prepared by leading researchers of the time and deals with the whole range of intentional reuses of wastewater, including treatment to produce potable water.

In assessing the potential health effects associated with reuse in agriculture, the vast majority of the research cited dealt with the survival of pathogens in wastewater, in the soil or on crops. Very little reliable epidemiological evidence was available with regard to the reuse of wastewater and related disease. The exceptions to this were a number of reports indicating a direct relationship between the use of untreated sewage in agriculture and the outbreak of the disease. As a result of this evidence the report recommends that any wastewater used in agriculture should be treated and suggests, as a minimum, primary treatment (essentially a sedimentation process providing between 2 and 4 hours retention) should be given to wastewater used for irrigation of crops not intended for direct human consumption – in other words, restricted irrigation.

Restricted use of treated wastewater for irrigation can create problems for farms where crop rotation is practised and where mixed fodder and vegetable crops are grown. Apart from the obvious management problems it creates, it assumes a level of health education and standards of personal hygiene among field workers that may not exist. The expert group, therefore, suggested that a very high degree of treatment and disinfection was needed if unrestricted use of wastewater for irrigation was intended.

Their approach was to recommend a bacteriological standard of not more than 100 coliform organisms/100 ml in 80% of samples, since this had been demonstrated to be technically feasible. It should be possible to achieve this by heavy chlorination (15–20 mg/l of chlorine with contact periods of 1 to 2 hours) of wastewater effluent from treatment processes that achieve biological stabilization. These include activated sludge processes, biological filtration and waste stabilization ponds.

The proportional reduction of coliform indicator organisms is believed to be representative of the bacterial pathogen reduction and some pathogens are known to be more sensitive to disinfection by chlorine. Such heavy chlorination as proposed by the report should, in addition, inactivate protozoan cysts and reduce the number of enteric viruses, though not to the same extent as the reduction of bacteria. Helminth ova receive relatively little attention in the report, except that note is made that some may be removed by primary sedimentation and that waste stabilization pond treatment of 5 to 7 days retention is likely to be more effective. In some circumstances, presumably where helminth infections are endemic, it suggests that sand filtration or equivalent polishing methods may be required.

There is clear concern, expressed by those responsible for the technical report, as to the lack of epidemiological evidence. Without this research evidence they could do little more than recommend guidelines based on pathogen survival and the efficiency of treatment processes in removing them.

Since 1973 more epidemiological studies have been undertaken to evaluate and attempt to quantify the health aspects of wastewater and excreta use in agriculture and aquaculture. Spurred on by the International Decade for Drinking Water Supply and Sanitation, a considerable amount of fundamental research has been done on water-related infections, their transmission and environmental control. In order to evaluate this new material properly, the World Health Organization, the World Bank, UNDP and IRCWD have sponsored major reviews to re-evaluate the state-of-the-art of this subject. The evidence from these was brought together in July this year at a meeting in Engelberg, culminating in the Engelberg Report (IRCWD, 1985).

The report suggests that some of the earlier approaches, including TR 517, require fundamental revision. In order of importance it attaches greatest concern to helminthic pathogens, specifically intestinal nematodes (*Ascaris, Trichuris*, and the hookworms), followed by excreted bacterial infections and finally excreted viral infections. To quote from the synopsis of the Report:

> The first quality criterion for both excreta and wastewater use in agriculture is the complete, or almost complete, removal of the eggs of intestinal nematodes (to a geometric mean of ≤ 1 viable nematode egg per litre). A major reduction (to a geometric mean of $\leq 10^3$ faecal coliforms per 100 ml) in the concentration

of excreted bacteria is recommended for unrestricted use of wastewater in agriculture. If these standards are met other pathogens, such as trematode eggs and protozoal cysts, are also reduced to undetectable levels. For wastewater, the required degree of purification is assured by a waste stabilization pond system of 4–5 cells and an overall retention time of 20 days. For excreta and wastewater use in aquaculture, it is possible that less stringent standards are acceptable.

The tentative microbiological quality guideline for treated wastewater reuse in agricultural irrigation given in the Engelberg Report is reproduced in *Table 5.4*.

Table 5.4 Tentative microbiological quality for treated wastewater reuse in agricultural irrigation[1]

Reuse process	*Intestinal nematodes[2] (geometric mean number of viable eggs per litre)*	*Faecal coliforms (geometric mean number/100 ml)*
Restricted irrigation[3]		
Irrigation of trees, industrial crops, fodder crops, fruit trees[4] and pasture[5]	$\leqslant 1$	Not applicable[3]
Unrestricted irrigation		
Irrigation of edible crops, sports fields and public parks[6]	$\leqslant 1$	<1000[7]

Source: IRCWD (1985)

[1] In specific cases, local epidemiological, sociocultural and hydrogeological factors should be taken into account, and these guidelines (which are based on an epidemiological evaluation of actual, rather than potential, health risks) modified accordingly.
[2] *Ascaris, Trichuris* and hookworms.
[3] A minimum degree of treatment equivalent to at least a 1 day anaerobic pond followed by a 5 day facultative pond or its equivalent is required in all cases.
[4] Irrigation should cease 2 weeks before fruit is picked, and no fruit should be picked off the ground.
[5] Irrigation should cease 2 weeks before animals are allowed to graze.
[6] Local epidemiological factors may require a more stringent standard for public lawns, especially hotel lawns in tourist areas.
[7] When edible crops are always consumed well-cooked, this recommendation may be less stringent.

The Report places great store on the value of waste stabilization ponds for the treatment of wastewater in tropical and subtropical countries. However, the extensive land requirements of ponds may limit their use and alternative treatment processes should be investigated with regard to their efficiency at achieving the microbiological guidelines proposed. The introduction of a guideline value for helminth eggs is both new and set at a stringent level. While research techniques are available for egg counts, the Engelberg group recognizes that it is imperative to 'develop as soon as possible a simple standardized test to quantify the concentration of viable eggs in all types of treated excreta and wastewaters'. It is perhaps a measure of their concern about the problem of helminthic disease associated with wastewater reuse that they propose such a criterion without the means being currently available to monitor it.

The Engelberg Report does not consider the reuse of sludges from municipal or industrial wastewater treatment processes, or chemical contaminants, nor reuses of wastewater and excreta other than in agriculture and aquaculture. The proposals contained within it are tentative and further research is still required. However, it is hoped that the Report will form the basis for a fundamental review of guidelines.

Treating wastewater for irrigation

The previous section discussed the guidelines that might be considered appropriate for specific irrigation reuse schemes. Both documents referred to suggest appropriate treatment processes for the guideline criteria they propose. While a detailed discussion of treatment processes is beyond the scope of this paper, there are a number of issues related to treatment that have a direct bearing on the health risk associated with reuse projects.

First, there is the need for reliability. A process that can only achieve its quality objectives when working at peak efficiency is unlikely to provide the necessary reserve of safety normally expected in the public health sector. Simplicity of operation, minimal need for operator intervention, simple maintenance and low operating costs are all likely to enhance reliability.

Second, the design of wastewater treatment plants has usually been based on the need to reduce organic and suspended solids loads to limit pollution of the environment. Pathogen removal was very rarely considered an objective. For reuse of effluents in agriculture this must now be of primary concern and processes should be selected and designed accordingly.

Finally, as with field workers and irrigation managers who will be using wastewater effluent, there is the need for training or retraining of treatment plant operators and supervisors so that they understand the new objectives for the treatment processes they control.

Conclusions

With adequate planning control and a complementary programme of training and health education, the reuse of treated wastewater for irrigation can be of substantial benefit to the community, particularly where natural water resources are limited.

Control should include a surveillance programme of the quality of raw wastewater and of the wastewater effluent used for irrigation. Crops should also be monitored, although the frequency of this will depend upon the potential for contamination from the wastewater supplied. It may also be necessary to include groundwater in the surveillance. Applying wastewater for irrigation gives much greater opportunity for infiltration, and thus the potential exists for contamination of the groundwater resource. In turn, this may threaten boreholes or wells used for potable supplies.

Waste stabilization ponds appear to be the most efficient treatment in terms of pathogen removal (see Chapter 8). However, the amount of land required is substantial and other treatment processes may have to be used. There is a need to evaluate these processes and particularly to identify the optimum techniques for upgrading existing treatment works to fulfil their new role.

Finally, at the planning stage there is the need for a sound, thorough background epidemiological study so that the impact of a wastewater reuse project on the health of the local community can be adequately assessed.

References

BRADLEY, D. J. (1978) Health aspects of water supplies in tropical countries. In *Water, Wastes and Health in Hot Climates,* edited by Feachem, R., McGarry, M. G. and Mara, D. D. Chichester: John Wiley

COHEN, J. and DAVIES, A. M. (1971) Epidemiological aspects of cholera El Tor outbreak in a non-endemic area. *Lancet,* **2 (7715),** 86

FAO (1979) *Irrigation and Drainage Paper No. 29, Water Quality for Irrigation.* Rome: FAO

IRCWD (1985) *Engelberg Report: Health Aspects of Wastewater and Excreta Use in Agriculture and Aquaculture.* Dubendorf, Switzerland. International Reference Centre for Wastes Disposal

PICKFORD, J. (1978) Water treatment in developing countries. In *Water, Wastes and Health in Hot Climates,* edited by Feachem, R., McGarry, M. G. and Mara, D. D. Chichester: John Wiley

SCHAEFER, W. J. (1985) *Health Aspects of Reuse of Treated Wastewater for Irrigation.* Background Document to Inter-Country Seminar on Health Aspects of Wastewater Reuse. Bahrain: WHO, EMRO

SHUVAL, H. I. (1977) Health considerations in water renovation and reuse. In *Water Renovation and Reuse,* edited by Shuval, H. I. New York: Academic Press

SHUVAL, H. I. (1981) Parasitic disease and wastewater irrigation. In *Sanitation in Developing Countries,* edited by Pacey, A. Chichester: John Wiley

WHO (1972) *Health Hazards of the Human Environment.* Geneva: WHO

WHO (1973) *Technical Report Series No. 517, Reuse of Effluents: Methods of Wastewater Treatment and Health Safeguards.* Report of a WHO Meeting of Experts, Geneva

WHO (1977) *Technical Report Series No. 612, Pesticide Residues in Food.* Report of the 1976 Joint FAO/WHO Meeting, Geneva

WHO (1979) *Technical Report Series No. 639, Human Viruses in Water, Wastewater and Soil.* Report of a WHO Scientific Group, Geneva

WHO (1984) *Guidelines for Drinking-Water Quality,* Vol. 1. Geneva: WHO

WHO-EURO (1981) *Report and Studies No. 54, The Risk to Health of Microbes in Sewage Sludge Applied to Land.* Report on a WHO Working Group, Copenhagen

WHO-EURO (1983) *Environmental Health Impact Assessment of Irrigated Agricultural Development Projects: Guidelines and Recommendations.* Report prepared for Regional Office for Europe.

Chapter 6

Environmental hazards of sewage and industrial effluents on irrigated farmlands in Japan

Toyoaki Morishita*

Introduction

The main islands of Japan are located on the rim of the Asian continent and have plentiful precipitation throughout the year, averaging between 1000 mm in low rainfall areas and 3000 mm in the higher rainfall areas. The quality of fresh water is, in general, much better than in many other countries, particularly in arid or semi-arid regions, because of the shorter flow distance of the rivers. The steeper slope of the river beds means faster flowing streams, which is the reason for the high quality of the water, although it does cause repeated floods in wide areas of the basins.

From the historical point of view, control and management of running river water has been supported by the farmers who used the river water for irrigated rice farming in the flood basin. No problems arose from their plentiful use of high quality water for irrigation. It is only in the last two or three decades that competition for this limited resource has developed among farmland irrigation, industrial water use and municipal water supply, and that the farmer has been requested to use irrigation water more efficiently. A proof of this may be that the money allotted for irrigation canal management is, even now in Japan, based not on the amount of irrigation water used but only on the area of farmland irrigated.

Table 6.1 Ministry approved water quality standards for rice irrigation in Japan

Item	pH	COD	SS	DO	T–N
Range	6–7.5	6 ppm	100 ppm	5 ppm	1 ppm
Item	EC	As	Zn	Cu	
Range	0.3 S	0.05 ppm	0.5 ppm	0.02 ppm	

Reflecting this particular situation in Japan, very high quality standards have been recommended for irrigation water, as shown in *Table 6.1*. This water quality standard, which was approved by the Ministry of Agriculture, Forestry and Fishery in Japan, is only a desirable guideline for rice irrigation and places no legal restriction for irrigation use. Another example is an upland farm irrigation project

* Institute of Applied Biochemistry, University of Tsukuba, Japan

in the Kasima district where brackish water flows into the canal water in the project area. The electrical conductivity and the chloride concentration were recommended not to exceed 0.5 siemens (S) and 150 ppm Cl^{-1}, respectively. These recommendations are based on criteria for 0% yield decrement for susceptible crop plants. The values are quite strict but ensure a better quality for irrigation water than in many other countries.

The use of sewage effluent for crop irrigation could become common practice in water-short areas. However, as long as the principle of 0% yield decrement for susceptible crops is kept (and it will be for at least some time), direct use of sewage or industrial effluents for crop irrigation will not be considered in the present situation in Japan. As a result there is only limited interest in this subject even at the research and experimental levels.

Meanwhile, the recent rapid growth of industrialization and urbanization has brought serious water pollution hazards to several areas in Japan. In fact, it is true that the water quality standards mentioned above have not been satisfied in numerous areas where there is now advanced water pollution, and the actual quality of the water in these areas is far lower than these quality standards. The following *Tables 6.2* and *6.3* show the actual situation regarding the water pollution hazard in Japan.

Table 6.2 Proportion of pollution-hazard affected area in representative river basins in Japan

River basin	*Total area (T)* (ha)	*Affected area (A)* (ha)	*A/T %*	*Pollutant*
Ishikari	135 114	9 494	7.0	SS (paper mill)
Kitagami	112 234	3 216	2.9	
Tone	226 180	9 822	4.3	
Shinano	211 505	9 542	4.5	
Jinzu	7 567	1 323	17.5	Heavy metals
Tenryu	40 615	103	0.3	
Yasaku	18 230	10 908	59.8	N and organic SS
Kiso	28 843	11 924	41.3	N and organic SS
Yodo	36 882	7 910	21.4	N and organic SS
Yoshino	12 051	1 898	15.7	
Onga	12 152	3 231	26.6	SS (coal mine)
Kuma	13 944	17	0.1	
Tikugo	43 349	961	2.2	
Total surveyed	1 682 473	119 697	7.7	

Table 6.3 Range of water quality in pollution-hazard affected areas in Japan

Item	*Quality standard*	*Range in polluted water*
pH	6.0–7.5	3.1–9.2
COD	below 6 ppm	0.5–53.6 ppm
DO	above 5 ppm	0.6–15.1 ppm
SS	below 100 ppm	0.0–147 ppm
T–N	below 1 ppm	0.0–50.7 ppm
EC	below 0.3 S	0.03–4.44 S

The data in *Table 6.2* from the survey by the Agricultural Structure Improvement Bureau indicate that the areas damaged by water pollution in irrigated farmlands in the main Japanese river basins occupy 7.1% of the total, ranging between the least at 0.1–0.3% in the Kuma and Tenryu rivers, to the largest at 41.3–59.8% in the Kiso and Yasaku rivers. Since this survey excluded most of the more heavily polluted small streams running through urban areas, the most recent serious situation of water pollution in suburban areas in Japan may not be revealed by this table.

Before the 1960s the main cause of water pollution in irrigated farmlands in Japan was wastewater drainage from various industrial factories, such as mining, metal smelting, paper milling, food processing, chemical industry, etc. With recent progressive urbanization the rapid change in lifestyle in rural areas, strengthening of the legal restrictions and improvement in wastewater disposal from these factories, domestic wastewater including some raw sewage and sewage effluent has taken the place of the former major pollution source, which was industrial wastewater. *Table 6.4* clearly shows this drastic change.

Table 6.4 Increasing areas and changing pattern of water pollution hazard in irrigated farmlands in Japan

Year	*Number of cases*	*Affected areas* (ha)	*Proportion of areas by pollution sources* (%)			
			Mine	*Factory*	*Domestic*	*Others*
1958	304	99 164	44	46	46	10
1965	898	126 711	27	54	15	4
1970	1526	194 322	16	39	34	11
1975	1349	157 325	13	22	59	6

Source: Survey Data. Agricultural Structure Improvement Bureau (1976)

In the 1965 survey, 27% of the pollution hazard in the irrigated farmland came from mining waste, 54% from factory waste and only 15% from domestic wastewater. After only 10 years, domestic wastewater accounted for 59% of the total, and the proportion of the mining and factory waste came down to 13% and 22%, respectively. Furthermore, about 70% of the newly added pollution hazard in 1975 was accounted for by domestic wastewater.

As indicated above, drainage from domestic wastewater into the irrigation water is responsible for the major part of the present pollution in the irrigated farmlands all over Japan. The future prospect is that sewage and industrial effluents will probably have to be used for farm irrigation in order to use the limited water resources more efficiently.

Before considering the beneficial effects of sewage application on farm crops, it is essential to evaluate strictly every probable environmental hazard to surface and groundwater, soil, atmosphere, plant, animal and human life, and to make use of this wastewater for irrigation only under proper control and management. Japanese experiences with serious water pollution should provide much valuable information and possible solutions.

The following sections of this chapter will discuss two major topics as they relate to experience in Japan: first, research projects on the land application of sewage and sewage effluent; and second, particular features of nitrogen and cadmium pollution in the irrigated farmlands.

Research projects on land application of sewage and sewage effluent in Japan

The principal strategy for water pollution control in Japan is the basin–unit sewage disposal system, where the collected sewage and industrial wastes from the greater part of a certain basin is treated in a centralized large-scale sewage disposal plant located near the river mouth. Such a system inevitably introduces two major conflicting problems: inefficiency and high cost of sewerage in the less populated rural areas; and inefficient utilization of limited water resources.

The major research projects on land application of sewage or treated effluent are conducted from the viewpoint of being a countermeasure for these contradictions. The two major research projects going on in Japan at present illustrate this point: they are (1) farmland application of sewage aiming at more efficient disposal of the wastewater in the rural area, conducted by the Ministry of Agriculture; and (2) forestland application of sewage effluent for recharge of groundwater, by the Ministry of Construction. The components of these two experimental projects are given in *Table 6.5*.

Table 6.5 Components of the research projects on land application of sewage effluent in Japan

(1) *Studies on the method of sewage treatment in rural areas*
(by the Ministry of Agriculture)

(a) Wastewater disposal in the rural area

(i) Pattern and source of analysis of waste disposal
(ii) Water use analysis
(iii) Traditional disposal method
(iv) Structure analysis of the rural community

(b) Particular treatment method for a rural community

(i) Land treatment by using the purifying ability of soil
(ii) Another treatment method without soil

(c) Land application technique

(i) Individual technical factors for land application
(ii) Effect on the properties of irrigated soil
(iii) Effect on the irrigated crop plant

(d) Establishment of the total design

(2) *Studies on forestland application of sewage effluent*
(by the Ministry of Construction)

(a) Effect of sewage application on the properties of forest soil
(b) Effect on growth and composition of tree and herb plants
(c) Effect on soil fauna
(d) Effect on the surrounding environments
(e) Behaviour analysis of the nitrogen in soil lysimeter
(f) Identification of the prediction model on flow and quality of water

Previously, in traditional rural communities in Japan, at least before the 1960s, most of the nightsoil and wastewater was returned for natural recycling in the ecosystem of the area, and the natural purifying capacity held the situation in control. The nightsoil was applied to farmland as a major nutrient source for crops, and the wastewater was dispersed to farmland or land around the farmer's own house. In many cases direct discharge of these wastes into the streams was prohibited by common law in the communities. The breakdown of the rural communities, the changing pattern of nightsoil disposal, the rapid increase in water use for daily life, all of these changing patterns in the lifestyle of rural communities have caused serious water pollution in the small streams which used to be kept clean and were used for rice irrigation without any trouble. The advanced state of pollution and the increasing breakdown in the sense of moral responsibility in the community is causing further water pollution.

One of the most effective measures would be to provide the whole of such areas with a complete sewerage system, but it would take a very long time and a vast amount of funds. This strategy of basin–unit sewerage systems adopted in the 1970s of rapid economic growth has turned into a heavy burden in the 1980s of steady growth. Some trial projects have been started to find a properly modified disposal system, especially for domestic waste in rural areas.

Furthermore, the programme for separating water courses between water for reuse and drainage water in this basin–unit sewerage system should also be effective for pollution control, but direct draining of the surface water into the sea, on the other hand, would diminish the potential use of the wastewater after purification by natural cyclic turnover.

Although Japan has sufficient per-area precipitation, as mentioned earlier, the per caput precipitation is now insufficient. In the former traditional wastewater disposal system, surface water was used again through one or other recycling method. The aim at present is to revive the use of forestland application of sewage effluent, but in a more sophisticated system, so that cyclic use can be made of the limited water resources.

A general conclusive evaluation of these research projects requires a greater accumulation of experimental data, particularly on the critical conditions in order

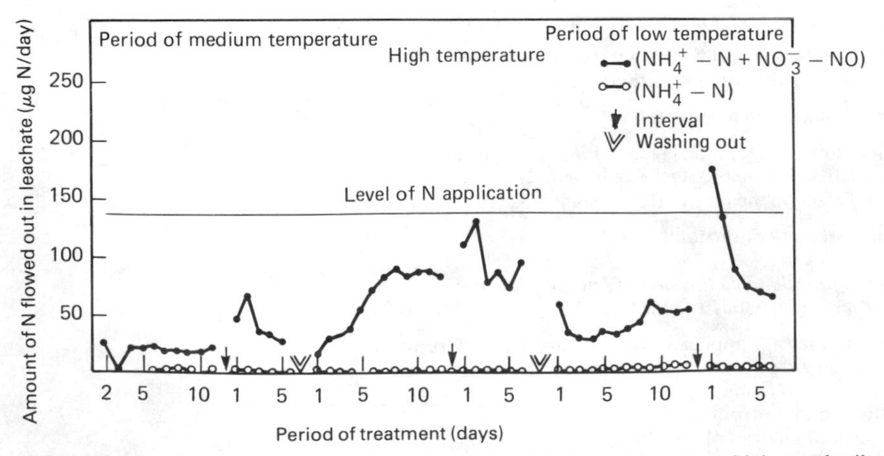

Figure 6.1 Changing flow pattern of NH_3–N and NO_3–N in the leachate from high speed soil treatment tank filled with the Mure soil

to avoid any kind of irreversible environmental damage. Major attention is being paid to chemical environmental hazards in these projects, and considerable data are now available on the water quality in the surrounding areas. Results of the field surveys and laboratory experiments probably mean that the critical pollutant from these sewage applications is the high level of nitrate nitrogen, ranging from 10–50 ppm N found in the surface and groundwater. *Figure 6.1* represents a typical pattern of chemical pollutants in the soil column.

Organic and ammonium nitrogen, phosphate, chemical oxygen demand (COD) materials and other pollutants in the sewage were effectively removed, except for nitrate nitrogen, and a low level of them was found in the leachate from the soil column. On the contrary, most of the nitrate nitrogen, including that generated by nitrification of organic and ammonium nitrogen, found its way into the leachate, and only a limited portion, for example in our experiment 26–36% of the total nitrogen, can be estimated to be removed by denitrification in the soil column.

Thus, although soil retention of the organic matter and ammonium nitrogen would provide enough time for their complete decomposition and nitrification in the soil, rapidly percolating nitrate nitrogen has only a poor chance of denitrification, mainly because of a lack of necessary organic carbon sources.

Particular features of nitrogen and cadmium pollution hazards

Among the chemical pollutants in sewage and industrial effluent, the most critical pollution hazards in irrigated famlands are nitrogen and cadmium. As mentioned earlier, excess supplies of nitrogen and cadmium to irrigated farmlands form two of the most important problems facing Japan recently. In this section some particular features and the critical conditions for these two pollutants will be investigated, based on field and laboratory experiments.

Excess supply of nitrogen in irrigated farmlands

Rice cultivation in Japan, where rice grows under submerged conditions, requires large amounts of irrigation water. The approximate figures for total water use are estimated as 1500–2000 mm on average; daily water depth decrement × days of irrigation (around 100–150 mm × 100 days = 1000–1500 mm) plus some additional amounts for seedling transplantation (about 500 mm). Accordingly, 5 ppm N in the irrigation water, for example, corresponds to around 75–100 kg N/ha of nitrogen

Table 6.6 Grade evaluation of irrigation water in relation to N concentration and effects on rice

Grade	*Range of N concentration*	*Effect on growth and yield of rice*
0	Below 1 ppm N	None
I	1–3	Slightly overluxuriant growth
II	3–5	Overluxuriant growth with occasional yield decrement
III	5–10	Appreciable yield decrement
IV	Above 10	Serious yield decrement

Source: Tokyo Municipal Agricultural Experimental Station (1967)

supplied, which is nearly equal to the standard application rate of fertilizer in Japan. Thus, irrigation with the nitrogen-enriched polluted water would supply a considerable excess amount of nutrient nitrogen to the growing rice plants, and result in a significant yield loss of rice through lodging, failure to ripen, and increased susceptibility to pests and diseases through over luxuriant growth.

A graded evaluation of irrigation water was given by the Tokyo Municipal Agricultural Experimental Station in 1967 and is listed in *Table 6.6*. The Ministry-approved water quality standard of below 1 ppm N just corresponds to the 0 grade, no detectable effect and 0% yield decrement for rice in this evaluation. A low level of nitrogen supplied by the nitrogen-enriched irrigation water would, in certain conditions, give some benefit by partially replacing applied fertilizer. However, the total evaluation is negative, at least in the recent situation in Japan, because it would be impossible to control the supply of nitrogen and accompanying organic waste and it would not be possible to control the growth pattern of the rice plant due to fertilizer. It would also cause harmful reduction of the paddy soil.

Particular features of the pollution hazard from cadmium

Repeated application of wastewater to farmland soil may contribute a significant quantity of heavy metals, such as mercury, copper, cadmium, and some others, that accumulate in soils. It is well recognized that these heavy metals tend to accumulate at the surface of the soil, and that a significant amount might enter the food chain through crops grown on farmland receiving wastewater.

Judging from its serious hazard to human health, its easy incorporation in the food chain and its general biotoxicity, cadmium is the most critical element. In fact, the greatest attention has been paid to cadmium contamination of rice grown on polluted paddy fields in Japan. Particular features of the cadmium pollution hazard in the soil–rice plant system are as follows: (1) the high absorption capacity of cadmium by the rice plant and the high tendency of translocation into the grain part of the crop; and (2) the tendency to accumulate and remain in the surface layer of soil.

Table 6.7 Absorption percentage of heavy metals by one cropping of rice plant

Metal	*Range (%)*	*Average (%)*
Zn	0.14–1.10	0.35
Pb	0.19–0.43	0.29
Cu	0.14–0.74	0.33
Cd	0.26–5.67	1.09
Hg	0.006–0.089	0.032

Source: Uptake of rice/metals in surface soil (%). From survey data in the Nifu and Annaka districts by T. Morishita

Cadmium is highly soluble in the soil solution, as reflected by the greater solubility constant of $Cd(OH)_2$, and higher absorption percentage by the rice plant which has been proved by field and greenhouse studies. *Table 6.7* gives a comparison of the absorption percentage by the rice plant of various heavy metals. The figures listed in the table, except for mercury, were obtained from a field survey in the Annaka district around a zinc smelter. Mercury is well known for its high biotic toxicity, but has a very low absorption percentage by rice. Only cadmium was shown as having a very high absorption percentage, ranging from 0.26 to 5.67% and with a maximum translocation percentage of 5–7%.

The solubility of Cd in soil and the uptake by the rice plant vary greatly, depending on the redox potential of the soil through reversible conversion between CdS and $Cd(SO_4)_2$. Conditions encouraging uptake by the plant may be consistent with those for translocation of Cd into the rice grain. Consequently, under these conditions, a very high proportion of soil Cd, sometimes in the range of 0.2–0.4%, was found in the brown rice grain.

The critical soil concentration of cadmium is also particularly low and the empirical potential criteria of soil Cd that will produce polluted rice grains were estimated in the range of either double or five times in natural conditions, as shown in *Table 6.8*. This means that non-polluted soil, having around 0.4 or 0.5 ppm Cd, may produce about 0.08 ppm Cd in brown rice, and only a little increase up to 0.82, 1.25, or 2.1 ppm of soil Cd has the potential to produce heavily polluted brown rice with 1.0 ppm Cd in any of the districts.

Another particular feature of Cd is its ability to remain in the surface soil. As shown in the irrigation–leachate equilibrium presented in *Figure 6.2*, most of the Cd in irrigated water, above 90% for 10 ppb and 98% for 100 ppb, is probably absorbed and remains in the soil above the turning point of equilibrium where the

Table 6.8 Empirical potential criteria of soil Cd in polluted brown rice (ppm Cd/dried soil)

District	*Potential criteria of Cd concentration of soils*	
	0.4 ppm Cd brown rice	*1.0 ppm Cd brown rice*
Fuchu, Toyama Prefecture	0.72	0.82
Kurobe, Toyama Prefecture	0.74	1.25
Annaka, Gunma Prefecture	1.3	2.1

Source: From field survey data by T. Morishita and T. Morishita and H. Anayama

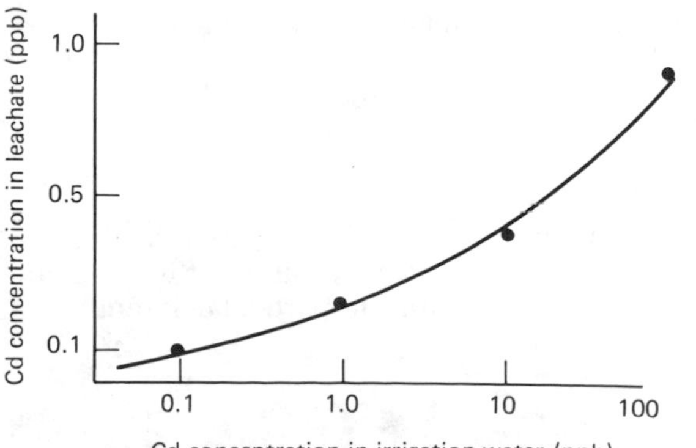

Figure 6.2 Schematized equilibrium of cadmium concentration between irrigation water and leachate in paddy soil (by T. Morishita)

leachate Cd become equal to the irrigation Cd. The turning point of 0.1 ppb Cd nearly corresponds to the natural background in river water.

An approximate evaluation of the balance sheet of Cd in an irrigated paddy rice field is given in *Table 6.9*. These figures were obtained by field survey and experiment in the Fuchu district in Toyama Prefecture. Without any supply of Cd from artificial sources, other than phosphate fertilizer, the balance is hardly maintained between the incoming and the total outflow. It is probable that application of fertilizer could increase the content of soil Cd because the rice stubble is usually not removed and remains in the field. Any additional supply of Cd through polluted irrigation water would cause a significant increase in soil Cd in the irrigated farmland, even at very low concentrations.

Table 6.9 Approximate estimation of cadmium balance in paddy rice field

Item	*Calculation*	*Amount of cadmium*
Incoming	Amount Cd concentration	(g/ha)
Irrigation	45 000 t × 0.16 ppb	7.2
Fertilizer	1.5 t × 8.4 ppm	12.6
Total		19.8
Outflowing		
Rice grain	5.2 t × 0.007 ppm	0.4
Straw	6.3 t × 0.63 ppm	3.9
Stubble	1.4 t × 4.9 ppm	6.9
Leachate	45 000 t × 0.10 ppb	4.5
Total		15.7

Source: From field survey and experiment by T. Morishita and H. Anayama

Under existing environmental quality standards in Japan it is recommended that Cd in river water should not exceed 10 ppb Cd. Approximate estimation gives the prediction that continuous irrigation with water containing 10 ppb Cd for one cropping of rice would cause an increase of 0.1 ppm Cd in the surface soil (0–15 cm). This suggests that heavily polluted rice with more than 1.0 ppm Cd could be produced only after several years' application of the river water if the environmental quality standard approved by the Japanese Environment Agency is maintained.

Conclusions

It is recommended that, because of the practical and financial difficulties in reclamation of polluted field soil, the use of sewage and industrial effluent should proceed under careful control and management, with the particular intention of avoiding irretrievable pollution by heavy metals.

Acknowledgements

The author wishes to express his sincere thanks to Professor Dr T. Yoshida and Y. Ohta of the University of Tsukuba for their kind and useful suggestions. He is also grateful to many Japanese scientists and official staff working on this subject, who have produced official periodicals which were very useful for writing this paper, but most of them have had to be omitted from the listed references because they were written only in Japanese.

Bibliography

FAO (1976) *Water Quality for Agriculture.* Irrigation and Drainage Paper 29, Food and Agriculture Organization of the United Nations, Rome

IIMURA, K. (1981) Heavy metals in soils. In *Heavy Metal Pollution in Soils of Japan,* edited by Kitagishi, K. and Yamane, I. Tokyo: Japan Scientific Societies Press (in English), pp. 19–50

IIMURA, K., ITO, H., CHINO, M., MORISHITA, T. and HIRATA, H. (1977) Behaviour of contaminant heavy metals in soil-plant system. *Proceedings of the International Seminar on Environmental and Fertilizer Management in Intensive Agriculture.* Tokyo, Japan (in English), pp. 357–368

MORISHITA, T. (1981) The Jintsu River Basin: contamination of soil and paddy rice with cadmium discharged from Kamiokia mine. In *Heavy Metal Pollution in Soils of Japan,* edited by Kitagishi, K. and Yamane, I. Tokyo: Japan Scientific Societies Press (in English), pp. 107–124

MORISHITA, T., KISHINO, K. and IDAKA, S. (1982) Mercury contamination of soils, rice plants and human hair in the vicinity of a mercury mine in Mie Prefecture, *Japanese Journal of Soil Science and Plant Nutrition,* **28,** 523–534 (in English)

MORISHITA, T. and MINAMI, Y. (1983a) Retention and outflow of soluble nutrient salts in high speed soil treatment of wastewater. *Japanese Journal of Soil Science and Plant Nutrition,* **54,** 117–123

MORISHITA, T. and MINAMI, Y. (1983b) Removal of organic and nitrogenous components in high speed soil treatment of wastewater. *Japanese Journal of Soil Science and Plant Nutrition,* **54,** 199–204 (with English summary)

STUDY GROUP ON AGRICULTURAL WATER USE (n.d.) *Water Supply in Japanese Agriculture (Nihon no Nogyo Yousui).* Tokyo: Chikyu-sha (in Japanese), pp. 127–165

Chapter 7

Discussion

Mr A. T. Shammas, Regional Agricultural Water Research Centre, Saudi Arabia suggested that the risk and the size of the problem in reusing effluent in irrigation should not be exaggerated. For example, salinity and sodium problems could be solved. Effluent water should not be compared with the good quality water in temperate regions or humid regions. In the Near East, under arid conditions, effluent is often a better quality water than the irrigation water and underground water. From experiments in Saudi Arabia, use of effluent is reducing salinity, and reducing the sodium content of the soil throughout the profile. So, in the process of technology transfer to this part of the world, the specific local conditions must be taken into account.

Professor M. B. Pescod accepted the examples given but indicated that the same situation did not apply everywhere. Sometimes wastewater contains a salinity level which prevents it from being used in irrigation. This occurs now in Bahrain where, as a result of infiltration of saline groundwater into the sewerage system, they have a very high salinity wastewater. Although the natural groundwater that they now use for irrigation is high in salinity, because the effluent is a resource which the authorities are going to produce rather than a natural resource which the farmers have been used to abstracting and using, with whatever consequences, they do not want the salinity in the effluent to be greater than about 2000 mg/l. At the moment it is 3000 mg/l and so they are having to invest in improving the sewerage system to try and prevent infiltration of saline groundwater. That is their approach because treatment to adjust salinity is not normally economic. The suggestion of lime treatment being a possibility is for where there are likely to be only marginal effects on crops and, by adjusting particularly the sodium balance with other cations, lime treatment might prove to be cost-effective. Very often, however, the wastewater is of a quality which is better than the groundwater that is being used. That is part of the total consideration of the local situation in relation to the quality of the effluent, the crops that are going to be produced with it and alternatives in the form of irrigation water.

Mr A. Gur, FAO Wastewater Expert, Saudi Arabia asked why we have to worry so much about heavy metals as long as wastewater is of domestic nature? He also referred to statements on page 28 (Chapter 3) and asked if this meant that we should not trust the soil–aquifer treatment capacity which had been documented in chapters in this Seminar? He wondered how Professor Pescod compared lime treatment and stabilization ponds with conventional types of treatment, using activated sludge or trickling filters. Also, as Professor Pescod finds the California bacteriological standards rather luxurious, what would be his suggestion for the bacteriological quality of effluents for some countries he has visited in the Middle East?

Professor Pescod indicated that, in his chapter addressing the broad issue of guidelines for effluent quality, he had tried to cover all possibilities. Certainly, if you are talking about primarily domestic sewage then you would not expect heavy metals to be a problem, but as Dr Morishita indicated, in industrialized areas, or even in sewered cities where there has been little thought given to industrial effluent control, these can build up in the sewage. In Riyadh, at least in the industrial area which has its own treatment plant, there may be high heavy metal potential, depending on the regulations governing the control of industrial effluents. With purely domestic sewage, heavy metal problems would not be expected but looking at the full range of potential we have to make sure that heavy metals do not build up in the soil over a matter of 20–30 years of irrigation.

On the second point, the treatment capacity of the soil, the intention was to suggest caution when effluent reuse for irrigation was to be introduced. A knowledge of the local hydrogeology is essential to be able to attempt to predict what the likely removal will be under the actual conditions with which you are dealing. In other words, you should not just trust to data which has been reported for a particular site; you have to compare the site conditions and the types of soil. That is why the site-specific nature of all these decisions was stressed, because the soils will vary, some may be more permeable than others, some may have fissured characteristics which allow short circuiting to the water table, and so on. Unless you look into the hydrogeology you should not rely on the soil treatment characteristics that you might normally expect. A knowledge of the type of soil, and an understanding of the potential for treatment in the soil, is something that we need to take account of. Some of the later papers on recharge present some data on the actual treatment that can be expected. Normally we can rely on soil treatment but caution is necessary in the case of heavy metals, which will be removed and the levels will increase in that soil over time.

Lime treatment in combination with ponds was discussed because ponds are sometimes eliminated as a treatment process because they take up too much land. Although under normal circumstances a long retention time is required, if you can precede them by some type of treatment, and lime was suggested because studies on lime have been carried out at Newcastle, then ponds might become more competitive with conventional techniques. Lime treatment before more conventional secondary treatment might affect treatment in those processes because we are talking about a very high pH. In the case of ponds there is a long retention time for the pH to be reduced by the absorption of CO_2 from the atmosphere. At the moment, lime treatment with conventional biological treatment of waste is not recommended, because costly pH adjustment thereafter will be necessary, but it might be viable in combination with ponds.

Californian standards are quite rigorous and they are being adopted in the Near East. In Bahrain, for example, planning for when improvement to the sewerage system will allow them to reuse their effluent from the treatment plant, they are now investing in ozonation after tertiary treatment by rapid gravity sand filters and chlorination. That is a luxury approach which perhaps the economy of Bahrain can tolerate, but if you look at a country like the Sudan or India or Pakistan it is just not feasible. The Engelberg Report approach is making sensible suggestions but we must interpret these in the context of the system of irrigation being used. As engineers, we should be looking for utilization of quite low grades of sewage or partially treated effluent, and minimizing the contact or exposure to the public through the choice of irrigation technique, the selection of crops and the marketing of crops. We must not place the public at risk but at the same time we do not have to rely on high investment in treatment and, more important perhaps, assume the control of difficult treatment processes to be 100% effective, which is not feasible in most developing countries.

Dr S. Al-Salem, Jordan asked how much can be achieved in reduction of pathogens and in retention time by using lime treatment with waste stabilization ponds.

Professor Pescod suggested there would be a reduction in retention time as a result of the increased kill of pathogens through lime treatment but could not quantify this at present. The results at Newcastle would have to be compared with what you might expect to achieve in ponds without pretreatment. It is expected that pond retention could be reduced to something of the order of 10 days, whereas if you are talking about pathogen reduction, or faecal coliform reduction, to a low level, you are probably talking about 40 days in ponds. Everything depends on the quality of the sewage and the impact of the lime and ponds compared with the impact of the alternative treatment. This is an area where more research is required; we have really just started on lime treatment studies. It should also be realized that lime is not always available at reasonable cost in all countries; some countries have to import lime and that would tend to make lime treatment uneconomical.

Mr A. Al-Nakib, Kuwait referred to Professor Pescod mentioning that precautions should be taken when using sewage effluent for agriculture and asked what possibility there was of groundwater contamination from irrigation water. What is the possibility for the heavy metals and toxic materials to precipitate in the soil when we use tertiary-treated wastewater?

Professor Pescod presumed that Mr Al-Nakib was thinking of the type of tertiary treatment that had been installed in Kuwait, secondary treatment in activated sludge tanks followed by prechlorination and postchlorination and rapid-gravity sand filtration. This is, of course, the sort of tertiary treatment approach to effluent reuse which has been taken in the United States. If activated carbon adsorption is added then a full tertiary-treatment system is obtained and this would normally produce a very high quality effluent. Trace organics, that might be a problem elsewhere, would not really be a problem in Kuwait with the level of industry there. Tertiary-treated effluent coming from that system will be a high quality water which could probably be used for potable supply and, therefore, once it is applied as irrigation water, should it reach the groundwater table, it would certainly not be contaminating. That is, apart from whatever agricultural fertilizers or soil conditioners that it might pick up and leach through the soil.

What might be the level of heavy metals and toxics precipitation in the soil when using this tertiary-treated effluent? Heavy metals or toxics would not be expected to be present in the sewage in any great amount in the Kuwait situation because industry is not highly developed and will not produce those types of waste, although there may be problems in some parts of Saudi Arabia where the petrochemicals industry is starting to develop. In the future, that might be a concern in Kuwait but the Shuaiba Development Authority is certainly monitoring the situation, predicting what might happen and planning for controlling the system.

Dr G. B. Shende, India appreciated Professor Pescod's view about the site specificity of wastewater disposal on land because the soils differ, the climates differ and, at the same time, overemphasis on the quality of the wastewater to be applied to the land is not necessary because soil itself is a very powerful purifying agent when wastewater is applied to the land. The data now available on this subject is mostly from temperate climates. When it is applied to the warmer climates of developing countries we feel that we should not be so sceptical about wastewater quality because in India the survival rate of pathogens and helminths in soils is very low because of the difference in climatic conditions.

He questioned Dr Hillman in respect of pathogens which are specific to animal or man hosts, but do not survive on plant species. Should we be so stringent with the microbiological quality of wastewater when it is to be reused for agricultural irrigation?

Mr P. J. Hillman interpreted Dr Shende's question as suggesting that we need not be so stringent in the quality of water applied to crops because pathogens do not survive for any length of time on crops. It all depends on where you are and on the circumstances. If you develop or adopt fairly simple precautions, such as stopping irrigation for a period before harvesting, then the desiccation effect and the ultraviolet light effect can, in fact, kill off pathogens, and therefore it may be possible to adopt a less stringent standard. It is for the individual and the individual authorities, knowing the circumstances with which they are faced, to make their own judgements.

Dr J. A. Aziz, Pakistan referred to Professor Pescod's paper, which mentions the management difficulties in applying relaxed standards and asking farmers not to grow certain crops. In Pakistan, at the moment, there is no pollution control legislation and all industries, and even municipalities, are discharging raw sewage into the rivers. On the other hand, before the raw sewage reaches the river, farmers take up this wastewater and buy it from the municipalities. They are so fond of this kind of wastewater, they buy it; they feel that the wastewater, since it contains a lot of nutrients, is giving them a higher crop yield. It is doubtful if the government could force them not to grow certain crops in certain seasons. So this proposal might give rise to management problems. Concerning lime treatment, facultative and maturation ponds with about 40 days' retention time are normally used for pathogen removal but will it be possible to eliminate maturation ponds and just have facultative ponds for treatment of the wastewater if lime treatment is adopted?

Professor Pescod replied first on the matter of controlling farm use of water and cropping pattern. He believed that administrative structures have to be set up very firmly at the planning stage but up to now had not seen too responsible an approach in most countries. In the Indian subcontinent, whenever a wastewater is collected, whether it is treated or not, farmers will use the effluent after it is discharged to a wadi or surface channel. That is what happens in Lahore; the sewage is collected and pumped into a canal and, from that point, the farmers use it indiscriminately. It is essentially raw sewage because there is no flow in the canal other than the sewage. If that is the level of control it is impossible to protect health, so it is essential to set up a proper management structure if the objective is to improve health control in the system. The health status of much of the population around Lahore is low and there is widespread transmission of parasites, for example, which cannot be entirely unrelated to indiscriminate reuse of raw sewage.

It is hoped that, in the simplest and cheapest way, system planning can be introduced, with the result that public health in a particular area can be protected. No matter how difficult management is, if it is necessary it should be applied. Maybe this is an altruistic approach, possible only in a country which is fairly well organized adminstratively and where the people are fairly law-abiding. Public education, which is all part of managing the system, is something governments have to realize they should invest in.

Coming to the last point made, whether lime treatment would eliminate maturation ponds, Professor Pescod pointed out that lime treatment does two things. It achieves some treatment beyond primary treatment, both in terms of a kill of pathogens and removal of helminth ova and protozoans, and it removes some solids and BOD. So the ponds which follow, perhaps the first one can be called a facultative pond and later ones maturation ponds, will have less to do. Whatever the ponds are called, the retention time required in the stabilization pond system would be reduced to something that might make it competitive with alternative, and more difficult to manage, conventional treatment systems.

Mr R. O. Cobham, Oxford, United Kingdom drew attention to Professor Pescod and other speakers having stressed the importance of selecting the type of crop which can be grown with the application of treated effluent. However, a statement in Professor Pescod's chapter, from experience seems to be an over-generalization and requires some clarification as to what led to the particular statement, namely, 'that timber crops can be grown with practically untreated wastewater'. Although that is a statement of fact, bcause it does happen, is it a dcsirable practice from the point of view of control of health? Even though the public is usually, one hopes, excluded from such areas, surely one has to pay equal attention to the health of the workers who are responsible for irrigation and maintenance operations within afforested areas?

Professor Pescod accepted Mr Cobham's point as valid criticism of the statement as it stands. The suggestion was that it might not be in the best interests of the country to think always of using effluent to irrigate crops that are food crops. In some places, where it is just impossible to achieve an acceptable level of health control, either through treatment or management of the designed system, then perhaps the simplest way of getting some productivity out of the effluent would be to irrigate forest land because you could use a very low level of effluent quality. Certainly you would have to be concerned about the health of workers and carefully choose the system of irrigation, but this level of organization should not be beyond the capacity of governments. In Kuwait, for example, the farm where they reuse treated effluent is off-bounds to the public and the farm workers are regularly monitored and protected in so far as is possible. It is not beyond the bounds of reality to manage a simple system like that in other countries and it is certainly made easier if the crops that are grown are not edible, eliminating any adverse public health impact.

Dr M. Abu-Zeid, Egypt said that, theoretically, it might be true that an integrated-system approach concerning effluent reuse is probably most efficient but there exist many practical, organizational, socioeconomic and administrative factors that make this approach very difficult to apply, especially in developing countries. So in some cases complete treatment with minimum downstream-user restrictions may be more practicable. Did Professor Pescod have data, cost data or evaluation data, which would support this integrated approach and help to sell the idea to governments? A second question related to groundwater contamination in places where no sewage collection system exists; can anything be done to minimize groundwater contamination in such cases?

Professor Pescod replied that the integrated approach which had been suggested is not only sensible, it is also ideal, as pointed out by Dr Abu-Zaid. It had not been adopted anywhere, and there are great practical difficulties in its implementation. However, it has potential anywhere to arrive at the best rational decision overall, and the decision must be based on cost and on feasibility in the context of the particular environment, both sociopolitical and technical. It may be that, in some places, the decision to invest in extensive, conventional tertiary treatment might be economically justifiable. That might sometimes be the right answer; in Bahrain, the decision-makers are convinced that is the best approach because of the high health status of the people. It is the sociopolitical climate there which has forced that decision and they, fortunately, have the resources to implement it. On the other hand, in Egypt it would be difficult to find the resources to adopt such a costly solution.

Professor Pescod admitted that at the moment he could not produce the cost data to support this integrated approach; he was hoping that the collaborative research programme which was planned for the next few years would not just look at the technical problems but would go into economic comparisons to try and attach costs to the different alternatives that might become available. Indeed, recently a Nigerian PhD student at Newcastle looked at the economics of crop selection for his region, assuming effluent was to be used for irrigation water. He came up with a model which eventually suggested a particular crop would be most economic. This was not obvious from the information available about which crops were salt-tolerant and what the relative productivities were, and so on. His predictions that a particular crop would be most economically produced using the effluent were specific to that area. That is the sort of planning governments should be involved in, looking at the total economics and the national impact of reuse in economic terms, rather than just adopting a policy of free use of the effluent by any farmer, investing heavily in treatment. There are better ways of managing the system.

The second point was concerned with groundwater contamination, where there is no wastewater collection system. That is always a problem and again it depends on the hydrogeology and on the type of sanitation systems. In the seminar paper presented by the participant from Mexico, mention is made of 80% of the population using the fields for sanitation, creating surface water contamination, as well as groundwater contamination. There is always a great potential for shallow groundwater pollution from on-site sanitation but, depending upon the soil conditions, there is a good chance of groundwater at any great depth being protected by the waste treatability capacity of the soil. Research, sponsored by the World Bank and the United Nations Environment Programme, is continuing on this subject.

Mr Chr. Photiou, Department of Agriculture, Cyprus accused Professor Pescod of being in favour of partial treatment before using the effluent for agricultural production. From another point of view the suggestion was to place much attention on the irrigation systems to be used, adopting the type of irrigation system which creates least danger. Such a system is likely to be a sub-surface method of trickle irrigation, or something similar. Do not partial treatment and safe methods of irrigation oppose each other, due to the high load of suspended solids or organic matter in the partially treated effluent?

Professor Pescod admitted it was pretty glib of him to say that we should be going for trickle subsurface irrigation with partial treatment when we have some way to go in the design of irrigation systems. He would like to think that we could do that without any treatment, without any investment in treatment, if that was feasible, but we have the physical constraints of the system to consider. A lot more research is required; for example, the Institute of Hydrology in Britain is conducting research in Mauritius on trickle irrigation, but only with normal irrigation water. This work is still at a research stage, attempting to supply the right amount of water at the right time, consistent with the crop requirements, so trickle irrigation is still a research system in the normal context of agricultural production. It certainly requires a lot more work before we can reliably suggest it for delivering partially treated effluent and it may be that we will have to consider biological aftergrowth in the system and how we might control that if we are going to lower the quality of the effluent that we supply.

Mr M. Lamtiri-Laarif, Ministre de l'Agriculture, Morocco asked if it was possible to irrigate citrus fruit and vegetables with untreated sewage and, if so, what safeguards and precautions must be taken.

Mr Hillman replied that the selection of an appropriate standard for treated wastewater to be used for irrigation depends on many factors. However, priority should be given to measures that minimize the risk to public health, and specifically to irrigation workers. Crops eaten raw should be particularly well protected. However, with citrus fruits it may be possible to limit the treatment to that which produces a virtually helminth-free effluent, providing spray irrigation is not used. It would seem sensible to stop irrigation with wastewater effluent 2 or 3 weeks before harvest and fallen fruit should not be picked from the ground. This assumes, of course, good management of both the irrigated area and field workers. The use of seasonal labour or immigrant workers presents a particular problem in this regard.

Vegetables need to be further protected and a bacteriological standard is required. The level at which this is set depends on local conditions and, as the Engelberg Report suggests, the guideline of $\leqslant$ 1000 faecal coliforms/100 ml might be relaxed a little if the crop is always consumed well-cooked. However, relaxing such a value implies a good standard of control during harvesting and marketing of crops.

Dr G. B. Shende put a question to Dr Morishita. In *Table 6.6* he had shown that at nitrogen concentrations in excess of 5 ppm N/l there is a decreasing yield of paddy. It is very difficult to assess from this statement how much nitrogen was applied to the crop. With this concentration of nitrogen it is unlikely that more than 100 kg or 120 kg can be supplied if the irrigation rate is normal. The problem of the decreased yield seems to be due to an unassessed supply of phosphates which are resulting in increased vegetative growth and depressed productivity.

Dr Morishita indicated that he was a plant nutritionist so, very often, he was using a high concentration of nutrient solution, even 100 ppm, but when irrigation water is supplied to submerged paddy fields most of the nitrogen accumulated in the top soil of the paddy field. The amount of total irrigation water was about 2 mm/cm^2, so an estimate of nitrogen supplied is 200 kg N. When 5 ppm N is applied in the irrigation water, about 100 kg N will be supplied which, under normal circumstances is nearly the same level of fertilizer that Japanese farmers apply in rice farming. However, one problem is uncontrolled supply. When supplied under control it may not cause a problem, but when the farmers do not watch the supply of nitrogen problems will arise. Another point is the accompanying organic material in the wastewater. As rice is grown under submerged conditions an excess of organics will create anaerobic conditions in the soil. For these two main reasons, Japanese rice will be damaged under such conditions.

Dr S. Al-Salem, Jordan commented on page 39 (Chapter 4). The medium-strength wastewater is typical of the type of wastewater found in developed countries such as Canada and the United States but BOD_5 and total suspended solids of 200 mg/l is not average for developing countries. For example, in Jordan, the average BOD_5 and suspended solids concentrations are about 500 mg/l, and the same apply in Syria and Egypt.

Dr A. Kandiah explained that he had taken the information from published data. As mentioned, some of these things are taken from specific examples and certainly will not be the same in other countries.

Mr Khalil I. Al-Salem, Kuwait refered to irrigation with treated sewage effluent in Kuwait where there had been no noticeable effect of excess nitrogen. The exception may be in storage life; potatoes cannot be kept for a long time and the effect appears clearly with corn. Although phosphorus is more than needed, when the farmers or the labourers started using this effluent they added phosphorus and the effect was that the plants appeared to be wilting. The plants were even wilting while the water was available. This effect was surprising and was discussed during a visit to Germany. Dr Tietjen gave the answer: phosphorus is no longer added to the plants and now the situation is excellent.

Dr M. Abu-Zeid drew attention to the example of cadmium control of one of the industrial companies at the end of Dr Morishita's discussion and wondered if some details of the programme could be given. What programme was applied to come up with these results? What was the cost of the programme, what volume of effluents are we talking about and what is the relationship between the effluents and the river water, assuming that there is a connection between the industrial water and the river water?

Dr Morishita replied that contact with company staff was maintained every year and there is an agreement between the native farmers' associations and the Mitsi Company so that they provide all kinds of data. For 3 or 4 years each checked the other. Samples of the same effluents and the same water are analysed separately. Thereafter the results are compared and a reliable relationship has developed. When valuable information is received, for example from this meeting, it is passed on to the Company. The Company staff also provide their data on pollution control and this is checked. For example, every year the cadmium concentration of the main discharge from the factories and the river water is checked. Although the Mitsi Company was previously the enemy of the vegetable farmers, more recently they are very good friends for pollution control.

Mr A. Gur directed a question to Mr Hillman asking if WHO was considering shedding light by initiating a study on the rather controversial subject of disinfecting agents to be used in wastewater disinfection, which is quite different from drinking water disinfection.

Mr Hillman was not aware of any work currently under way in this regard. Literature does exist on the efficiency of other disinfection processes, such as ozonation for the production of potable waters. However, as mentioned in the new WHO Guidelines for Drinking Water Quality, the relative simplicity, low cost and efficiency of chlorine, plus the residual that can be left, suggests it will remain the most common disinfection process for potable waters. There is no doubt, however, that high organic load in wastewater makes chlorination less efficient and alternatives deserve careful study.

Mr Cobham, Oxford, UK asked Mr Hillman if he could say a little more about the aerosol distribution dangers associated with sprinkler irrigation. To what extent, in his view and experience, do these dangers relate at all to the use of tertiary-treated effluent and, secondly, can these dangers be remedied to any extent if irrigation is confined to night-time periods?

Mr Hillman admitted that tertiary treatment will further reduce the number of viable virus units in wastewater effluent, some more than others. However, apart from some high-technology, high-cost processes, tertiary treatment was unlikely to remove all viruses. The suggestion of night-time sprinkler irrigation would presumably greatly reduce the risk to field workers but would have less influence on the hazard to nearby communities. The difficulty is in assessing just what the danger really is from aerosols and, as indicated in Chapter 5, more epidemiological studies are needed.

Mr A. Bouzaidi, Tunisia asked, regarding environmental hazards, why Dr Morishita had only mentioned cadmium. Were there any other hazardous substances in effluents?

Dr Morishita pointed out that he had failed to mention one important matter, the recent wide scattering of cadmium-containing products in our daily life. One Japanese scientist checked the wastewater from his house and found large amounts of zinc. He followed this up and found in hair-conditioners or skin-conditioners many zinc compounds and in these compounds, there is often some cadmium. Many products incorporate cadmium; cadmium batteries, cadmium paints, televisions, and many other items. There is a clear tendency for increasing cadmium levels in our domestic wastewater or domestic sludge. In Japan sometimes more than 5 ppm of cadmium is found in domestic sewage sludge. Mining company discharges were previously the main source of cadmium pollution in Japan but, more recently, our daily life has become the main cause of cadmium pollution. The situation will become the same in any country.

Dr A. Hamdy, Mediterranean Institute, Italy referred to the chapters by Dr Kandiah and Professor Pescod (4 and 3). The most important thing to be considered, when using sewage effluent in irrigation, is water quality, crops, soil type, all of these. What are the authors' thoughts on the possibility of using untreated sewage effluent by modifying the systems of irrigation which we are now using and adapting crop rotation to be applied under different soil conditions ? Can a strategy for using sewage without any pretreatment be arrived at or not?

Professor Pescod suggested that irrigation with raw wastewater was feasible under strict management control and with careful choice of crops. In the case of a normal domestic sewage there should be no adverse effects on the soil and, apart from the possibility of high nitrogen levels promoting excessive leaf growth, crop production should not suffer. The main risk from irrigating with raw sewage is the transmission of disease. In those countries now allowing raw sewage irrigation of any crop, there is considerable endemic infection in the population. It should be an objective of the Seminar to improve on such practices in the future, not only by costly and sophisticated tertiary treatment but also through integrated planning of the overall reuse system, which will include the lowest level of sewage treatment and proper management of effluent application and crop processing.

Dr Kandiah agreed, in principle, with what Professor Pescod said. As mentioned by one of the participants, it is known that effluents are not all the time, in most cases, high in salts and sodium. Sewage effluent is often of better quality, in terms of salt and sodium, than groundwater. So it is possible, from the point of view of chemical constitutents, that sewage effluent can be used with minimum treatment. Our major concern is with health. In fact, we are already criticized for using conventional irrigation waters because it is said that mosquito-carried diseases and bilharzia are spread by normal irrigation. Also, as mentioned by Dr Shende from India, depending on climatic conditions, temperature and soils, a lot of these health hazards are minimized. It is possible, and all have to work towards it, to adopt a strategy of using sewage effluent with minimum treatment. No treatment or minimum and economically viable treatment which is very simple and low cost should be applicable. However, from the point of view of chemical standards, most effluents are quite usable and often of better quality than groundwater sources now being used for irrigation.

Mr Shammas expressed the opinion that the cadmium case was giving the impression that the problem was the result of the use of sewage effluent. It should be kept in mind that the source of this cadmium was not sewage, it was due to industrial activity, mining for zinc, and cadmium is often associated with zinc. It was in the river water, it was coming to the fields, so it was not sewage. It was untreated industrial waste going into the river and causing this 'itai-itai' disease, not the cadmium in the rice but the cadmium taken in the water. The river water was used as a source of drinking water, and for cooking it was potable water. It should be clear to us that this was not a problem that arose from the use of sewage effluent. It was because the effluent from industry was not treated. Normally, sewage effluent will be almost free of heavy metals due to the metals precipitating out in the sludge in sewage treatment.

Dr Morishita answered the first question by suggesting that the wastewater from the mining activity might have been the major souce of cadmium pollution. There is some discussion about which is the major cause of cadmium intake, rice or water. Some scientists calculated the cadmium intakes by rice and water and found the major part of the cadmium intake came from the river water. The Mitsi Company purchase cadmium from all over the world and they must discharge it. However, the cadmium concentration in the river water never reaches the necessary level to cause 'itai-itai' disease. Our conclusion now is that the major source of cadmium in 'itai-itai' patients is rice and other crops.

Dr P. Economides, Department of Veterinary Services, Cyprus asked if, instead of trying to form criteria based on pathogen indicators in treated effluents used for irrigation, it would be better to find mechanisms which will enable a responsible authority to define and impose conditions for the planning, establishment and operation of treatment plants and the safe use of the effluents produced.

Mr Hillman pointed out that both the WHO (1973; see p. 59) technical report and the recent Engelberg Report (IRCWD, 1985; see p. 60) suggest treatment processes that should, with the normal degree of operational control and maintenance, achieve reliably the quality criteria proposed. It would seem reasonable, therefore, for responsible authorities to prescribe treatment processes and procedures for the use of effluent and crops to be grown, rather than set effluent quality standards. However, this may lead to a certain degree of conservatism in process design and in the procedures proposed. More importantly, it will lead to stereotype designs which will be less easily adapted to the specific circumstances of particular locations and effluent loads. Nevertheless, in terms of simplicity of control and in order to minimize public health hazards, this may be the most appropriate approach for some countries.

Section II

Treatment of sewage

Chapter 8

Waste stabilization ponds: the production of high quality effluents for crop irrigation

D. Mara*

Introduction

Waste stabilization ponds are a low-cost, low-technology, but highly efficient method of wastewater treatment. They are shallow basins with earth embankments into which wastewater continuously flows and from which treated effluent is discharged. Because ponds are an entirely natural process unaided by any mechanical devices, they have long hydraulic retention times (weeks, rather than hours, as in many other wastewater treatment processes), and hence a large land area requirement. There are three main types of ponds, and these are usually arranged in the following sequence to form a series of ponds capable of treating any biodegradable wastewater to any required quality standard:

- *Anaerobic ponds* (2–5 m deep), which receive raw wastewater, are most advantageously fed with strong organic wastes (biochemical oxygen demand [BOD_5] > 300 mg/l) or those with a high suspended solids content (SS > 300 mg/l); the BOD_5 loading (100–400 g/m^3 d) is so high that the ponds are completely devoid of dissolved oxygen, and the SS settle to the bottom of the pond (usually 2–5 m deep) where they undergo vigorous anaerobic digestion at temperatures > 15°C.
- *Facultative ponds* (1–2 m deep), which receive either raw wastewater (primary faculative ponds) or settled wastewater (secondary facultative ponds) have a lower anaerobic zone and an upper aerobic zone where oxygen for bacterial metabolism is largely provided by the photosynthetic activity of microalgae which grow profusely to give the pond liquid a deep green coloration.
- *Maturation ponds* (1–2 m deep), in which the facultative pond effluent is treated further, principally to reduce the number of excreted pathogens and nutrients (N and P) to the desired level, although there is some additional removal of BOD; like facultative ponds they are photosynthetic ponds, and their size and number control the final effluent quality.

Pond systems comprising a properly designed series of anaerobic, facultative and maturation ponds have a number of advantages over other, more conventional wastewater treatment processes:

* Professor of Civil Engineering, University of Leeds, UK

- They are usually the cheapest form of treatment, both in construction and in operation and maintenance terms, and do not require any input of external energy (other than solar energy).
- They are able to reduce the concentration of excreted pathogens to very low levels (for example, a six-log unit reduction of faecal coliforms is easily obtainable, whereas most other treatment processes can only achieve a two-log unit reduction at best), and this is of prime importance when the effluent is to be reused in agriculture or aquaculture or where the transmission rate of excreta-related diseases is high, as in most developing countries.
- They are well able to absorb hydraulic and organic shock loads.
- They can tolerate high concentrations of heavy metals (up to around 30 mg/l).
- Their operation and maintenance is very simple; floating scum on facultative and maturation ponds must be removed, the grass on the embankments cut regularly and anaerobic ponds desludged every 3–5 years when they are half full of sludge.

As a result of these very significant advantages ponds are widely used throughout the world wherever land is available at reasonable cost. Although pond efficiency increases with ambient temperature, ponds are used as far north as Alaska and as far south as New Zealand, where the city of Auckland has over 5 km^2 of ponds. One-third of all wastewater treatment facilities in the United States is ponds, and in Europe they are widely used in France, Germany and Portugal; and they are common in the Middle East, Africa, Asia and Latin America.

Chapter 9 gives a more extensive review of stabilization ponds technology.

Pond design procedures

There are many ways in which a series of anaerobic, facultative and maturation ponds can be designed (Mara, 1976). The approach used in this chapter is based on the following principles.

Anaerobic ponds

A conservative, empirical approach is used based on the permissible volumetric BOD_5 loading (λ_v, g/m^{3}d). Meiring *et al.* (1968) recommended that λ_v be in the range 100–400 in order to maintain anaerobic conditions and prevent odour nuisance. In Bavaria, a design loading of 100 is used, with satisfactory results even in winter (Bucksteeg, 1983). This value is adopted as the design loading for temperatures below 10°C. In order to allay designers' fears of odour release in summer, a design loading of 300 (rather than 400) is adopted for temperatures above 20°C. In the range 10–20°C the following linear interpolation equation is used:

$$\lambda_v = 20T - 100 \qquad (1)$$

where T is the temperature, °C. The BOD_5 removal in anaerobic ponds is conservatively taken as 40% below 10°C and 60% above 20°C, and with linear interpolation in between:

$$R = T + 20 \qquad (2)$$

where R is the percentage BOD_5 removal. The use of such equations may appear simplistic, but in the absence of good field data they are as reasonable assumptions as can be made.

Facultative ponds

For primary facultative ponds (that is, those that receive raw wastewater), the following modification of McGarry and Pescod's (1970) equation is used:

$$\lambda_s = 15T - 50 \quad (3)$$

where λ_s is the design areal BOD_5 loading, kg/had. Experience in both Germany (Bucksteeg, 1983) and France (Agence de Bassin Loire-Bretagne and Centre Technique du Genie Rural des Eaux et Forêts, 1979) indicates that a primary facultative pond operates well, even in winter, at a loading of $100\,kg\,ha^{-1}d^{-1}$; this corresponds to a temperature of 10°C in equation (3), and so this loading is used for temperatures of 10°C and below and equation (3) for temperatures above 10°C.

In primary facultative ponds, up to 30% of the influent BOD_5 is removed in the sludge layer in the form of methane gas (Marais, 1970a). In secondary facultative ponds (that is, those that receive settled wastewater, for example the effluent from an anerobic pond), this cannot occur as there is essentially no sludge layer. Thus a factor of 0.7 is applied to equation (3) to give the following design equation for secondary facultative ponds:

$$\lambda_s = 10.5T - 35 \quad (4)$$

Maturation ponds

BOD is not used as a measure of the efficiency of maturation ponds, which are designed on the basis of faecal coliform removal. This is usually a more appropriate basis for design, especially when the effluent is to be reused in agriculture or aquaculture (*Table 8.1*) since effluents with low faecal coliform numbers will normally have an acceptable BOD_5 concentration. Thus BOD_5 removal in facultative and maturation ponds is not considered in this model, only the removal of faecal coliforms. Marais (1970b) proposed the following equations for the removal of faecal coliforms in a series of anaerobic, facultative and maturation ponds:

$$N_e = N_i/[1 + k_T t_a)(1 + k_T t_f)(1 + k_T t_m)^n] \quad (5)$$

$$k_T = 2.6(1.10)_T^{-20} \quad (6)$$

where N_e and N_i are the concentrations of faecal coliforms/100 ml of the final effluent and raw wastewater, respectively; k_T is the first order rate constant for faecal coliform removal at T°C, d^{-1}; t_a, t_f and t_m are respectively the mean hydraulic retention times (= volume/flow) in the anaerobic, facultative and maturation ponds, and n is the number of maturation ponds (which are assumed to be the same size; this is desirable (Marais, 1970b) but not always topographically possible). Equation (6) has recently been shown to be applicable in northern France (Demillac and Baron, 1982).

Equation (5) is usually solved by trial and error as it contains two unknowns, t_m and n – t_a and t_f are known by this stage in the design; N_i is either known or taken as, for example, 1×10^8; and N_e is stipulated by the local regulatory agency or estimated from Table 5.4 (page 61). For example, one might try two ponds at 7 days, or three ponds at 5 days, and so on until the equation gives an acceptable value of N_e. However, examination of equation (5) shows that it is better to have a larger number of smaller ponds, rather than a smaller number of larger ponds. Marais (1970b) recommends that the minimum hydraulic retention time in maturation ponds (t_m^{min}) should be around 3 d in order to minimize shortcircuiting.

Thus, if t_m^{min} is taken as 3 d, then if a single maturation pond of, say, 6 days retention gives an acceptable value of N_e, it will usually be better to have two ponds each with 3 days' retention. This provides a factor of safety at only a small additional cost. In general terms, designers should, therefore, investigate maturation pond retention times in the range:

$$t_m^{min} \leqslant t_m < t_m^{min} [(n + 1)/n] \quad (7)$$

starting initially with $n = 1$ and increasing n as necessary in order to meet the required value of N_e. This procedure is tiresome when done manually but is ideally suited to solution by a microcomputer, which produces a much more efficient design in terms of land area requirement (Gambrill *et al.*, 1986).

Increased summer populations

In Mediterranean Europe (for example) winter temperatures are low, but in summer, when tourism may increase the population to be served by a factor of 2–20 (occasionally more) (Drakides and Calignon, 1983), temperatures are higher and so the ponds are able to treat greater volumes of wastewater. If the population increase factor (PIF) is defined as the maximum summer (resident + tourist) population divided by the winter (resident) population, then winter conditions control the design if the PIF is equal to or less than the permissible summer loading divided by the permissible winter loading. For example, in the case of primary facultative ponds, winter conditions control the design if, from equation (3):

$$\text{PIF} \leqslant (3T_s - 10)/(3T_w - 10) \quad (8)$$

where T_s and T_w are the summer and winter temperature, respectively. Similar relationships apply to anaerobic and secondary facultative ponds.

Pond performance in northeast Brazil

It is clear from equation (5) that any pond system can be easily designed to produce any required effluent standard in terms of faecal coliform concentration. The design procedure given above is, of course, conservative; higher performance will generally be attained in practice. For example, in northeast Brazil (Mara, Pearson

Table 8.1 Performance of a series of five waste stabilization ponds in northeast Brazil (mean pond temperature: 26°C)

Sample	*Retention time* (days)	*BOD₅* (mg/l)	*Suspended solids* (mg/l)	*Faecal coliforms* (/100 ml)	*Intestinal nematode eggs* (/litre)
Raw wastewater	–	240	305	4.6×10^7	804
Effluent from:					
anaerobic pond	6.8	63	56	2.9×10^6	29
facultative pond	5.5	45	74	3.2×10^5	1
maturation pond 1	5.5	25	61	2.4×10^4	0
maturation pond 2	5.5	19	43	450	0
maturation pond 3	5.8	17	45	30	0

Source: Mara, Pearson and Silva (1983)

and Silva, 1983) a series of five ponds, with an overall retention time of 29.1 days and operating at a temperature of 26 ± 1°C, reduced the faecal coliform concentration from 4.6×10^7/100 ml of raw sewage to 30/100 ml in the final effluent – a reduction of 99.99993% or slightly more than six log units (*Table 8.1*). The removal of helminth ova and protozoan cysts was 100% – complete removal. No other sewage treatment process can achieve such efficiency at the same cost.

Pond effluent reuse for crop irrigation in Alexandria, Egypt: a case study*

By the year 2005 Alexandria is expected to have a population of 5.6 millions. Domestic sewage production will be some 840 000 m^3/d and industry will double this to around 1 840 000 m^3/d. Water for irrigating the Nile Delta region, which has been an area of intense agricultural activity for some 3000 years or more, comes principally from the Nile and there is a highly developed canal system for supplying irrigation water and removing drainage water. However, less water is now available from the Nile than before and, as a result, it is not possible to develop fully the agricultural potential of desert areas hitherto not irrigated. Reuse of treated wastewater effluent is clearly a partial solution to this problem and, from an agronomic point of view, it is better to treat the sewage in a series of ponds rather than to discharge it to waste in the Mediterranean sea.

Assuming that the total wastewater flow from Alexandria has 600 mg l^{-1} of BOD_5 and 1×10^9 faecal coliforms per 100 ml, and that the winter and summer temperatures are 15 and 25°C, respectively, some 97 km^2 of ponds would be required to treat the wastewater to a standard of <1000 faecal coliforms per 100 ml. The effluent (some 1 236 000 m^3/d, allowing for 2 m evaporation per annum) could then be discharged into the canal system and reused for irrigation. Alternatively, the wastewater could be treated to a less stringent standard and reused for the irrigation of industrial or fodder crops (*Table 8.1*). On the assumption that each hectare of irrigated land requires 10 000 m^3 of water per annum, reuse of Alexandria's wastewater would enable some 45 000 hectares of currently unused (and, due to lack of water, currently unusable) desert to be put into full agricultural production. The crops most likely to be irrigated are sugar beet in summer and alfalfa, maize and broad beans in rotation in winter.

References

AGENCE DE BASSIN LOIRE-BRETAGNE AND CENTRE TECHNIQUE DU GENIE RURAL DES EAUX ET FORÊTS (1979) Lagunage Naturel et Lagunage Aere: Procèdes d'Epuration des Petites Collectivités. Orleans: Agence de Bassin Loire-Bretagne

BUCKSTEEG, K. (1983) Experiencc des étangs de stabilisation non aerés en Bavarie: domaine d'utilisation, dimensionement, traitement des eaux usées dilués, construction. *La Tribune du Cebedeau,* **36(481),** 533–540

---

* The information presented herein was kindly provided by Dr G. Parry and Dr H. W. Pearson of the Environmental Advisory Unit of the Department of Botany, University of Liverpool, England, which is advising the Governorate of Alexandria on wastewater reuse through a contract with the Overseas Development Administration of the Foreign and Commonwealth Office, London. The pond design figures were obtained by the author using the microcomputer program referred to above; they include a factor of 1.2 applied to the total pond mid-depth to allow for embankments, etc.

DEMILLAC, R. and BARON, D. (1982) *Inactivation Comparée des Virus Hydriques et des Bactéries Indicatrices de Contamination Fécale en Bassins de Lagunage.* Rennes: École National de la Santé Publique.

DRAKIDES, M. and CALIGNON, M. (1983) *Lagunes à Charge Estivale.* Pierre-Benite: Agence de Bassin Rhone-Mediterranée-Corse.

GAMBRILL, M. P., MARA, D. D., ECCLES, C. R. and BAGHAEI-YAZDI, N. (1986) Microcomputer-aided design of waste stabilization ponds in tourist areas of Mediterranean Europe. *Public Health Engineer,* **14(2),** 39–41

McGARRY, M. G. and PESCOD, M. B. (1970) Stabilization pond design criteria for tropical Asia. In: *Proceedings of the Second International Symposium for Waste Treatment Lagoons,* edited by R. E. McKinney. Laurence: University of Kansas, pp. 114–132

MARA, D. D. (1984) *Sewage Treatment in Hot Climates.* Chichester: John Wiley

MARA, D. D., PEARSON, H. W. and SILVA, S. A. (1983) Brazilian stabilization pond research suggests low-cost urban applications. *World Water,* **6(7),** 20–24

MARAIS, G. V. R. (1970a) Dynamic behaviour of oxidation ponds. In: *Proceedings of the Second International Symposium on Waste Treatment Lagoons,* edited by R. E. McKinney. Laurence: University of Kansas, pp. 15–46

MARAIS, G. V. R. (1970b) Faecal bacterial kinetics in waste stabilization ponds. *Journal of the Environmental Engineering Division, American Society of Civil Engineers,* **100(EE1),** 119–139

MEIRING, P. G., DREWS, R. J., van ECK, H. and STANDER, G. J. (1968) A guide to the use of pond systems in South Africa for the purification of raw and partially treated sewage. *CSIR Special Report WAT 34.* Pretoria: National Institute for Water Research

Chapter 9

Design, operation and maintenance of wastewater stabilization ponds

M. B. Pescod* and D. D. Mara†

Introduction

The application of stabilization ponds for wastewater treatment depends on investment in a collection system. In developing countries, sewerage systems are not common in urban areas, nor are they necessarily the most appropriate sanitation technology, but where collection can be justified stabilization ponds will generally be the treatment system of first choice. The exceptions will be where land costs (or opportunity costs) are very high and in areas of karst geology with highly permeable strata. However, even where there is no sewerage system, nightsoil collection in tankers will allow stabilization ponds to be adopted for treatment. In addition, organic industrial effluents, particularly from agro-industry, will be amenable to treatment in ponds either separately or in combination with domestic sewage.

Given the centralized collection of organic wastewater and reasonable land cost, stabilization pond treatment will not only be competitive in total cost but will have distinct advantages in terms of the simplicity of construction, operation and maintenance over conventional wastewater treatment processes. The absence of mechanical and electrical equipment will be a particular benefit of pond systems in developing countries. Under hot climatic conditions, high ambient temperatures will produce high rates of reactions and biomass growth in stabilization ponds.

Terminology

The collective term 'stabilization ponds' has tended to supersede alternatives such as 'oxidation ponds' or 'treatment lagoons' to describe shallow, manmade lagoons which may be adopted to treat organic wastewaters without equipment in a once-through mode of operation. More specific terminology applies to differentiate between particular forms of pond, sometimes indicating the type of action predominating.

Anaerobic ponds are primarily used to treat strong organic wastewaters and are devoid of oxygen as a result of the high organic loading applied. *Facultative ponds* operate at lower organic loadings and dissolved oxygen persists in the water column

* Professor and Head, Department of Civil Engineering, University of Newcastle upon Tyne, UK
† Professor of Civil Engineering, University of Leeds, UK

throughout the day, mainly due to the presence of algae. *Maturation ponds* are aerobic lagoons used as a polishing stage after facultative ponds in the treatment of wastewaters containing faecal material, and their principal function is pathogen removal. *Aerated lagoons* are basically activated sludge units without sludge recycle and operate at low mixed liquor suspended solids levels (MLSS 200–500 mg/l) and relatively long retention times. *High-rate aerated ponds* are very shallow lagoons usually incorporating mechanical aeration and are designed to optimize the growth of algae for harvesting and reuse. Macrophyte ponds are maturation ponds in which submerged or emergent aquatic weeds are allowed to grow, with the intention of upgrading the quality of effluent.

Pond combinations

Wastewater stabilization pond systems are designed to achieve different forms of treatment in up to three stages in series, depending on the organic strength of the input waste and the effluent quality objectives. For ease of maintenance and flexibility of operation, at least two trains of ponds in parallel are incorporated in any design. Strong wastewaters, with BOD_5 concentration in excess of about 300 mg/l, will frequently be introduced into first-stage anaerobic ponds which achieve a high volumetric rate of removal. Weaker wastes or, where anaerobic ponds are environmentally unacceptable, even stronger wastes (say up to 1000 mg/l BOD_5) may be discharged directly into *primary* facultative ponds. Effluent from first-stage anaerobic ponds will overflow into *secondary* facultative ponds which comprise the second stage of biological treatment. Following primary or secondary facultative ponds, if further pathogen reduction is necessary, maturation ponds will provide tertiary treatment. Typical pond system configurations are given in *Figure 9.1* (World Health Organization, 1986).

Anaerobic ponds

Anaerobic ponds are very cost-effective for the removal of BOD, when it is present in high concentration. Normally, a single anaerobic pond in each treatment train is sufficient if the strength of the influent wastewater, L_i, is less than 1000 mg/l BOD_5. For high-strength industrial wastes, up to three anaerobic ponds in series might be justifiable, but the retention time, t_{an}, in any of these ponds should not be less than 1 day (McGarry and Pescod, 1970).

Anaerobic conditions in first-stage stabilization ponds are created by maintaining a high volumetric organic loading, certainly greater than 100 g BOD_5/m^3 d. Volumetric loading, λ_v, is given by:

$$\lambda_v = \frac{L_i Q}{V} \tag{1}$$

where
L_i = influent BOD_5, mg/l,
Q = influent flow rate, m^3/d, and
V = pond volume, m^3.

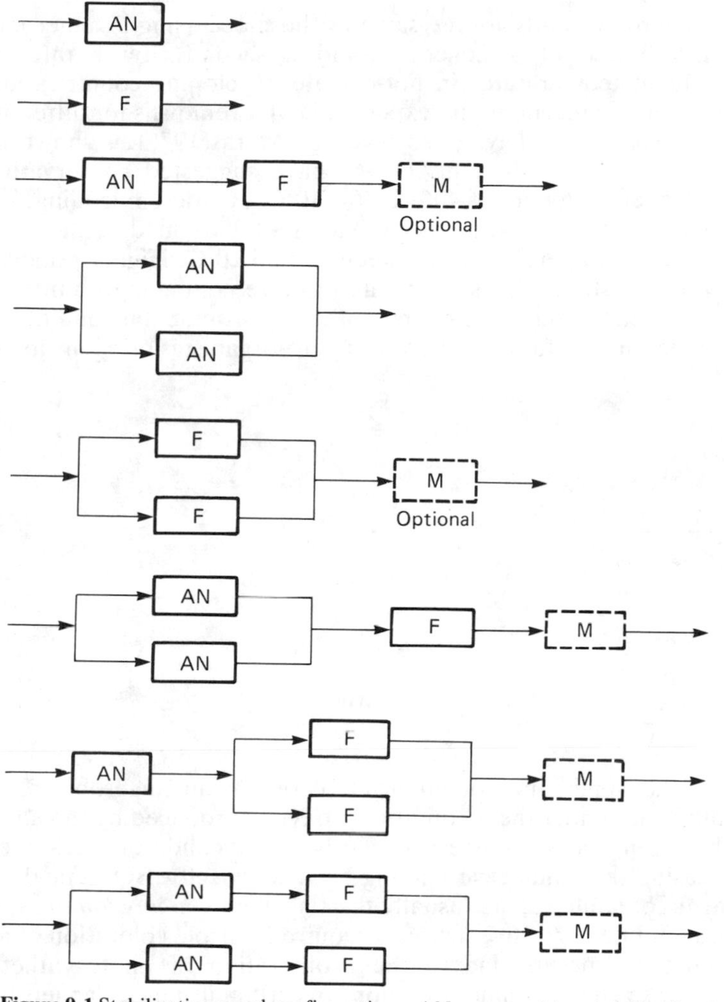

Figure 9.1 Stabilization pond configurations. AN = anaerobic pond; F = facultative pond; M = maturation pond

Or, since $V/Q = t_{an}$, the retention time:

$$\lambda_v = \frac{L_i}{t_{an}} \tag{2}$$

Very high loadings, up to 1000 g BOD_5/m^3d, achieve efficient utilization of anaerobic pond volume, but with wastewater containing sulphate concentrations in excess of 100 mg/l the production of H_2S is likely to cause odour problems. In the case of typical municipal sewage, it is generally accepted that a maximum anaerobic pond loading of 400 g BOD_5/m^3d will prevent odour nuisance (Meiring *et al.*, 1968).

Anaerobic ponds normally have a depth between 2 and 5 cm and function as open septic tanks with gas release to the atmosphere. The biochemical reactions which take place in anaerobic ponds are the same as those occurring in anaerobic digesters, with a first phase of acidogenesis and a second slower rate of methanogenesis. Ambient temperatures in hot-climate developing countries are conductive to these anaerobic reactions and expected BOD_5 removals for different retention times in treating sewage have been given by Mara (1976) as shown in *Table 9.1*. More recently, Gambrill *et al.* (1986) have suggested conservative removals of BOD_5 in anaerobic ponds as 40% below 10°C, at a design loading, λ_v, of 100 g/m^3 d, and 60% above 20°C, at a design loading of 300 g/m^3 d, with linear interpolation for operating temperature of between 10 and 20°C. Higher removal rates are possible with industrial wastes, particularly those containing significant quantities of organic settleable solids. Of course, other environmental conditions, particularly pH, must be suitable for the anaerobic micro-organisms bringing about the breakdown of BOD.

Table 9.1 BOD removals in anaerobic ponds loaded at 250 g BOD_5/m^3 d

Retention, t_{an} days	*BOD_5 removal %*
1	50
2.5	60
5	70

During operation, the appearance of an algal bloom in an anaerobic pond indicates underloading and, under these conditions, oxygen produced by the algae can inhibit the methanogenic phase of digestion. Truly anoxic conditions must then be restored by increasing the volumetric loading to at least 100 g BOD_5/m^3 d. A very thin surface layer containing algae, usually the flagellate *Chlamydomonas*, is not generally harmful and no corrective action is required. Purple coloration of an anaerobic pond sometimes occurs due to the proliferation of photosynthetic bacteria but this is likewise an acceptable condition. In certain instances, anaerobic ponds become covered with a thick scum layer, which is thought to be beneficial but not essential, although it may give rise to increased fly breeding.

Facultative ponds

Primary

The effluent from anaerobic ponds will require some form of aerobic treatment before discharge, and facultative ponds will often be more appropriate than conventional forms of secondary treatment for application in developing countries. Primary facultative ponds will be designed for the treatment of weaker wastes and in sensitive locations where anaerobic pond odours would be unacceptable. Solids in the influent to a facultative pond and excess biomass produced in the pond will settle out forming a sludge layer at the bottom. This benthic layer will be anaerobic and, as a result of anaerobic breakdown of organics, will release soluble organic products to the water column above.

Organic matter dissolved or suspended in the water column will be metabolized by heterotrophic bacteria with the uptake of oxygen, as in conventional aerobic biological wastewater treatment processes. However, unlike in conventional processes, the dissolved oxygen utilized by the bacteria in facultative ponds is replaced through photosynthetic oxygen production by microalgae, rather than by aeration equipment. Especially in treating municipal sewage in hot climates, the environment in facultative ponds is ideal for the proliferation of microalgae. High temperature and ample sunlight create conditions which encourage algae to utilize the carbon dioxide (CO_2) released by bacteria in breaking down the organic components of the wastewater and take up nutrients (mainly nitrogen and phosphorus) contained in the wastewater. The symbiotic relationship illustrated in *Figure 9.2* contributes to the overall removal of BOD in facultative ponds, described diagrammatically by Marais (1970a) as in *Figure 9.3*.

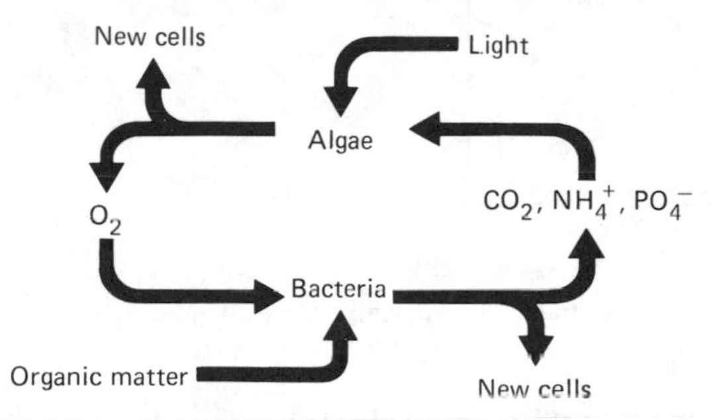

Figure 9.2 Symbiosis of algae and bacteria in stabilization ponds

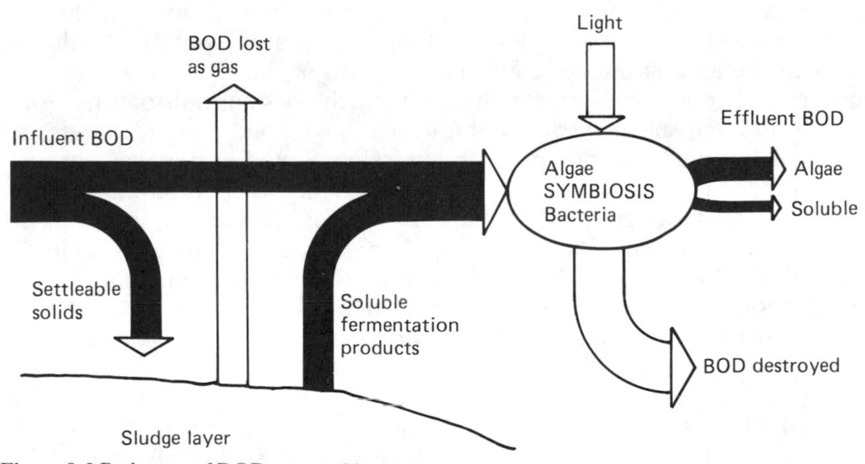

Figure 9.3 Pathways of BOD removal in primary facultative ponds (from Marais, 1970a)

To maintain the balance necessary to allow this symbiosis to persist the organic loading on a facultative pond must be strictly limited. Even under satisfactory operating conditions, the dissolved oxygen concentration (DO) in a facultative pond will vary diurnally as well as over the depth. Maximum DO will occur at the surface of the pond and will usually reach supersaturation in tropical regions at the time of maximum radiation intensity, as shown in *Figure 9.4*. From that time until sunrise, DO will decline and may well disappear completely for a short period. For a typical facultative pond depth, D_f, of 1.5 m the water column will be predominantly aerobic at the time of peak radiation and predominantly anaerobic at sunrise. As illustrated in *Figure 9.4*, the pH of the pond contents will also vary diurnally as algae utilize CO_2 throughout daylight hours and respire, along with bacteria and other organisms, during the night.

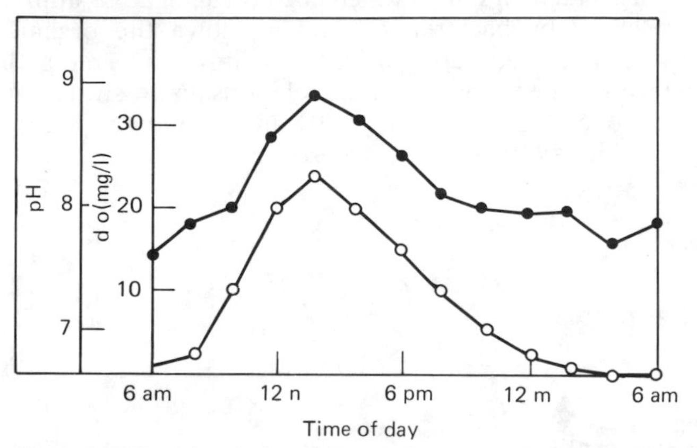

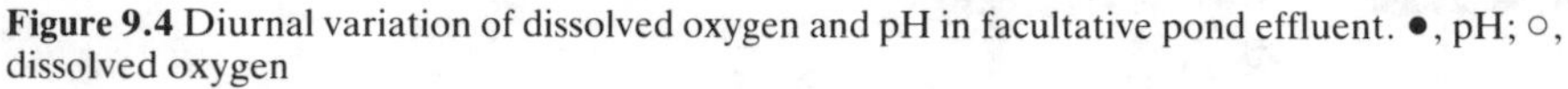

Figure 9.4 Diurnal variation of dissolved oxygen and pH in facultative pond effluent. ●, pH; ○, dissolved oxygen

Wind is considered important to the satisfactory operation of facultative ponds by mixing the contents and helping to prevent short-circuiting. Intimate mixing of organic substrate and the degrading organisms is important in any biological reactor but in facultative ponds wind mixing is also essential to prevent thermal stratification causing anaerobiosis and failure. Facultative ponds should be orientated with the longest dimension in the direction of the prevailing wind.

Although completely-mixed reactor theory with the assumption of first-order kinetics for BOD removal can be adopted for facultative pond design (Marais and Shaw, 1961), such a fundamental approach is rarely adopted in practice. Instead, an empirical procedure based on operational experience is more common. The most widely adopted design method currently being applied wherever local experience is limited is that introduced by McGarry and Pescod in 1970. A regression analysis of operating data on ponds around the world relating maximum surface organic loading, λ_s in lb/acre d, to the mean ambient air temperature, T in °F, of the coldest month resulted in the following equation:

$$\lambda_{s(max)} = 10(1.054)^T \quad (3)$$

Mara (1976) modified this formula and introduced a generous factor of safety in suggesting the following linear design equation (with λ_s in kg/ha d and T in °C):

$$\lambda_s = 20T - 120 \quad (4)$$

Subsequently, Arthur (1983) suggested that best agreement with available operating data, including a factor of safety of about 1.5, is represented by the relationship:

$$\lambda_s = 20T - 60 \quad (5)$$

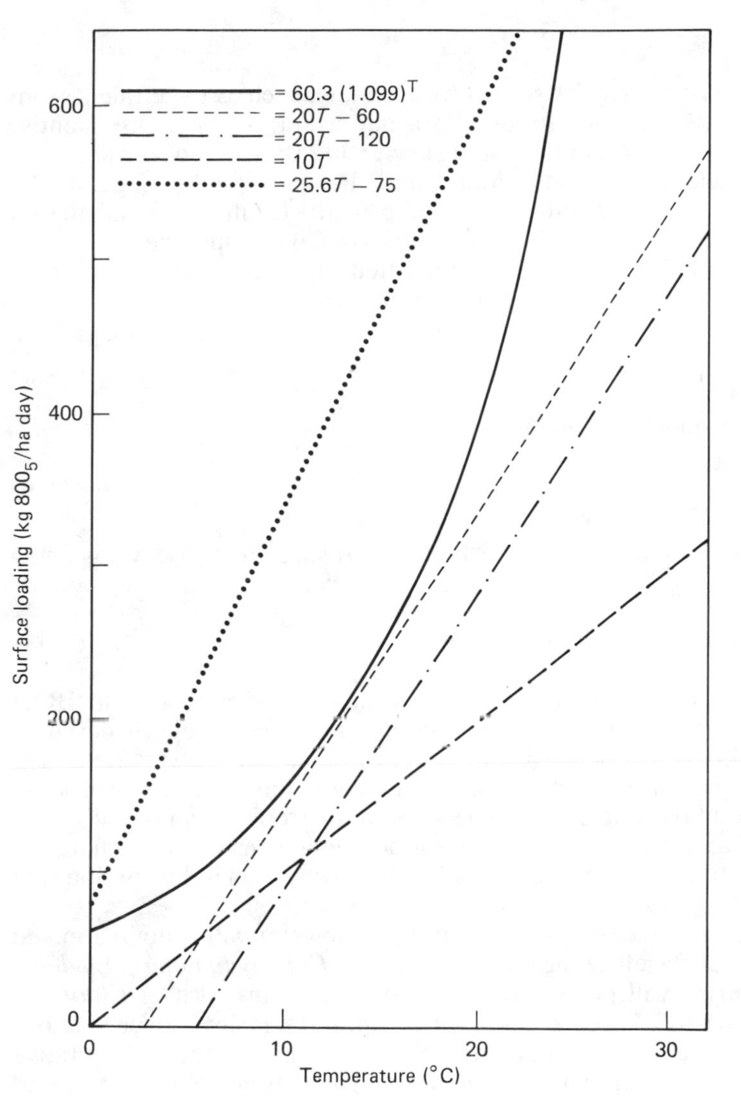

Figure 9.5 Comparison of alternative design equations for facultative stabilization ponds

Figure 9.5 shows a comparison of various formulations for primary facultative pond design, including the following:

$$\lambda_s = 60.3\,(1.099)^T \tag{6}$$

which is the °C hectare (ha) version of the original formulation of McGarry and Pescod (1970),

$$\lambda_s = 10T \tag{7}$$

which is the formulation now recommended by Mara and Pearson (1988) for use in Mediterranean Europe, and

$$\lambda_s = 25.6T + 75 \tag{8}$$

the formula first suggested by Arthur (1981).

Secondary

Secondary facultative ponds receive settled wastewaters, such as the effluent from an anaerobic pond or 'small bore' sewage (Otis and Mara, 1985). Consequently, they will not have such an active or deep sludge layer and BOD removal will be less than in primary facultative ponds. Mara and Pearson (1986) suggest the introduction of a factor of 0.7 into whichever primary facultative pond design equation is adopted to adapt it for the design of secondary facultative ponds.

Surface (or areal) BOD loading can be translated into a mid-depth facultative pond area (A_f in m^2) requirement using the formula:

$$A_f = \frac{10\ L_i\ Q}{\lambda_s} \tag{9}$$

Thus, adopting the loading equation (5):

$$A_f = \frac{L_i\ Q}{2T - 6} \tag{10}$$

and the mean hydraulic retention time in the facultative pond (t_f in days) is given by:

$$t_f = \frac{A_f\ D_f}{Q} = \frac{L_i\ D_f}{2T - 6} \tag{11}$$

The removal of BOD in facultative ponds (λ_f in kg/ha d) is related to BOD loading and usually averages 70–80% of λ_s. Retention time in a properly designed facultative pond will normally be 20–40 days and, with a depth of about 1.5 m, the area required will be significantly greater than for an anaerobic pond. The effluent from a facultative pond treating municipal sewage in the tropics will normally have a BOD_5 between 50 and 70 mg/l as a result of the suspended algae. On discharge to a surface water this effluent will not cause problems downstream if the dilution is of the order of 8:1.

Efficiently operating facultative ponds treating wastewater will contain a mixed population of flora but flagellate algal genera such as *Chlamydomonas, Euglena, Phacus* and *Pyrobotrys* will predominate. Non-motile forms such as *Chlorella, Scenedesmus* and various diatom species will be present in low concentrations unless the pond is underloaded. Algal stratification often occurs in facultative ponds, particularly in the absence of wind-induced mixing, as motile forms respond to changes in light intensity and move in a band up and down the water column. The relative numbers of different genera and their dominance in a facultative pond vary from season to season throughout the year but species diversity generally decreases with increase in loading. Sometimes, mobile purple sulphur bacteria appear when facultative ponds are overloaded and sulphide concentration increases, with the danger of odour production. High ammonia concentrations also bring on the same problem and are toxic to algae, especially above pH 8.0.

Maturation ponds

The effluent from facultative ponds treating municipal sewage or equivalent input wastewater will normally contain at least 50 mg/l BOD_5, and if an effluent with lower BOD_5 concentration is required it will be necessary to use maturation ponds.

For sewage treatment, two maturation ponds in series, each with a retention time of 7 days, have been found necessary to produce a final effluent with BOD_5 <25 mg/l when the facultative pond effluent had a $BOD_5 < 75$ mg/l.

A more important function of maturation ponds is the removal of excreted pathogens to achieve an effluent quality which is suitable for its downstream reuse. Although the longer retention in anaerobic and facultative pond systems will make them more efficient than conventional wastewater treatment processes in removing pathogens, the effluent from a facultative pond treating sewage will generally require further treatment in maturation ponds to reach effluent standards imposed for reuse. Faecal coliform bacteria are commonly used as indicators of excreted pathogens and maturation ponds can be designed to achieve a given reduction of faecal coliforms (FC). Protozoan cysts and helminth ova are removed by sedimentation in stabilization ponds and a series of ponds with overall retention of 11 days or more will produce an effluent containing $\nless$ 1 nematode egg/litre (Mara and Silva, 1986).

Reduction of faecal coliform bacteria in any lagoon (anaerobic, facultative and maturation) has been found to follow first-order kinetics:

$$N_e = \frac{N_i}{1 + K_b t} \tag{12}$$

Where

N_e = number of faecal coliforms/100 ml of effluent,
N_i = Number of faecal coliforms/100 ml of influent,
K_b = first-order rate constant for FC removal, d^{-1}, and
t = retention time in any pond, d.

For N ponds in series, equation (12) becomes:

$$N_e = \frac{N_i}{(1 + K_b t_{an})\,(1 + K_b t_{fa})\,(1 + K_b t_{m_1})\,(1 + K_b t_{m_n})} \tag{13}$$

Where, t_{m_n} = retention time in the nth maturation pond.

The value of K_b is extremely sensitive to temperature and was shown by Marais (1970b) to be given by:

$$K_{b(T)} = 2.6(1.9)^{T} - 20 \tag{14}$$

Where, $K_{b(T)}$ = value of K_b at T°C

A suitable design value of N_i in the case of sewage treatment is 1×10^8 faecal coliforms/100 ml, which is slightly higher than average practical levels.

The value of N_e should be obtained by substituting the appropriate levels of variables in equation (13), assuming a retention time of 7 days in each of two maturation lagoons (for sewage). If the calculated value of N_e does not meet the reuse effluent standard, the number of maturation ponds should be increased, say to three or more each with retention time 5 days, and N_e recalculated. A more systematic approach is now available whereby the optimum design for maturation ponds can be obtained using a simple computer program (Gambrill *et al.*, 1986).

Maturation ponds will be aerobic throughout the water column during daylight hours and the pH will rise above 9.0. The algal population of many species of non-flagellate unicellular and colonial forms will be distributed over the full depth of a maturation pond. Large numbers of filamentous algae, particularly blue-greens, will emerge under very low BOD loading conditions. Very low concentrations of algae in a maturation pond will indicate excessive algae predation by zooplankton such as *Daphnia* sp. and this will have a deleterious effect on pathogen die-off, which is linked to algal activity.

Aerated lagoons

Aerated lagoons may be *partially mixed* or *completely mixed* depending on the power input of the aerators. Aerators in a completely mixed lagoon provide enough energy (about 20 watts/m^3) to maintain the solids in suspension and this is more than that required to satisfy the oxygen demand for aerobic oxidation at the low MLSS levels prevailing in aerated lagoons. In a partially mixed aerated lagoon, a power input of about 4 watts/m^3 will transfer sufficient oxygen to meet the needs of the bacteria but will allow a sludge layer to form at the bottom of the lagoon in quiescent locations. Careful siting of many small aerators, rather than a few large ones, will minimize solids settlement and prevent the formation of troublesome sludge banks.

The design of aerated lagoons can be based on completely mixed reactor theory with the assumption of first-order kinetics for BOD removal. Adopting this theoretical approach, the rate of oxidation of organic matter is approximated by:

$$L_e = \frac{L_i}{1 + k_1 t_a} \tag{15}$$

Where

L_e = effluent BOD_5, mg/l,
L_i = influent BOD_5, mg/l,
k_1 = first-order rate constant for BOD_5 removal, d^{-1}, and
t_a = retention time in the aerated lagoon, d.

Retention time in an aerated lagoon is usually between 2 and 6 days, 4 days being a common value. A typical design value for k_1 in sewage treatment lagoons is 5 d^{-} at 20°C, with values at other temperatures being estimated from:

$$k_{1(T)} = 5(1.035)^{T-20} \tag{16}$$

The rate constant for organic industrial wastewaters must be determined by conducting laboratory and/or pilot plant studies. From an estimate of the BOD removal, the oxygen requirements can be calculated and, allowing for the difference between field conditions and the standard conditions of the manufacturer's aerator rating assessment, the aerator capacity decided.

In practice, however, more empirical procedures are adopted and typically a 4-day retention time is taken to achieve 70–90% BOD_5 removal, depending on operating temperature, in a 3–4 m deep partially mixed aerated lagoon with a power input of 4 watts/m^3. At tropical operating temperatures, BOD_5 removal can be expected to approach 90% in treating sewage in such an aerated lagoon and faecal coliform reduction might reach 96% (Arthur, 1983). Aerated lagoons should always be followed by settling ponds to remove effluent solids and the effluent will require maturation pond treatment if faecal coliform reduction is important. The fact that aerated lagoons take up less land than faculative ponds provides a mechanism for converting overloaded facultative pond systems to efficient treatment plants by the introduction of aeration equipment. Thus, aerated lagoons have a role to play both in reducing the initial land requirements of new pond treatment systems and in extending the capacity of existing ponds without increasing land requirements.

High-rate aerobic ponds

Facultative ponds used for wastewater treatment are designed to achieve BOD removal and algae are incidental, although necessary, to this process. Long retention time in 1.5 m deep ponds is optimal for wastewater treatment but not for algal growth. However, the high protein content of algae (40–50%) and their potential for bioconversion to methane make their harvesting from wastewater ponds of possible commercial value. With this as the goal, the yield of algae from a pond becomes the primary design objective. Ponds which are optimized for algal biomass production are termed high-rate aerobic ponds. A reasonable estimate of productivity from such ponds is 12 g/m^2 d (50 MT/ha year) of dry-weight, ash-free algae (Benemann *et al.*, 1977).

High-rate aerobic ponds are still at the experimental stage because of the difficulty of harvesting the algae economically from the relatively dilute suspension in pond effluent. The problems of marketing dried algae as a food supplement for animals (pigs, cattle and chickens) and methane fermentation costs also act as deterrents. Design of aerobic algal culture ponds can only be based on empirical procedures drawing on experience, and research in Thailand (McGarry, Lin and Merto, 1973) provides data of relevance to hot climate developing countries. Pilot-scale aerobic ponds, 450 mm deep were broom-mixed manually once a day to prevent sludge deposits. BOD_5 loadings up to 350 kg/ha d produced algal concentrations between 150 and 250 mg/l but artifical shading of ponds reduced algal concentration to as low as 50 mg/l. It was found that the noon-time concentration of algae, C_c mg/l, was related to the total radiant energy received per unit volume of water, S_a in kcal/l, by the equation:

$$C_c = 57.4 + S_a - 0.0225\, S_a \quad (17)$$

S_a is obtained from:

$$S_a = \frac{S\, t_a}{100\, d} \quad (18)$$

where
S = total incident radiation, cal/cm^2 d.
t_a = retention time, d, and
d = pond depth, m.

The yield of algae, Y_c in kg/ha d, was found to be:

$$Y_c = \frac{10\, d\, C_c}{t_a} \quad (19)$$

$$\text{or } Y_c = 0.5S - \frac{574\, d}{t_a} - \frac{2.25\,(10^{-5})\, t_a S^2}{d} \quad (20)$$

Thus, to maximize yield in any location, the retention time should be minimized and the depth increased (up to a maximum of 600 mm for aerobic ponds). Unfortunately, this would reduce algal concentration and increase the difficulties of algal harvesting.

For reuse as animal feed or bioconversion to methane, the algae must be separated from the pond liquid effluent and their size (predominantly less than 20 μm in diameter) makes this a considerable engineering problem. Research on techniques to control algal species so that larger varieties predominate in aerobic ponds is proceeding in the United States (Benemann *et al.*, 1977). Size-selective recycle of harvestable species and nutritional limitation are the two methods under investigation but these are unlikely to be put into practice in developing countries in the near future. If larger algae (such as *Oscillatoria* or *Microactinium*) can be produced, then simpler types of solid/liquid separation techniques, such as microstraining, sedimentation and filtration, will be feasible. However, under the normal conditions prevailing in aerobic ponds, more complex and costly processes, for example alum flocculation and flotation or centrifugation, are necessary to concentrate the algae. In addition, for reuse as livestock feed, vacuum filtration or heat drying of the algal concentrate will normally be required.

Macrophyte ponds

Maturation ponds which incorporate floating, submerged or emergent aquatic plant species are termed *macrophyte ponds* and these have been used in recent years for upgrading effluents from stabilization ponds. Macrophytes take up large amounts of inorganic nutrients (especially N and P) and heavy metals (such as Cd, Cu, Hg and Zn) as a consequence of their growth requirements and decrease the concentration of algal cells through light shading by the leaf canopy and, possibly, adherence to gelatinous biomass which grows on the roots.

Emergent species such as *Phragmites communis* and *Scirpus lacstris* have been used in Europe in 0.5 m deep macrophyte ponds but require regular cutting and harvesting, attract animals and birds and encourage fly and mosquito breeding. Floating species, with their larger root systems, are more efficient at nutrient stripping. Although several genera have been used in pilot schemes, including *Salvinia, Spirodella, Lemna* and *Eichornia* (O'Brien, 1981), *Eichornia crassipes* (water hyacinth) has been studied in much greater detail. In tropical regions, water hyacinth doubles in mass about every 6 days and a macrophyte pond can produce more than 250 kg/ha d (dry weight). Nitrogen and phosphorus reductions up to 80% and 50% have been achieved. In Tamil Nadu, India, studies have indicated that the coontail, *Ceratophyllum demersum,* a submerged macrophyte, is very efficient at removing ammonia (97%) and phosphorus (96%) from raw sewage and also removes 95% of the BOD_5. It has a lower growth rate than *Eichornia crassipes,* which allows less frequent harvesting.

Fly and mosquito breeding is also a problem in floating macrophyte ponds but this can be partially alleviated by introducing larvae-eating fish species such as *Gambusia* and *Peocelia* into the ponds. It should be recognized that pathogen die-off is poor in macrophyte ponds as a result of light shading and lower dissolved oxygen and pH than in algal maturation ponds. In their favour, macrophyte ponds can serve a useful purpose in stripping pond effluents of nutrients and algae and at the same time produce a harvestable biomass. Floating and submerged macrophytes are fairly easily collected by floating harvesters. The harvested plants might be fed to cattle, used as a green manure in agriculture, composted aerobically to produce a fertilizer and soil conditioner, or can be converted into biogas in an anaerobic digester, in which case the residual sludge can then be applied as a fertilizer and soil conditioner (UN Economic and Social Commission for Asia and the Pacific, 1981).

Aquaculture in waste disposal

Mention has already been made of the difficulty of harvesting algae from high-rate aerated ponds for algae reuse. An alternative approach is to use fish to harvest the algae and convert plant protein to animal protein. Species such as carp and tilapia can be grown in maturation ponds or in fish ponds fertilized with stabilization pond effluent or stored nightsoil. Yields up to 28 000 kg/ha year have been achieved in fertilized ponds where a broader polyculture of fish is adopted. With the long experience of fish farming in many developing countries, particularly in Asia, and the widespread need for animal protein, this form of wastewater reuse has the greatest potential for application throughout the world. A recent review of recycling organic wastes into fish in the tropics emphasized the attraction of this form of treatment for nightsoil recovery in rural and urban areas (Edwards, 1980).

Physical design of ponds

When the process design for stabilization pond systems has been selected it must be translated into a physical design. Actual pond dimensions, suited to the available site, must be selected, embankments and pond inlet/outlet structures designed and decisions taken on preliminary treatment, pond lining, pipework, fencing, etc.

Pond location

Ponds should be located at least 200 m from the community they serve, mainly to discourage people visiting the ponds. Odour release, even from anaerobic ponds, is most unlikely to be a problem in a well-designed system, but the public may need assurance about this at the planning stage. A minimum distance of 200 m normally allays any fears on this score.

There should be vehicle access to the ponds and, in order to minimize earthworks, the site should be flat or gently sloping; the soil must be suitable. Ponds should not be located within 2 km of airports as the birdlife – especially seagulls – attracted to the ponds may constitute a risk to air navigation.

Preliminary treatment

Preliminary wastewater treatment, such as screening and/or grit removal, is sometimes provided prior to treatment in ponds. A manually raked bar screen is all that need be provided in small pond systems (serving less than around 2000 people) if maintenance can be guaranteed. Otherwise, it is better to introduce raw sewage into the primary pond. For larger populations, mechanically raked screens and mechanical grit separators may be considered but these are the most vulnerable parts of the system if maintenance is suspect. The design of such preliminary treatment facilities should follow conventional recommended practice. Adequate provision must be made for the disposal of screenings.

Immediately after preliminary treatment, if any, there should be a stormwater overflow, if the sewer system is not of the separate type, and a Parshall or Venturi flume to measure wastewater flow. Automatic flow recorders are advisable for large flows but too troublesome in small pond installations. A flow measurement device is essential since pond performance cannot otherwise be assessed.

Pond geometry

There has been little rigorous work done on determining optimal pond shape. The most common shape is rectangular, although there is much variation in the length-to-breadth ratio. Clearly, optimal pond geometry, which includes not only the shape of the pond but also the relative positions of inlet and outlet, will minimize hydraulic shortcircuiting.

In general, anaerobic and primary facultative ponds should be rectangular with length-to-breadth ratios of less than 3, in order to avoid sludge banks forming near the inlet. Secondary facultative and maturation ponds should, wherever possible, have higher length-to-breadth ratios (up to 20 to 1) so that they better approximate plug flow conditions; high length-to-breadth ratios may also be achieved by placing baffles in the pond. Ponds do not need to be true rectangles, but may be gently curved if necessary or if desired for aesthetic reasons. A single inlet and outlet are usually sufficient, and these should be located in diagonally opposite corners of the pond.

In order to facilitate wind-induced mixing the pond should be located so that its longest dimension (diagonal) lies in the direction of the prevailing wind. If this is seasonally variable, then the summer wind direction should be used as this is when thermal stratification is most likely to occur. In order to minimize hydraulic shortcircuiting, the inlet should be located such that the wastewater flows into the pond against the prevailing wind.

The areas calculated by process design procedures are mid-depth areas and the dimensions calculated from them are thus mid-depth dimensions. These need to be corrected for the slope of the embankment, as shown in *Figure 9.6*. The dimensions and levels which the contractor needs to know are those of the base and the top of the embankment; the latter includes the effect of the freeboard (F).

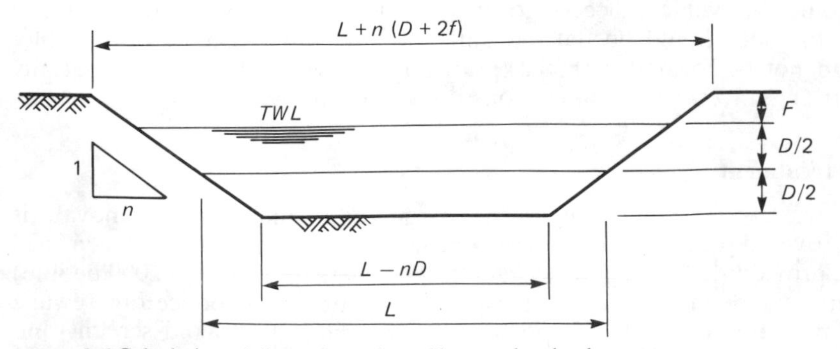

Figure 9.6 Calculation of dimensions of pond base and embankment top

The minimum freeboard that should be provided is decided on the basis of preventing waves, induced by the wind, from overtopping the embankment. For small ponds (under 1 ha in area) 0.5 m freeboard should be provided; for ponds between 1 and 3 ha the freeboard should be 0.5–1 m, depending on site considerations. For larger ponds the freeboard may be calculated from the equation (Oswald, 1975):

$$F = (\log A)^{1/2} - 1 \quad (21)$$

where

F = freeboard, m
A = pond area, m^2

Pond liquid depths are commonly in the following ranges:

anaerobic ponds	2–5 m
facultative ponds	1–2 m
maturation ponds	1–2 m

The depth chosen for any particular pond depends on site considerations (presence of shallow rock, minimization of earthworks). In primary facultative ponds, especially those with high length-to-breadth ratios, it is often advantageous to provide a deeper zone (2–5 m) near the inlet for sludge settlement and digestion; this is especially useful in coastal or desert areas for small pond systems treating wastewater with a high grit load when no grit removal facilities are included.

For any pond installation, but especially for pond systems treating flows in excess of around 5000 m^3/day, it is sensible, in order to increase operational flexibility, to have two or more series of ponds in parallel. The available site topography may in any case necessitate such a subdivision. Usually the series are equal, that is to say they receive the same flow, and arrangements for splitting the raw wastewater flow into equal parts after preliminary treatment must be made; this is best done by providing penstock-controlled flumes ahead of each series.

Geotechnical considerations

The principal objectives of a geotechnical investigation are to ensure correct embankment design and to determine whether the soil is sufficiently permeable to require lining. The following properties of the soils at the proposed pond location must be measured:

- Particle size distributions.
- Maximum dry density and optimum moisture content (modified Proctor test).
- Atterberg limits.
- Organic content.
- Coefficient of permeability.

Organic and plastic soils and medium to coarse sands are not suitable for embankment construction. If there is no suitable local soil with which at least a stable and impermeable embankment core can be formed, then it must be brought to the site at extra cost, and the local soil (if suitable) used for the embankment slopes. If the local soil is totally unsuitable, construction costs will be very high and ponds may not be the most economic treatment system.

Ideally, embankments should be constructed from the soil excavated from the site, and there should be a balance between cut and fill, although it is worth noting that ponds constructed completely in cut may be a cheaper alternative, especially if embankment construction costs are high. The soil used for embankment construction should be compacted in 150–250 mm layers to 90% of the maximum dry density determined by the modified Proctor test. Shrinkage of the soil occurs during compaction (10–30%) and excavation estimates must take this into account. After compaction, the soil should have a coefficient of permeability of $<10^{-7}$ m/s (measured *in situ*). Embankment design should allow for vehicle access to facilitate maintenance.

Embankment slopes are commonly 1–3 internally and 1–1.5–2 externally. Steeper slopes may be used if the soil is suitable; slope stability should be ascertained according to standard soil mechanics procedures for small earth dams. Embankments should be planted with grass to increase stability; slow growing rhizomatous species, such as *Cynodon dactylon* (Bermuda grass), should be used to minimize maintenance.

External embankments should be protected from stormwater erosion by providing adequate drainage. Internally, embankments require protection against erosion by wave action, and this is best achieved by precast concrete slabs or stone riprap at top water level. Such protection also prevents vegetation growing down the embankment into the pond and so providing a suitably shaded habitat for mosquito breeding.

Seepage losses must be at least smaller than the inflow less net evaporation, in order to maintain the water level in the pond. The maximum permissible permeability of the soil layer making up the pond base can be determined from d'Arcy's law:

$$k = (86\,400\,Q_s/A(\Delta l/\Delta h) \qquad (22)$$

where,

k = maximum permissible permeability, m/s
Q_s = maximum permissible seepage flow, m^3/d
A = base area of pond, m^2
Δl = depth of soil layer below pond base to aquifer or more permeable stratum, m
Δh = hydraulic head (= pond depth + Δh), m

If the permeability of the soil is more than the maximum permissible, the pond must be lined. A variety of lining materials is available and local costs dictate which should be used. Satisfactory lining has been achieved by stabilizing the soil with ordinary portland cement (8 kg/m^2), by plastic membranes and by introducing a 150 mm layer of low permeability soil.

Inlet and outlet structures

There is a wide variety of designs for inlet and outlet structures and provided basic concepts are followed their precise design is relatively unimportant. First of all, they should be simple and inexpensive; while this should be self-evident, it is all too common to see unnecessarily complex and expensive structures. Second, they should permit samples of the pond effluent to be taken with ease. The inlet to anaerobic and primary facultative ponds should discharge below the liquid level so as to minimize shortcircuiting (especially in deep anaerobic ponds) and reduce the quantity of scum (which is important in facultative ponds). Inlets to secondary facultative and maturation ponds can discharge either above or below the liquid level. Some simple inlet designs are shown in *Figure 9.7*.

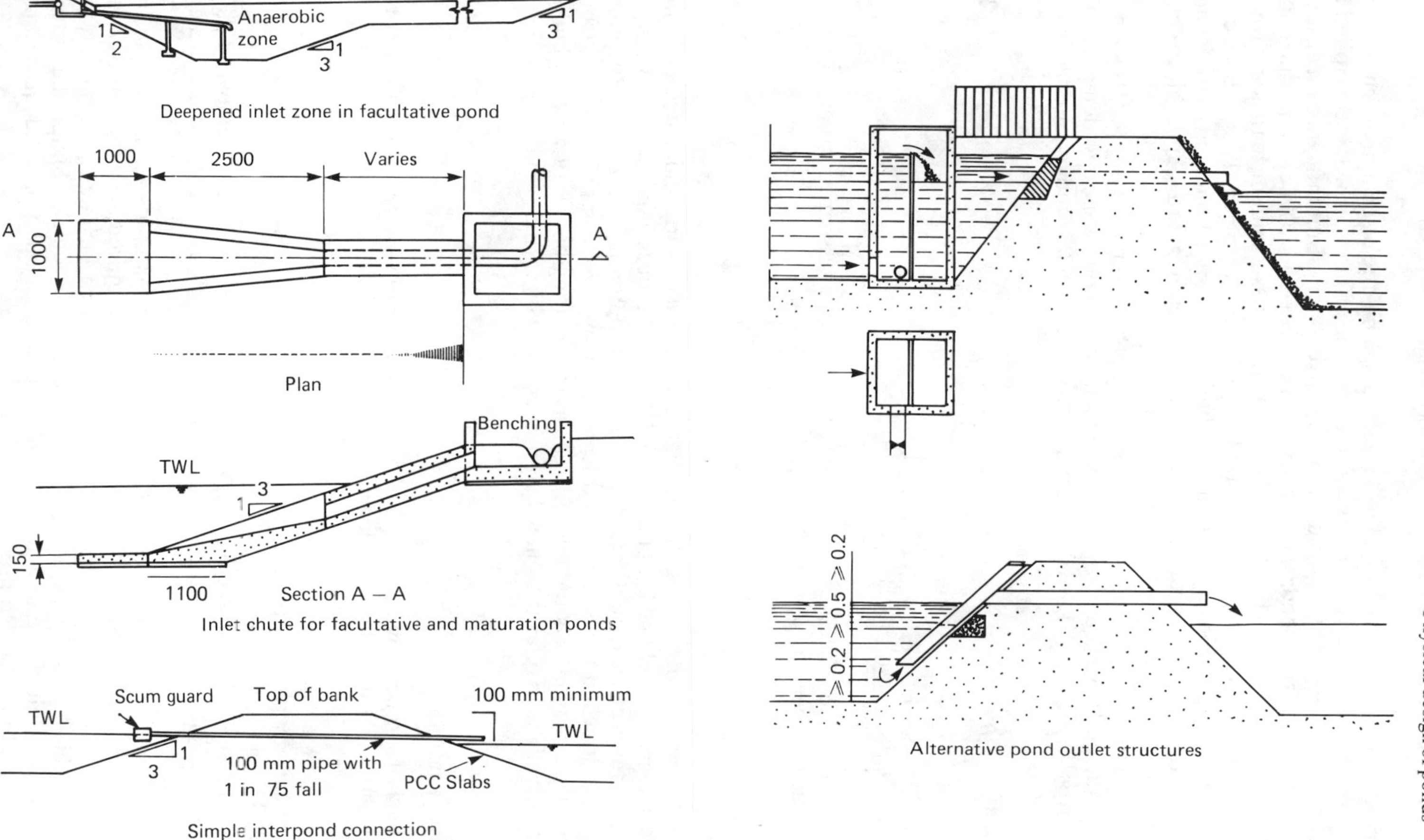

Figure 9.7 Inlet and outlet structures

The outlet of all ponds should be protected against the discharge of scum by the provision of a scum guard. The takeoff level for the effluent, which is controlled by the scum guard depth, is important as it can have a significant influence on effluent quality. In facultative ponds the scum guard should extend just below the maximum depth of the algal band when the pond is stratified, so as to minimize the quantity of algae (and hence BOD) leaving the pond; in practice a scum guard depth of 60 cm is adequate. In anaerobic and maturation ponds (where algal banding is irrelevant) the takeoff should be near the surface, and a scum guard depth of 15 cm is recommended.

Some simple designs for outlet structures are also shown in *Figure 9.7.* If a weir is used in the outlet structure the following formula should be used to determine the head over the weir and, knowing the pond depth, to calculate the required height of the weir:

$$q = 0.0567\, h^{3/2} \qquad (23)$$

where
q = flow per metre length of weir, l/s
h = head of water above weir, mm

The weir need not necessarily be strictly linear; often a U-shaped structure is more economical, especially at high flow rates.

Security

Ponds (other than in very remote installations) should be surrounded by a chainlink fence and gates should be kept padlocked. Warning notices, advising that the ponds are a wastewater treatment facility and therefore potentially hazardous to health, should be attached to the fence to discourage people visiting the ponds, which, if properly maintained, should appear as pleasant, inviting bodies of water. Children are especially at risk as they may be tempted to swim in the ponds.

Operator facilities

The facilities to be provided for pond operators depend partly on their number but should normally include the following:

- First aid kit.
- Strategically placed lifebuoys.
- Washbasin and toilet.
- Storage space for protective clothing, grass-cutting and scum-removal equipment, screen rake and other tools, sampling boat (if provided) and life jackets.

With the exception of the lifebuoys, these can be accommodated in a simple (for example wooden) building. This can also house, if required, sample bottles and a refrigerator for sample storage. For small systems such facilities are generally not provided but they should be available in the service vehicle. Simple laboratory facilities and a telephone should be provided at larger installations. Adequate space for car parking is also required.

Operation and maintenance of ponds

Start-up procedures

Pond systems should be commissioned during the hottest period of the year in order to establish as quickly as possible the necessary microbial populations to effect waste stabilization. Prior to commissioning, all ponds must be free from vegetation.

Anaerobic ponds should be filled with raw sewage and seeded with digesting sludge from, for example, an anaerobic digester at a conventional sewage treatment plant or with septage from local septic tanks. The ponds should then be gradually loaded up to the design loading rate over the following week (or month, if the ponds are not seeded); care should be taken to maintain the pond pH above 7 to permit the development of methanogenic bacteria. It may be necessary during the first month or so to dose the pond with lime or soda ash. If, due to an initially low rate of sewer connections in newly sewered towns, the sewage is weak or its flow low, it is best to bypass the anaerobic pond until the sewage strength and flow is such that a loading of at least 100 g $BOD_5/m^3 d$ can be applied to it.

It is preferable to fill facultative and maturation ponds with fresh water (from a river, lake or well; mains water is not necessary) so as to permit the gradual development of the algal and heterotrophic bacterial populations. Primary facultative ponds may advantageously be seeded in the same way as anaerobic ponds. If freshwater is not available, primary facultative ponds should be filled with raw sewage and left for 3–4 weeks to allow the microbial population to develop; some odour release is inevitable during this period.

Routine maintenance

The maintenance requirements of ponds are very simple, but they must be carried out regularly otherwise there will be serious odour, fly and mosquito nuisance. Routine maintenance tasks are as follows:

- Removal of screenings and grit from the preliminary works.
- Cutting grass on the embankment (this is necessary to prevent the formation of mosquito breeding habitats; the use of slow growing grasses minimizes this task).
- Removal of floating scum from the surface of facultative and maturation ponds (this is required to maximize photosynthesis and obviate fly breeding).
- Spraying the scum on anaerobic ponds (which should not be removed as it aids the treatment process), as necessary, with a suitable insecticide to prevent fly breeding).
- Removal of accumulated solids from the inlets and outlets.
- Repairing any damage to the embankments caused by rodents or other animals.
- Repairing any damage to external fences and gates.

The operators must be given precise instructions on the frequency at which these tasks should be done, and their work must be regularly inspected. Operators may also be required to take samples and carry out some routine measurements.

Solids in raw wastewater, as well as biomass produced, settle out in first-stage anaerobic ponds and they require desludging when half full of sludge. This occurs every n years:

$$n = V/2Ps \qquad (24)$$

where

V = volume of anaerobic pond, m^3
P = population served
s = sludge accumulation rate, m^3/person year

A suitable sludge accumulation design value is probably around 0.05 m^3/person year. Thus, at a design loading of 100 g BOD_5/m^3 and assuming a BOD_5 contribution of 50 g/person day, desludging would be required every 5 years. Sludge removal can be readily achieved by using a raft-mounted sludge pump which discharges into either an adjacent sludge lagoon or tankers which transport the sludge to a landfill site, central sludge treatment facility or other suitable disposal location.

Operator requirements

The number of operational staff required for pond systems is small by comparison with conventional sewage treatment works, and depends on:

- The size of the system.
- The type of preliminary treatment (manually or mechanically operated).
- The local cost and quality of labour.

In developing countries, for systems less than 8 ha, serving up to about 20 000 people, one full-time operator is sufficient; a half-time operator could be employed if grass-cutting, embankment repairs and other major maintenance work were done by a visiting service crew. Two full-time operators will be required for pond systems between 8 and 20 ha, serving up to 50 000 population. Operator training is essential if even such a simple wastewater treatment system is to perform satisfactorily on a continuous basis.

Monitoring and evaluation

Once a waste stabilization pond system has been commissioned, a routine monitoring and evaluation programme should be established so that its real, as opposed to design, performance can be determined and the quality of its effluent known. Routine monitoring of the final effluent quality of a pond system permits a regular assessment to be made of whether or not the effluent is complying with the local discharge or reuse standards. Moreover, should a pond system suddenly fail or its effluent start to deteriorate, the results of such a monitoring programme often give some insight into the cause of the problem and may indicate what remedial action is required.

Effluent quality monitoring programmes should be simple but provide reliable data. Representative samples of the raw wastewater and the final effluent should be taken at least monthly, although for small installations quarterly samples usually suffice. Since pond effluent quality shows a significant diurnal variation (although this is less pronounced in maturation ponds than in facultative ponds), 24 h flow-weighted composite samples are required for most parameters, although grab samples are satisfactory for some.

The evaluation of pond performance and behaviour, although a much more complex procedure than the routine monitoring of effluent quality (Pearson, Mara and Bartone, 1986), is nonetheless extremely useful as it provides information on how underloaded or overloaded the system is, and thus by how much, if any, the loading on the system can be safely increased as the community it serves expands, or whether further ponds (either in parallel or in series) are required. It also indicates how the design of future pond installations in the region might be improved to take account of local conditions.

Table 9.2 Economic cost comparison of waste stabilization pond system with other wastewater treatment processes

(1) Waste stabilization pond system (46 ha)			
			(US $ million)
	Costs:	Capital cost (including land)	5.68
		Operating cost	0.21
			5.89
	Benefits:	Irrigation income	0.43
		Pisciculture income	0.30
			0.73
Net present value: US $ 5.16 million			
(2) Aerated lagoon system (50 ha)			
			(US $ million)
	Costs:	Capital cost (including land)	6.98
		Operating cost	1.28
			8.26
	Benefits:	Irrigation income	0.43
		Pisciculture income	0.30
			0.73
Net present value: US $ 7.3 million			
(3) Oxidation ditch system (20 ha)			
			(US $ million)
	Costs:	Capital cost (including land)	4.80
		Operating cost	1.49
			6.29
	Benefits:	Irrigation income	0.43
Net present value: US $ 5.86 million			
(4) Biological filter system (25 ha)			
			(US $ million)
	Costs:	Capital cost (including land)	7.77
		Operating cost	0.86
			8.63
	Benefits:	Irrigation income	0.43
Net present value: US $ 8.20 million			

Source: Arthur (1983); see text for design assumptions and objectives.

Comparative cost of pond systems

Arthur (1983) has given the most recent and most detailed economic cost comparison between stabilization ponds, aerated lagoons, oxidation ditches and biological filters. He made certain assumptions (for example, use of maturation ponds following the aerated lagoon and chlorination of oxidation ditch and biological filter effluents) to ensure that the effluents from the four processes were of similar bacteriological quality, for irrigation or aquaculture reuse. Data for the comparison were taken from Sana'a in the Yemen Arab Republic and a design population of 250 000, with 120 l/capita day and 40 g BOD_5/capita day, were assumed. Influent faecal coliform level was taken as a 2×10^7/100 ml and an effluent standard of 1×10^4 faecal coliform/100 ml, and BOD_5 of 25 mg/l was required.

Assuming an opportunity cost of capital of 12% and land value of US\$5/m^2, Arthur calculated land area requirements and total net present value of each system as shown in *Table 9.2*. The waste stabilization pond option was cheapest under the conditions assumed but the minimum cost solution is very sensitive to the price of land. In most developing country locations, a land value of US\$5/m^2 would more than cover existing land costs and stabilization ponds would normally be the preferred option.

References

AD HOC PANEL OF THE US NATIONAL RESEARCH COUNCIL (1981) Food, Fuel and Fertilizer from Organic Wastes. Washington DC: National Academy Press

ARTHUR, J. P. (1983) Notes on the design and operation of waste stabilization ponds in warm climates of developing countries, *World Bank Technical Paper No. 7.* Washington DC: World Bank

ARTHUR, J. P. (1981) The development of design equations for facultative waste stabilization ponds in semi-arid areas. *Proceedings of the Institution of Civil Engineers,* **71(2),** 197–213

BENEMANN, J. R., WEISSMAN, J. C., KOOPMAN, B. L. and OSWALD, W. J. (1977) Energy production by microbial photosynthesis. *Nature,* **268,** 19–23

EDWARDS, P. (1980) A review of recycling organic wastes into fish with emphasis on the tropics. *Aquaculture,* **21,** 261–279

GAMBRILL, M. P., MARA, D. D., ECCLES, C. R. H. and BAGHAEI-YAZDI, N. (1986) Microcomputer-aided design of waste stabilization ponds in tourist areas of Mediterranean Europe. *Public Health Engineer,* **14(2),** 39–41

McGARRY, M. G. and PESCOD, M. B. (1970) Stabilization pond design criteria for tropical Asia. *Proceedings of the Second International Symposium on Waste Treatment Lagoons,* Kansas City, pp. 114–132

McGARRY, M. G., LIN, C. D. and MERTO, J. L. (1973) Photosynthetic yields and byproduct recovery from sewage oxidation ponds. *Advances in Water Pollution Research,* **4,** 521–531

MARA, D. D. (1976) *Sewage Treatment in Hot Climates.* Chichester: John Wiley

MARA, D. D. and PEARSON, H. W. (1986) Artificial freshwater environments: waste stabilization ponds. In: *Biotechnology, 8,* Weinheim: VCH Verlagsgesellschaft, pp. 177–208

MARA, D. D. and PEARSON, H. W. (1988) *Waste Stabilization Ponds: Design Manual for Mediterranean Europe.* Copenhagen: World Health Organization Regional Office for Europe

MARA, D. D. and SILVA, S. A. (1986) Removal of intestinal nematode eggs in tropical waste stabilization ponds, *Journal of Tropical Medicine and Hygiene,* **89 (2),** 71–74

MARAIS, G. V. R. (1970a) Dynamic behaviour of oxidation ponds. *Proceedings of the Second International Symposium on Waste Treatment Lagoons,* Kansas City, pp. 15–46

MARAIS, G. V. R. (1970b) Faecal bacterial kinetics in stabilization ponds, *Journal of the American Society of Civil Engineers, Environmental Engineering Division,* **100 (EEI),** 119–139

MARAIS, G. V. R. and SHAW, V. A. (1961) A rational theory for design of sewage stabilization ponds in Central and South Africa. *Transactions of the South African Institute of Civil Engineering,* **3,** 205

MEIRING, P. G., DREWS, R. M., van ECK, H. and STANDER, G. J. (1968) A guide to the use of pond systems in South Africa for the purification of raw or partially treated sewage. *CSIR Special Report WAT 34.* Pretoria: National Institute for Water Research

O'BRIEN, W. J. (1981) Use of aquatic macrophytes for wastewater treatment. *Journal of the American Society of Civil Engineers, Environmental Engineering Division,* **107, (EE4),** 681–698

OSWALD, W. J. (1975) *Waste Pond Fundamentals.* Unpublished document, Washington DC: World Bank

OTIS, R. J. and MARA, D. D. (1985) *The Design of Small Bore Sewer Systems.* TAG Technical Note No. 14, Washington DC

PEARSON, H. W., MARA, D. D. and BARTONE, C. R. (1986) Guidelines for the minimum evaluation of the performance of full-scale waste stabilization pond systems. *Water Research,* **21 (9),** 1067–1076

UN ECONOMIC AND SOCIAL COMMISSION FOR ASIA AND THE PACIFIC (1981) *Renewable Sources of Energy,* Vol. II, Bangkok: Biogas ECDC-TCDC Directory

WORLD HEALTH ORGANIZATION (1987) *Wastewater Stabilization Ponds; Principles of Planning and Practice. WHO-EMRO Technical Publication No.10.* Alexandria: Regional Office for the Eastern Mediterranean

Chapter 10

Groundwater recharge as a treatment of sewage effluent for unrestricted irrigation

Herman Bouwer*

Introduction

Sewage effluent is almost always a nuisance, even after it is treated. It may pollute surface water, it can present odour problems, it may spread disease, or it can aggravate mosquito or other insect problems. In dry areas, people are tempted to use it for irrigation, which can be a danger to public health when the crops are consumed by humans, directly or indirectly. Irrigation, however, is one of the best uses for sewage effluent because irrigation does not require top quality water, and the effluent contains nutrients (nitrogen, phosphorus, potassium and minor elements) that are of value to crops. The problem with using sewage effluent for irrigation, of course, is that it also contains disease-causing organisms and sometimcs undesirable chemicals if the effluent is high in industrial wastes. These problems can be overcome by proper treatment and by source control in the case of industrial wastes. Specifically, where sewage effluent is used for irrigation it should be treated so that (a) it meets public health requirements; (b) it meets chemical quality requirements for irrigation water, from a crops and soils standpoint; and (c) it is aesthetically acceptable to the public.

Quality requirements

The quality requirements from a public health standpoint depend on the crops that are irrigated and how they are consumed by humans. If only non-edible crops are grown (fibre and seed crops, for example), primary treatment may suffice. At the other extreme, if the sewage effluent is used for general irrigation, including vegetables consumed raw by people or crops brought raw into the kitchen where they can contaminate other food, the effluent must be treated to remove all disease-causing organisms (bacteria, viruses, protozoa and eggs of parasitic worms). This usually requires primary and secondary treatment of the effluent, followed by filtration and disinfection. An example of the various classes of crops or plants that can be irrigated with sewage effluent and the corresponding water quality requirements for public health considerations is shown in *Table 10.1*. Where sewage treatment is limited and only non-edible crops (fibre, seed or forage crops)

* Director and Research Hydraulic Engineer, US Water Conservation Laboratory, Agricultural Research Service, US Department of Agriculture

Table 10.1 Standards for irrigation with reclaimed wastewater in Arizona

	Crop and land use category							
	A	*B*	*C*	*D*	*E*	*F*	*G*	*H*
pH	4.5–9	4.5–9	4.5–9	6.5–9	4.5–9	4.5–9	4.5–9	4.5–9
Faecal coliforms (CFU/100 ml)								
Geometric mean (five-sample minimum)	1000	1000	1000	1000	1000	200	25	2.2
Single sample not to exceed	4000	4000	4000	4000	2500	1000	75	25
Turbidity (NTU)	–	–	–	–	–	–	5	1
Enteric virus (PFU/40 l)	–	–	–	–	–	–	125	1
Entamoeba histolytica	–	–	–	–	–	–	–	ND
Ascaris lumbricoides (roundworm eggs)	–	–	–	–	–	–	ND	ND
Common large tapeworm	–	–	ND	ND	–	–	–	–

Source: Matters (1981)

CFU colony-forming unit
NTU nephelometric turbidity units
PFU plaque-forming units
ND none detectable, using correct samples and methods and qualified personnel

A Orchards
B Fibre, seed and forage crops
C Pastures
D Livestock watering
E Processed food crops
F Landscaped areas, restricted access
G Landscaped areas, open access
H Crops to be consumed raw

can be irrigated, a strict programme of inspection and enforcement is necessary to make sure that farmers do not use the sewage effluent to grow some vegetable crops on the side.

The chemical quality requirements for using sewage effluent for irrigation are the same as those for regular irrigation water. These requirements are formulated to avoid adverse effects on the soil (primarily deflocculation of clay and decline of soil structure as caused by excessive sodium concentrations) and on the plant (salinity effects, excessive nitrogen and specific ion toxicities, particularly of trace elements). Examples of chemical quality requirements of irrigation water are shown in *Tables 10.2* and *10.3*. Most sewage effluents will meet these requirements. If the salinity of the effluent is high, salt-tolerant crops should be grown and special management techniques (adequate leaching, use of high quality water for germination of the crop, planting on the side of the ridges instead of on top of the ridges in furrow-irrigated crops) should be used to avoid undesirable effects due to salinity. Excessive concentrations of certain trace elements are usually due to industrial waste discharges in the sewer system. These can be reduced by proper source control and blending with other water is also a possibility. The nitrogen concentration requirements in *Table 10.2* are for general crop irrigation in warm, dry climates where water applications may be of the order of 1–2 m per season or per year. For cooler climates or supplemental irrigation, or where only grasses or other forage crops are grown, much higher nitrogen concentrations can be allowed.

The aesthetic aspects of irrigation with sewage effluent are of most concern in densely populated areas, but they are also important for the farmers and workers handling the effluent water. The aesthetic aspects involve the appearance of the effluent and its odour. Thus, the concentrations of suspended solids and of biodegradable organic matter should be minimized so that the water looks clear and does not smell.

Table 10.2 Guidelines for interpretation of water quality for irrigation

Problems and quality parameters	*No problems*	*Increasing problems*	*Severe problems*
Salinity effects on crop yield			
Total dissolved-solids concentration (mg/l)	<480	480–1920	>1920
Deflocculation of clay and reduction in K and infiltration rate			
Total dissolved-solids concentration (mg/l)	>320	<320	<128
Adjusted sodium adsorption ratio (SAR)	<6	6–9	>9
Specific ion toxicity			
Boron (mg/l)	<0.5	0.5–2	2–10
Sodium (as adjusted SAR) if water is absorbed by roots only	<3	3–9	>9
Sodium (mg/l) if water is also absorbed by leaves	<69	>69	
Chloride (mg/l) if water is absorbed by roots only	<142	142–355	>355
Chloride (mg/l) if water is also absorbed by leaves	<106	>106	
Quality effects			
Nitrogen in mg/l (excess N may delay harvest time and adversely affect yield or quality of sugar beets, grapes, citrus, avocados, apricots, etc.)	<5	5–30	>30
Bicarbonate as HCO_3 in mg/l (when water is applied with sprinklers, bicarbonate may cause white carbonate deposits on fruits and leaves)	<90	90–520	>520

Source: Ayers (1975)

Soil–aquifer treatment

Where sewage effluent is to be used for unrestricted irrigation (category H in *Table 10.1*), a fairly intensive treatment technology is required to meet the quality requirements. Typically, the treatment would have to consist of primary (settling) treatment, secondary (biological) treatment, coagulation, sedimentation, sand filtration and chlorination. The primary and secondary treatment steps could be replaced by lagooning or oxidation ditch treatment (carousel systems). Where hydrogeological conditions are favourable for groundwater recharge with infiltration basins, however, the necessary treatment can be obtained very simply by the filtration process as the sewage percolates through the soil and the vadose zone, down to the groundwater and then some distance through the aquifer. This 'soil–aquifer treatment' (SAT) process converts the sewage effluent into 'renovated water', which can be collected from the aquifer by wells or drains. Under the proper soil and hydrological conditions, the SAT process removes essentially all suspended solids, micro-organisms (bacteria, viruses, protozoa, parasitic ova, etc.) and phosphorus, and significantly reduces the concentrations of nitrogen and heavy metals. Thus, the renovated water from an SAT groundwater

Table 10.3 Recommended maximum limits in mg/l for trace elements in irrigation water

	Permanent irrigation of all soils	*Up to 20 year irrigation of fine-textured, neutral to alkaline soils (pH 6–8.5)*
Aluminium	5	20
Arsenic	0.1	2
Beryllium	0.1	0.5
Boron-sensitive crops	0.75	2
Boron-semitolerant crops	1	
Boron-tolerant crops	2	
Cadmium	0.01	0.05
Chromium	0.1	1
Cobalt	0.05	5
Copper	0.2	5
Fluoride	1	15
Iron	5	20
Lead	5	10
Lithium: citrus	0.075	0.072
Lithium: other crops	2.5	2.5
Manganese	0.2	10
Molybdenum	0.01	0.05*
Nickel	0.2	2
Selenium	0.02	0.02
Vanadium	0.1	1
Zinc	2	10

* For acid soils only

Source: National Academy of Sciences and National Academy of Engineering (1972)

recharge system will generally meet the public health, chemical and aesthetic requirements for unrestricted irrigation (unless, of course, the salt concentration of the sewage effluent is very high or there are toxic substances, such as boron, in the effluent at high concentrations that move through the soil and aquifer without much change in concentration). The effluent should have had primary treatment (plus secondary treatment if available) or lagooning or oxidation ditch treatment before it is used for infiltration into the SAT system. Pretreatment processes that leave high algal concentrations in the effluent should be avoided, however, because algae can severely clog the soil of infiltration basins.

While the renovated water from an SAT system is of much better quality than the sewage effluent going into the soil, it could be of lower quality than the native groundwater. Thus, SAT systems should be designed and managed to avoid encroachment into the native groundwater and to use only a portion of the aquifer for renovation of wastewater. This can be achieved with the systems shown in *Figure 10.1,* where the vadose zone below the infiltration basins and the portion of the aquifer between the infiltration basins and the collection facilities (wells or drains) is dedicated to SAT of partially treated sewage effluent. The distance between infiltration basins and wells or drains should be as large as possible (usually at least 50–100 m) to give adequate SAT. If the groundwater is deep and/or the soils are relatively fine, flow distances through the aquifer can be reduced.

If the groundwater table is high, the renovated water can be intercepted with open or closed drains (*Figure 10.1,* top). If the groundwater is deep, the renovated water must be collected using wells. Two configurations are possible. In one (*Figure 10.1,* centre), the infiltration basins are located in a cluster and are surrounded by a circle of wells. The wells then tend to pump a mixture of renovated water and of native groundwater from the aquifer outside the SAT system. This gives dilution of the renovated water with native groundwater but it is not desirable where pumping of native groundwater needs to be minimized or where the renovated water needs to undergo additional treatment. In the other layout (*Figure 10.1,* bottom), the infiltration basins are arranged in two parallel strips with the wells located on a line midway between the strips. Here, the wells deliver essentially 100% renovated water. Movement of renovated water into the aquifer outside the SAT system is avoided by monitoring groundwater levels with

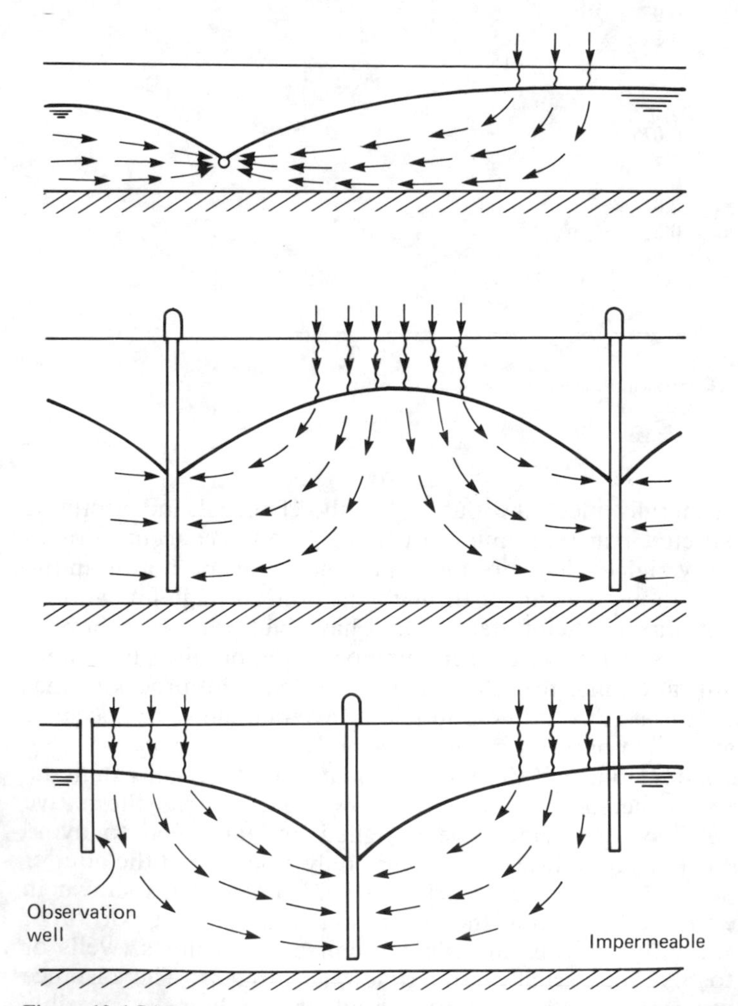

Figure 10.1 SAT systems with renovated-water recovery by drains (*top*), infiltration basins in centre surrounded by a circle of wells for pumping renovated water out of aquifer (centre), and infiltration basins in two parallel strips with wells midway between strips for pumping renovated water (*bottom*)

observation wells at the periphery of the infiltration strips (*see Figure 10.1,* bottom) and managing infiltration and pumping rates so that the water levels in the monitoring wells do not rise higher than the groundwater table in the aquifer outside the SAT system. This prevents hydraulic gradients and, hence, water movement from the SAT system to the native groundwater outside the system.

A variation of the system in *Figure 10.1* (top), is used in humid areas to reduce pollution of streams or lakes by sewage effluent or similar wastewater. Instead of discharging such wastewater directly into the stream or lake, it is applied to infiltrate basins a few hundred metres or so away from the stream or lake. After infiltration and movement to the underlying groundwater, the resulting renovated water then moves through the aquifer to the stream or lake, where it causes much less pollution than direct discharge of wastewater into surface waters. These systems require that the groundwater table is higher than the water level in the stream or lake, so that the groundwater flows to the surface water.

SAT systems require surface soils that are fine enough to give good purification of the sewage water, coarse enough to give high infiltration rates, and deep enough to give good treatment of the water before it enters the aquifer or other coarse-textured formations. Sandy loams to loamy sands several metres deep are very good soil profiles for wastewater renovation by SAT.

Phoenix plans

The city of Phoenix in south-central Arizona, Unites States, is interested in renovating part of its sewage effluent by SAT and in exchanging the renovated water with a nearby irrigation district for high quality groundwater from that district which the city would then use to augment its municipal water supply. There are two major sewage treatment plants in the Phoenix area: the 91st Avenue treatment plant (activated sludge, chlorination, capacity about 450 000 m^3/day or 120 M gal.) and the 23rd Avenue treatment plant (activated sludge, chlorination, capacity about 150 000 m^3/day or 40 M gal.). The SAT system for the Phoenix area would consist of a series of infiltration basins arranged in two parallel strips with wells on a line midway between the strips (*see Figure 10.1,* bottom). The feasibility of SAT in the Phoenix area was studied with two experimental systems, a small test project installed in 1967 and a larger demonstration project installed in 1975. The latter will be part of a future operational project (*Figure 10.2*) that will have a basin area of 48 ha and a projected capacity of about 50 million m^3/year.

As an example of the hydraulic capacities and the treatment that can be obtained with SAT systems, the results of these two test projects will be summarized in the remainder of this chapter.

Results of SAT test projects

Project descriptions and hydraulic loading rates

The first project was the Flushing Meadows project (Bouwer, Lance and Riggs, 1974; Bouwer *et al.*, 1980). This was an experimental project installed in the Salt River bed in 1967. It consisted of six parallel, long, narrow infiltration basins of about 0.13 ha each. The soil consisted of about 1 m of loamy sand underlain by sand and gravel layers. The groundwater table was at a depth of around 3 m. Monitoring

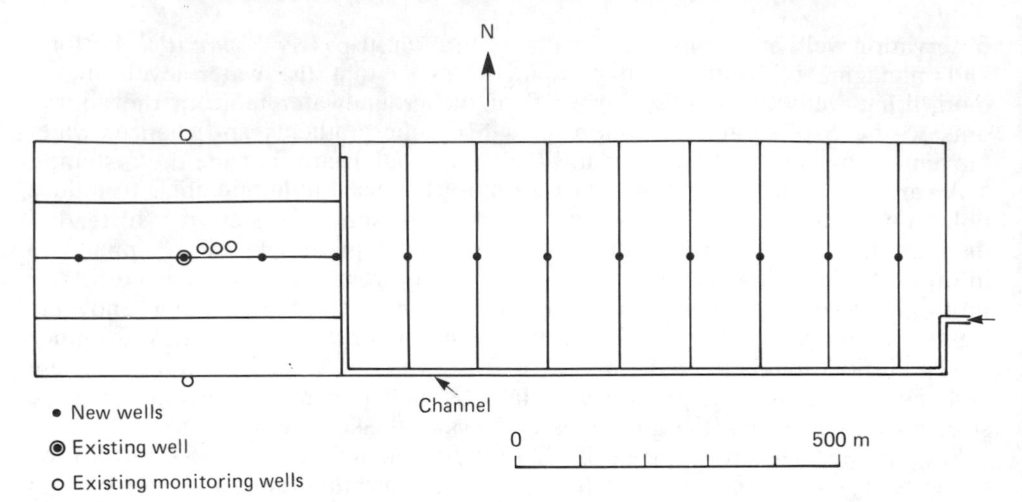

Figure 10.2 Sketch of 23rd Avenue project with 16 ha lagoon (*left*) that was split into four infiltration basins, and 32 ha lagoon (*right*) that can be split into nine infiltration basins to increase capacity of system to that of plant outflow. The 12 wells for pumping renovated water are shown as dots. Also shown is the bypass-supply channel

wells 6–9 m deep were installed at various points between the basins and away from the basins. This made it possible to sample renovated wastewater from the aquifer below the basins and after the water had moved laterally for some distance through the aquifer.

The second project was the 23rd Avenue project (Bouwer and Rice, 1984). This is a demonstration and future operational project installed in 1975 on the north side of the Salt River bed. It consists of a 16 ha lagoon split lengthwise into four infiltration basins of 4 ha each (*see Figure 10.2*). The soil lacks the loamy sand top layer of the Flushing Meadows project. Thus, the soil profile consists mostly of sand and gravel layers. The water table depth in the study period ranged between 5 m and 25 m and was mostly around 15 m. Monitoring wells for sampling renovated wastewater were installed in the centre of the project at depths of 18, 24 and 30 m, and on the north side of the basin complex at depths of 22 m (*see Figure 10.2*). In addition, a large production well (capacity about 10 000 m^3/day or 2000 gal/min) was drilled in the centre of the project with the casing perforated from 30 to 54 m depth.

Flooding and drying schedules were mostly 9 days flooding–12 days drying at the Flushing Meadows project and 14 days flooding–14 days drying at the 23rd Avenue project. Water depths in the basins were about 15–20 cm. During flooding, infiltration rates typically were between 0.3 and 0.6 m/day yielding a total infiltration or hydraulic loading of between 60 and 120 m/year for Flushing Meadows and about 100 m/year for the 23rd Avenue project. For the latter, the effluent from the treatment plant initially flowed through a 32 ha lagoon (*see Figure 10.2*) before it entered the infiltration basins. This gave problems of soil clogging in the infiltration basins due to heavy growth of algae in the lagoon, especially in the summer. The unicellular algae *Carteria klebsii* were particularly troublesome. In addition to forming a 'filter cake' on the bottom of the infiltration basin, the algae removed CO_2 from the wastewater for photosynthesis, which raised the pH and, in turn, caused precipitation of $CaCO_3$, which further aggravated the soil clogging.

Algal growth and the resulting soil clogging were avoided by constructing a bypass canal around the 32 ha lagoon (*see Figure 10.2*), reducing the detention time of the effluent from a few days to about half an hour. After the bypass channel was put into operation, hydraulic loading rates for the infiltration basins increased from 21 m/year to almost 100 m/year.

At a hydraulic loading rate of 100 m per year, 1 ha of infiltration basin can handle $100 \times 10\,000 = 10^6$ m^3 of wastewater/year. Thus, the 150 000 m^3/day of effluent from the 23rd Avenue wastewater treatment plant would require 55 ha (or about 135 acres) of infiltration basins, at 40 M gal./d of effluent. Almost all of this area can be obtained by converting the 32 ha lagoon east of the present infiltration system into infiltration basins (*see Figure 10.2*). This would give a total basin area of about 48 ha (120 acres), that could handle 48 million m^3/year (39 000 acre feet/year or 35 M gal./d) of effluent. The wells for pumping the renovated water from the aquifer could be located on the centreline through the project (*see Figure 10.2*). At a capacity of 10 000 m^3/day (about 2000 gal./min.) per well, 12 wells would be needed to pump renovated water out of the aquifer at the same rate as it infiltrates as wastewater in the basins, thus creating an equilibrium situation.

Quality improvements

For both projects, most of the quality improvement of the wastewater occurred in the vadose zone (this is the zone between soil surface and groundwater table). The quality improvements for these projects will be summarized in the following paragraphs. For additional details, reference is made to H. Bouwer, Lance and Riggs (1974); Bouwer *et al.* (1980) and Bouwer and Rice (1984), and to E. J. Bouwer *et al.* (1984).

Suspended solids

The suspended solids content of the renovated water at the Flushing Meadows project was less than 1 mg/l. From the 23rd Avenue project it averaged about 1 mg/l for the large production well. Most of these solids probably were fine aquifer particles that entered the well through the perforations in the casing. The suspended solids content of the secondary effluent at the 23rd Avenue project averaged about 11 mg/l.

Total dissolved solids

The total salt content of the renovated water increased slightly as it moved through the SAT system (from 750 to 790 mg/l at the 23rd Avenue project). Evaporation from the basins (including from the soil during drying) should increase the TDS content by about 2%. The rest of the increase probably was due to mobilization of calcium carbonate due to a pH drop from 8 to 7 as the effluent moved through the vadose zone.

Nitrogen

At the Flushing Meadows project, nitrogen removal from the effluent as it seeped through the vadose zone to become renovated water was about 30% at maximum hydraulic loading (100–120 m/year), but 65% of the loading rate was reduced to about 70 m/year, by using 9-day flooding–12-day drying cycles, and by reducing the water depths in the basins from 0.3 to 0.15 m. The form and concentration of nitrogen in the renovated water sampled from the aquifer below the basins were slow to respond to the reduction in hydraulic loading (Bouwer *et al.*, 1980). In the 10th year of operation (1977), the renovated water contained 2.8 mg/l of ammonium nitrogen, 6.25 mg/l nitrate nitrogen and 0.58 mg/l organic nitrogen, for a total nitrogen content of 9.6 mg/l. This was 65% less than the total nitrogen of the secondary sewage effluent, which averaged 27.4 mg/l (mostly as ammonium) in that year. At the 23rd Avenue project, the total N content in the secondary sewage effluent averaged about 18 mg/l, of which 16 mg/l was as ammonium. The 2-week flooding–drying cycles must have been conducive to denitrification in the vadose zone, because the total N content of the renovated water from the large centre well averaged about 5.6 mg/l, of which 5.3 mg/l was as nitrate, 0.1 mg/l as ammonium, 0.1 mg/l as organic and 0.02 mg/l as nitrite. The nitrogen removal thus was about 70%. This removal was the same before and after the secondary effluent was chlorinated, indicating that the low residual chlorine of the effluent by the time it infiltrated into the ground apparently had no effect on the nitrogen transformations in the soil.

The flooding and drying sequence that maximize denitrification in the vadose zone depends on various factors and must be evaluated for each particular system. Pertinent factors include the ammonium and carbon contents of the effluent entering the soil, infiltration rates, cation-exchange capacity of soil, exchangeable ammonium percentage, depth of oxygen penetration in the soil during drying and temperature. The combined laboratory and field data from the Flushing Meadows experiments showed that, to achieve high nitrogen removal percentages, the amount of ammonium nitrogen applied during flooding must be balanced against the amount of oxygen entering the soil during drying. Flooding periods must be long enough to develop anaerobic conditions in the soil. Infiltration rates must be controlled to the appropriate level for the particular effluent, soil and climate at a given site. Most of the nitrogen transformations in the Flushing Meadows studies occurred in the upper 50 cm of the vadose zone.

Phosphate

Phosphate removal increased with increasing distance of underground movement of the sewage effluent. After 3 m of downward movement through the vadose zone and 6 m through the aquifer, phosphate removal at the Flushing Meadows project was about 40% at high hydraulic loading and 80% at reduced hydraulic loading. Additional lateral movement of 60 m through the aquifer increased the removal to 95% (that is, to a concentration of 0.51 mg/l phosphate phosphorus versus 7.9 mg/l in the effluent). After 10 years of operation and a total infiltration of 754 m of secondary effluent, there were no signs of a decrease in phosphate removal. At the 23rd Avenue project, phosphate phosphorus concentrations in the last few years of the research averaged 5.5 mg/l for the secondary effluent and 0.37 mg/l for the renovated water pumped from the centre well. The shallower wells showed a higher phosphate content, indicating that precipitation of phosphate continued in the aquifer. For example, renovated water sampled from the 22 m deep north well showed phosphate phosphorus concentrations that averaged 1.5 mg/l. Most of the phosphate removal was probably due to precipitation of calcium phosphate.

Fluoride

Fluoride removal paralleled phosphate removal, indicating precipitation as calcium fluoride. At the Flushing Meadows project, fluoride concentrations in 1977 were 2.08 mg/l for the effluent, 1.66 mg/l for the renovated water after it had moved 3 m through the vadose zone and 3–6 m through the aquifer and 0.95 mg/l after it had moved an additional 30 m through the aquifer. At the 23rd Avenue project, fluoride concentrations averaged 1.22 mg/l in the secondary effluent and 0.7 mg/l in the renovated water from the centre well.

Boron

Boron was not removed in the vadose zone or the aquifer and was present at concentrations of 0.5 to 0.7 mg/l in both effluent and renovated water. The lack of boron removal was due to the absence of significant amounts of clay in the vadose zone and aquifer.

Metals

At the Flushing Meadows project, movement of the secondary effluent through 3 m of vadose zone and 6 m of aquifer reduced zinc from 193 to 35 μg/l, copper from 123 to 16 μg/l, cadmium from 7.7 to 7.2 μg/l and lead from 82 to 66 μg/l (Bouwer, Lance and Riggs, 1974). Cadmium thus appeared to be the most mobile metal.

Faecal coliforms

The secondary effluent at the Flushing Meadows project was not chlorinated and contained 10^5–10^6 faecal coliforms/100 ml. Most of these were removed in the top metre of the vadose zone but some penetrated to the aquifer, especially when a new flooding period was started. The deeper penetration of faecal coliforms at the beginning of a flooding period was attributed to less straining of bacteria at the soil surface because the clogged layer had not yet developed. Also, the activity of native soil bacteria at the end of a drying period was lower, producing a less antagonistic environment for the faecal coliforms in the soil when flooding was resumed. Faecal coliform concentrations in the water after 3 m of travel through the vadose zone and 6 m through the aquifer were 10–500/100 ml when the renovated water consisted of water that had infiltrated at the beginning of a flooding period and 0–1/100 ml after continued flooding. Additional lateral movement of about 100 m through the aquifer was necessary to produce renovated water that was completely free of faecal coliforms at all times.

At the 23rd Avenue project, faecal coliform concentrations in the secondary sewage effluent entering the infiltration basins were 10 000/100 ml prior to November 1980, when the effluent was not yet chlorinated and was first passed through a 32 ha lagoon. This concentration increased to 1.8×10^6/100 ml when the unchlorinated effluent was bypassed around the lagoon and flowed directly into the infiltration basins. It then decreased to 3500/100 ml after the effluent was chlorinated and still bypassed around the lagoon. The corresponding faecal coliform concentrations in the water pumped from the large centre well from a depth of 30–54 m averaged 2.3, 22 and 0.27/100 ml, respectively. The corresponding ranges were 0–40, 0–160 and 0–3/100 ml, respectively. Considerable faecal coliform concentrations were observed in the renovated water from the shallower wells, especially when the faecal coliform concentration of the infiltrating effluent was 1.8×10^6/100 ml. At that time, water from the 18 m deep well showed coliform peaks after a new flooding period was started that regularly exceeded 1000/100 ml and at one time even reached 17 000/100 ml. Thus, a considerable number of faecal coliforms passed through the vadose zone. However, chlorination of the effluent and resulting reduction of the faecal coliform concentration to 3500/100 ml prior to infiltration, and additional movement of the water through the aquifer to the centre well, produced renovated water that was essentially free of faecal coliforms.

Viruses

At the Flushing Meadows project, the virus concentrations of unchlorinated secondary effluent averaged 2118 plaque-forming units (PFU)/100 l (average of six bimonthly samples taken for 1 year). They included poliovirus, echovirus, coxsackie and reoviruses. No viruses could be detected in renovated water sampled after 3 m of movement through the vadose zone and 3–6 m movement through the aquifer. At the 23rd Avenue project, virus concentrations in the renovated water from the centre well averaged 1.3 PFU/100 l before chlorination of the secondary effluent and 0 PFU/100 l after chlorination of the secondary effluent. The combined effects of chlorination and SAT thus apparently resulted in complete removal of the viruses.

Organic carbon

At the Flushing Meadows project, the biochemical oxygen demand (BOD_5) of the effluent water after moving 3 m through the vadose zone and 6 m through the aquifer was essentially zero, indicating that almost all biodegradable carbon was mineralized. However, the renovated water still contained about 5 mg/l total organic carbon (TOC), as compared to 10–20 mg/l of TOC in the secondary effluent. At the 23rd Avenue project, the TOC concentration of the secondary effluent averaged 12 mg/l where it entered the infiltration basins and 14 mg/l at the opposite ends of the basins. This increase was probably due to biological activity in the water as it moved through the basins. The renovated water from the 18 m well (intake about 5 m below the bottom of the vadose zone) had a TOC content of 3.2 mg/l and that from the centre well (which pumped from 30 to 54 m depth) had a TOC content of 1.9 mg/l, indicating further removal of organic carbon as the water moved through the aquifer. The TOC removal in the SAT system was the same before and after chlorination of the secondary effluent, indicating that chlorination had no effect on the microbiological processes in the soil.

The concentration of organic carbon in the renovated water of 1.9 mg/l was higher than the 0.2–0.7 mg/l typically found in unpolluted groundwaters. The latter are mostly due to humic substances, such as fulvic and humic acids (Thurman, 1979). The renovated water from the SAT process could thus contain a number of synthetic organic compounds, some of which could be carcinogenic or otherwise toxic.

Removal of trace organic compounds in the vadose zone

The nature and concentration of trace organics in the secondary sewage effluent and in the renovated water from the various wells of the 23rd Avenue project were determined by Stanford University's Environmental Engineering and Science Section, using gas chromatography and mass spectrometry. The studies were carried out for 2 months with unchlorinated effluent and then for 3 months with chlorinated effluent, taking weekly or biweekly samples. As could be expected, the results showed a wide variety of organic compounds, including priority pollutants (many in concentrations of the order of μg/l, see E. J. Bouwer *et al.* 1984, and H. Bouwer and Rice, 1984).

Chlorination had only a minor effect on the type and concentration of organic compounds in the sewage effluent. Of the volatile organic compounds, 30–70% were lost by volatization from the infiltration basins. Soil percolation removed 50–99% of the non-halogenated organic compounds, probably mostly by microbial decomposition. Concentrations of halogenated organic compounds decreased to a lesser extent with passage through the soil and aquifer. Thus, halogenated organic compounds (including the aliphatic compounds chloroform, carbon tetrachloride, trichloroethylene and 1,1,1-trichloroethane and the aromatic dichlorobenzenes, trichlorobenzenes and chlorophenols) were more mobile and refractory in the underground environment than the non-halogenated compounds, which included the aliphatic nonanes, hexanes and octanes, and the aromatic xylenes, C3-benzenes, styrene, phenanthrene and diethylphthalate.

Other organic micropollutants
In addition to the aliphatic and aromatic compounds mentioned, other compounds tentatively identified in organic extracts of the samples of secondary sewage effluent and renovated water using gas chromatography–mass spectrometry were: fatty acids, resin acids, clofibric acid, alkylphenol polyethoxylate carboxylic acids (APECs), trimethylbenzene sulfonic acid, steroids, *n*-alkanes, caffeine, Diazinon, alkylphenol polyethoxylates (APEs) and trialkylphosphates. Several of the compounds were detected only in the secondary effluent and not in the renovated water. A few others – Diazinon, clofibric acid and tributylphosphate – decreased in concentration with soil passage but were still detected in the renovated water. The APEs appeared to undergo rather complex transformations during ground filtration. They appeared to be completely removed by soil percolation during the prechlorination period but, after chlorination, two isomers were found following soil passage, while others were removed.

The results of these studies showed that SAT is effective in reducing concentrations of a number of synthetic organic compounds in the sewage effluent but that the renovated water still contains a wide spectrum of organic compounds, albeit at very low concentrations. Thus, while the renovated water is suitable as such for unrestricted irrigation and recreation, recycling it for drinking would require additional treatment, such as activated carbon filtration, to remove the remaining organic compounds. The water would also have to be disinfected and reverse osmosis may be desirable.

Conclusions

The results of the Phoenix studies show that the renovated water from the SAT projects meets the public health, agronomic and aesthetic requirements for unrestricted irrigation, including vegetable crops that are consumed raw (Bouwer, 1982). The water also meets the standards for lakes with primary contact recreation (Matters, 1981). Potable use of the renovated water will require additional treatment, for example, activated carbon adsorption, reverse osmosis and disinfection. Such treatment, however, will be more effective and economical for renovated water from a soil–aquifer treatment system than for effluent from a conventional sewage treatment plant.

In the Phoenix studies, secondary effluent was used because that was what the treatment plants provided. The secondary (biological) treatment step, however, is not necessary because the SAT system can handle relatively large amounts of organic carbon. Thus, where sewage effluent is to be used for a rapid-infiltration system, primary treatment may suffice (Carlson *et al.,* 1982; Lance, Rice and Gilbert, 1980; Leach, Enfield and Harlin, 1980; Rice and Gilbert, 1978; Rice and

Bouwer, 1983). Some additional clarification or lime precipitation of the primary effluent may be desirable, however, to reduce suspended solids and to improve the quality of the primary effluent.

The cost of placing partially treated wastewater underground and pumping it from wells as renovated water is low and SAT systems do not require highly trained operators. Thus, where land availability and hydrogeological conditions are favourable for groundwater recharge by surface spreading of wastewater, SAT can play an important role in the reuse and recycling of municipal wastewater.

References

AYERS, R. S. (1975) Quality of water for irrigation, *Proceedings of the Irrigation and Drainage Division, Speciality Conference,* American Society of Civil Engineers, Logan, Utah, 13–15 August, pp. 24–56

BOUWER, E. J., McCARTY, P. L., BOUWER, H. and RICE, R. C. (1984) Organic containment behaviour during rapid infiltration of secondary wastewater at the Phoenix 23rd Avenue project. *Water Research,* **18(4),** 463–472

BOUWER, H. (1982) Wastewater reuse in arid areas. In: *Water Reuse,* edited by E. J. Middlebrooks. Ann Arbor, Ann Arbor Science Publishers, Inc., pp. 137–180

BOUWER, H., LANCE, J. C. and RIGGS, M. S. (1974) High rate land treatment. II. Water quality and economic aspects of the Flushing Meadows project. *Journal Water Pollution Control Federation,* **46(5),** 844–859

BOUWER, H., RICE, R. C., LANCE, J. C. and GILBERT, R. G. (1980) Rapid-infiltration research – The Flushing Meadows project, Arizona. *Journal Water Pollution Control Federation,* **52(10),** 2457–2470

BOUWER, H. and RICE, R. C. (1984) Renovation of wastewater at the 23rd Avenue rapid-infiltration project. *Journal Water Pollution Control Federation,* **56(1),** 76–83

CARISON, R. R., LINDSTEDT, K. D., BENNETT, E. R. and HARTMAN, R. B. (1982) Rapid infiltration treatment of primary and secondary effluents. *Journal Water Pollution Control Federation,* **54(3),** 270–280

LANCE, J. C., RICE, R. C. and GILBERT, R. G. (1980) Renovation of wastewater by soil columns flooded with primary effluent. *Journal Water Pollution Control Federation,* **52(2),** 381–388

LEACH, L. E., ENFIELD, C. G. and HARLIN, C. C. Jr (1980) Summary of long-term rapid infiltration system studies. *Report No. EPA-600/2-80-165,* Ada, Oklahoma: US Environmental Protection Agency, p. 48

MATTERS, M. F. (1981) Arizona rules for irrigating with sewage effluent. In: *Proceedings of Sewage Irrigation Symposium,* Phoenix, Arizona: US Water Conservation Laboratory, pp. 6–12

NATIONAL ACADEMY OF SCIENCES AND NATIONAL ACADEMY OF ENGINEERING (1972) *Report No. 5501-00520, Supt. of Documents*

RICE, R. C. and GILBERT, R. G. (1978) Land treatment of primary sewage effluent: water and energy conservation. In: *Hydrology and Water Resources in Arizona and the Southwest.* Tucson, Arizona: University of Arizona Press, pp. 33–36

RICE, R. C. and BOUWER, H. (1984) Soil-aquifer treatment using primary effluent. *Journal Water Pollution Control Federation,* **56(1),** 84–88

THURMAN, E. M. (1979) *Isolation, characterization and geochemical significance of humic substances from groundwater.* PhD thesis, Department of Geological Sciences, University of Colorado, Boulder, Colorado

Chapter 11

UK experience in the groundwater recharge of partially treated sewage: potential for irrigation purposes

H. A. C. Montgomery*

Introduction

About 300 Ml/d of sewage effluent are recharged to the ground in the United Kingdom, half of it to the Chalk and much of the remainder to the Triassic sandstones. Effects on groundwater quality have been studied at nine of the recharge sites, as shown in *Figure 11.1*. Until recently, groundwater recharge was

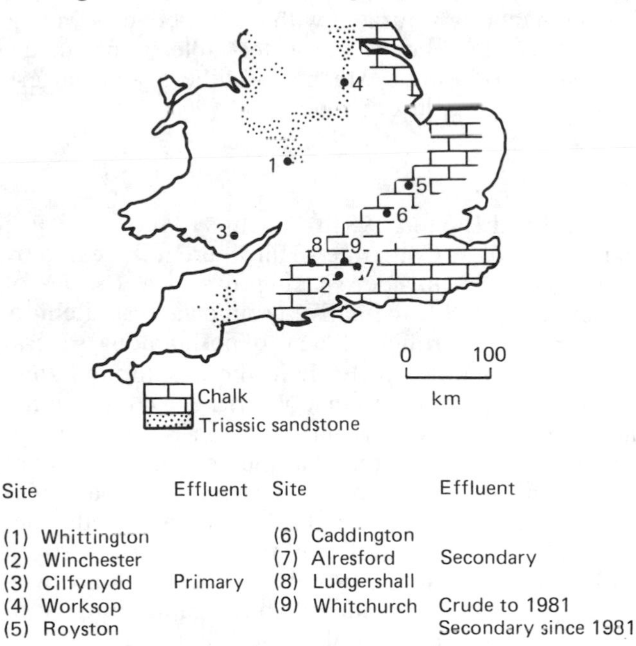

Site	Effluent	Site	Effluent
(1) Whittington		(6) Caddington	
(2) Winchester		(7) Alresford	Secondary
(3) Cilfynydd	Primary	(8) Ludgershall	
(4) Worksop		(9) Whitchurch	Crude to 1981
(5) Royston			Secondary since 1981

Figure 11.1 Location of recharge sites investigated

regarded in the United Kingdom merely as a convenient way of disposing of sewage, but it is now arousing interest as a method of conserving water resources, and a paper reviewing the UK studies from that point of view has been presented by Montgomery, Beard and Baxter (1984a). The present chapter discusses the

* Consultant Water Scientist, UK

same studies with regard to the possible use of aquifers to store sewage effluents intended as irrigation waters as, for example, in the Dan Region Project (Idelovitch, Terkeltoub and Michail, 1980).

Results for organic matter, nitrogen, phosphorus and microbiological contaminants are summarized in this chapter. Heavy metals were investigated at four of the sites but the results were too complex to be summarized in a short review; in general they show some enhancement of iron and manganese during recharge, whereas dissolved copper, lead and zinc were partially removed by recharge through chalk and through Triassic sandstone cemented with calcite.

The improvements in water quality reported in this chapter were attributable to biological, chemical and physical interactions with the soil and the aquifer rock, and have been distinguished from the further improvements which occurred downgradient of most of the recharge sites as a result of dilution of the effluents by the native goundwater.

Chalk sites

The Chalk is a soft, porous, Cretaceous limestone with many microfissures. It consists of almost pure calcium carbonate with a small admixture of clay. Most of the void space of about 40% is permanently saturated with water above as well as below the water table. Movement of water below the water table occurs in the 1–2% of the volume which consists of fissures. Much of southern and eastern England relies on the Chalk aquifer for supplies of water.

Winchester

In Winchester (Montgomery, Beard and Baxter, 1984b), about 11 Ml/d of sewage from the city is pumped to near the top of St Catherine's Hill where it is treated by sedimentation only and then descends the hillside, soaking away via a series of trenches and lagoons. The effluent mixes with the native groundwater and enters the River Itchen by diffuse upward flow through alluvial deposits along several kilometres' length of the river valley. Removal of BOD in the unsaturated zone (which ranges from 4 m to 36 m in depth) is similar to that occurring in conventional treatment by biological filtration. Removal of inorganic nitrogen is mostly more than 50% and sometimes as much as 90%, suggesting that the descending effluent traverses alternate aerobic and anaerobic microzones where nitrification is followed by denitrification. Further purification takes place in the saturated zone, and careful measurements have failed to show any effect of the effluent on nitrogen or phosphate levels in the River Itchen. Coliform bacteria and viruses are well removed. Dichlorobenzene (DCB) is the principal identifiable organic compound, definitely derived from the sewage, to escape removal in the Chalk.

It has now become necessary to decide the future of the Winchester works because a road to be built across the site will remove one-third of the recharge area including the lagoons. Experiments have been started to determine the soakage capacity of adjacent land by both ditch and surface irrigation; the use of ditches looks highly promising at the time of writing. Subject to a successful outcome to this investigation it is probable that the principle of primary effluent recharge will be retained, since the advantages over the option of secondary treatment and discharge to the adjacent river are seen as:

- Extensive purification which offers total protection from pollution risk to the water supply intake on the river 3 km below Winchester.
- No expenditure will be incurred in constructing or operating biological treatment plant.

Ludgershall

About 0.7 Ml/day of biologically treated sewage effluent is released from orifices in a distribution channel across a shallow porous topsoil directly overlying the chalk (Montgomery, Beard and Baxter, 1984b). The hydraulic performance is excellent, a loading of the order of 10^4 mm/day being applied. Chemically, the changes occurring in the unsaturated zone of 10–20 m comprise removal of most of the residual BOD, phosphate and ammonium, and slight, localized denitrification. Coliform bacteria are mostly removed in the unsaturated zone and are almost completely absent after a few hundred metres travel. No other changes were observed except those attributable to simple dilution by the native groundwater.

Alresford

About 0.6 Ml/d of biologically treated effluent are discharged to an unsaturated zone of approximately 20 m through a French drain network embedded in gravel (Beard and Montgomery, 1981). The investigation was primarily concerned with the removal of bacteria. As at Ludgershall, excellent removal of coliform bacteria was observed, by better than four orders of magnitude, a few hundred metres from the recharge site, with the unsaturated zone being more effective than the saturated zone per unit of distance traversed. Chemical changes were slight, apart from phosphate removal which was virtually complete. The gravel pack surrounding the French drains gradually becomes blocked at this site and major maintenance is required at intervals of the order of 10 years.

Whitchurch

Crude sewage (approximately 1.3 Ml/d) was discharged into trenches at this site until 1981, when a biological treatment works discharging to a French drain system was commissioned (Montgomery, Beard and Baxter, 1984b). Anomalies in the hydrogeology made the results more difficult to interpret than at the sites discussed previously. Nevertheless, there was a clear pattern, as at Winchester, of good removal of BOD, TOC and bacteria during the crude sewage discharge regime. Methane gas was detected in the unsaturated zone, proving the existence of anaerobic biological activity, and this probably accounted for the removal of 26–43% of the total dissolved nitrogen applied prior to 1981 (Baxter *et al.*, 1981). Dichlorobenzene was again the only trace organic compound, unambiguously of sewage origin, to persist, and 1 µg/l was present in the groundwater 400 m from the recharge area.

Since the changeover to biological treatment late in 1981 there has been a gradual improvement in groundwater quality at boreholes close to the soakaway area, with BOD values falling and ammonia being replaced by nitrate. There has been little change in water quality at the more remote boreholes. Concentrations of nitrogen are almost unchanged as the removal previously caused by denitrification in the ground is now effected by sedimentation in the treatment plant.

Caddington

About 1 Ml/d of biologically treated sewage effluent is discharged through a French drain system (Baxter and Edworthy, 1979). The effluent is chlorinated as a precaution because there is a borehole for public water supply 2.5 km down-gradient. Chemically the results were very similar to those described above for Ludgershall. Despite the chlorination, the overall reduction in bacterial numbers was no better than that observed at Ludgershall or Alresford, and traces of chlorinated hydrocarbons were detected in the groundwater.

Royston

Until 1979, when a treatment works discharging to surface waters was provided, this discharge consisted of 1.8 Ml/d of partially settled sewage of high chloride content which was recharged through lagoons (Tester and Harker, 1981). The results of groundwater quality investigations were very similar to those already described above for Winchester and Whitchurch.

Triassic sandstone sites

The Triassic sandstone is the second major aquifer in the United Kingdom. It is more variable than the Chalk, ranging from unconsolidated sands to cemented, fractured sandstones.

Whittington

This site is in the West Midlands and has received an average dry weather flow of some 12 Ml/d of mixed primary and industrial effluent for the last 90 years (Baxter and Clark, 1984; Baxter, 1981). Surface irrigation occurs from 120 to 150 chambers located across the site, connected by four buried carrier mains. Sluice gates in these chambers enable the area of effluent recharge to be varied to suit the needs of the farmer. The unsaturated zone is 25–30 m thick and the fractured sandstones have a transmissivity of between 500 and 1000 m^2/d. There is a negligible removal of total nitrogen over the saturated zone and similar nitrate contamination of Triassic sandstone groundwater has also been recorded at a second effluent recharge site just to the north of Whittington (Baxter, 1981). The cause of this reduced nitrogen removal, compared with the Chalk, is not known but may be related to greater oxygen penetration preventing the development of the anoxic conditions necessary for denitrification.

The removal of enteric bacteria and viruses observed beneath the recharge area is complete, unlike the Chalk sites where almost complete removal is only found 300–500 m down-gradient.

Whole rock analysis has shown that effluent recharge has leached calcite from the aquifer itself.

Worksop

About 7 Ml/day of settled sewage were applied to permanent pasture at this site by surface irrigation for many years, until a treatment works discharging to the adjacent river was opened in 1976 (Lucas and Reeves, 1980). Concern about rising nitrate levels at local boreholes led to a detailed investigation which showed that the groundwater was contaminated with nitrate down-gradient of the recharge site, up to 50 mg/l (as N) being observed at the borehole closest to the site. The implication was that most of the nitrogen content of the sewage entered the groundwater as nitrate and that nitrogen removal was negligible, as at Whittington but unlike the Chalk sites receiving settled or crude sewage.

Alluvial gravel site

Of all effluent recharged in the United Kingdom, 22% enters minor aquifers. Although such aquifers are not of strategic importance for water resource purposes they may nevertheless be locally important. Only one such site has been investigated in detail.

Cilfynydd

This site in South Wales (Baxter and Clark, 1984; Joseph, 1977) has no water resource significance and the investigation was designed to determine effects of effluent recharge upon groundwater quality in an alluvial gravel aquifer (actually containing sands, silts and clays in addition to gravel). About 2.2 Ml/d of primary effluent, of mixed domestic and industrial origin, has been allowed to flow over a series of fields on the river terraces of the River Taff for the last 90 years. Infiltration occurs through three main ditches, which are in continuous use, each leading to a series of secondary ditches. The locations of these secondary ditches vary across the recharge area, a new channel being cut when an existing one becomes blocked.

Considerable treatment of the recharged effluent is achieved, thereby reducing the effect of the effluent on the quality of the River Taff. Much of the effectiveness of this system would seem to be due to the changes made in the recharge area, which allows the unsaturated zone to 'rest' and regenerate its treatment capacity. Nearly 60% of the inorganic nitrogen and 30–50% of the organic carbon are removed. Phosphate and heavy metals, other than iron and manganese, are removed within the first 0.1 m of the soil zone. Concentrations of iron and manganese, however, increase markedly as a result of being leached from the soil/alluvium.

Discussion

The average natural infiltration to unconfined aquifers in the United Kingdom is typically 200–400 mm/a. The flow of sewage of, say, 1 Ml/d from a typical village is therefore equivalent in volume to the natural recharge to an area of the order of 1 km^2. Considerations of this kind imply that abundant groundwater is normally available in the United Kingdom for dilution of recharged effluents and this was confirmed during the studies reported. Such dilution would not normally be available in arid countries where, in general, recharged effluent would be expected to contain relatively little natural groundwater when abstracted for reuse. The quality of the effluent in storage is particularly important if it is to be reused with little or no dilution.

The studies described in this paper have confirmed observations in the United States and elsewhere (see, for example, Bouwer *et al.*, 1980; Hartman *et al.*, 1980; Idelovitch, Terkeltoub and Michail, 1980; Baxter and Clark, 1984) that artificial recharge is a remarkably effective method of removing organic matter, ammonia, bacteria and viruses. This is to be expected, by analogy with the biological filtration process of sewage treatment. Nitrogen was partially removed by recharge of primary effluent into the chalk and alluvial gravel aquifers, residual nitrogen being present as nitrate; although nitrogen was not removed by the Triassic sandstone it is thought that the adoption of a different method of operating the recharge sites, with deliberate use of alternate active and resting periods, would probably have promoted nitrogen removal by nitrification and denitrification (Metcalf and Eddy Inc., 1977). In general, the unsaturated zone was more effective than the saturated zone, per unit distance of flow, in promoting improvements in water quality.

Whether nitrogen removal should be encouraged or not at a new recharge site would depend on (a) how much of the nitrogen content of the sewage was required as a crop nutrient during irrigation; and (b) whether the water was also to be used for potable supply, in which case as much nitrogen as possible should be removed. The studies reviewed here have shown that nitrogen may be conserved by giving full biological treatment before recharge.

Phosphate is removed by recharge into calcareous rocks, presumably by precipitation as calcium phosphate. In these circumstances the value of phosphate as a potential plant nutrient is lost.

Iron and manganese are both deficient in sewage effluent as a medium for crop growth (Winfield and Bone, 1981) and might have to be supplemented if also deficient in the soil being irrigated. However, there was evidence of leaching of iron and manganese into the recharged effluents, especially at Cilfynydd, and such leaching might suffice to offset any deficiency. Toxic heavy metals tend to be associated with the suspended solids of sewage (El-Nennah and El-Kobbia, 1983) and to that extent would not be recharged (unless solubilized by anaerobic action) and further, partial removal may occur in the aquifer. Probably of more concern is the mineral content of sewage effluent. Calcium, magnesium, sodium, chloride and sulphate all tend to be conserved during artificial recharge and could restrict the use of recharged effluent for irrigation in arid climates in the long term.

The method of applying the effluent to the recharge area needs careful consideration. At the sites investigated in the United Kingdom, surface flow over the thin topsoil at Ludgershall was found to be the most satisfactory method of recharging a biologically treated effluent but was less satisfactory for primary effluent because of the growth of 'sewage fungus' on the land surface (Beard, M. J., personal communication). With primary effluent, trenches gave good hydraulic performance and were easy to maintain. Lagoons are less easily maintained than trenches. Subsurface tile drains (French drains) performed well with biologically treated effluent but required complete renovation after about 10 years, by which time the surrounding gravel pack had become clogged with organic solids. In an arid climate the need to restrict evaporation would seem to favour trenches or French drains.

Any site being considered for the recharge of sewage effluent should be investigated geologically beforehand. In particular, the presence of large fissures or karstic conditions would reduce the effectiveness of an aquifer for purifying sewage and might even permit the transmission of dangerous microbial contamination. An unsaturated zone of at least 10 m is desirable to take full advantage of the potential of an aquifer to treat sewage.

References

BAXTER, K. M. (1981) The effects of discharging a primary sewage effluent on the Triassic Sandstone aquifer at a site in the English Midlands. *First International Conference on Groundwater Quality Research,* Houston, Texas

BAXTER, K. M. and CLARK, L. (1984) Effluent recharge. The effects of effluent recharge upon groundwater quality. *WRC Report TR199.* Medmenham: Water Research Centre

BAXTER, K. M. and EDWORTHY, K. J. (1979) The impact of sewage effluent recharge on groundwater in the Chalk for an area in South East England. *International Symposium on Artificial Groundwater Recharge,* Dortmund, Germany

BAXTER, K. M., EDWORTHY, K. J., BEARD, M. J. and MONTGOMERY, H. A. C. (1981) Effects of discharging sewage to the Chalk. *Science of the Total Environment,* **21,** 77–83

BEARD, M. J. and MONTGOMERY, H. A. C. (1981) Survival of bacteria in a sewage effluent discharged to the Chalk. *Water Pollution Control,* **80,** 34–41

BOUWER, H., RICE, R. C., LANCE, J. C. and GILBERT, R. G. (1980) Rapid infiltration research at Flushing Meadows project, Arizona. *Journal Water Pollution Control Federation,* **52,** 2457–2470

EL-NENNAH, N. and EL-KOBBIA, T. (1983) Evaluation of Cairo sewage effluents for irrigation purposes. *Environmental Pollution (Series B),* **5,** 233–245

HARTMAN, R. B., LINSTEDT, K. D., BENNETT, R. F. and CARLSON, R. R. (1980) Treatment of primary effluent by rapid infiltration. *US EPA Report* EPA-600/2-80-207

IDELOVITCH, E., TERKELTOUB, R. and MICHAIL, M. (1980) The role of groundwater recharge in wastewater reuse: Israel's Dan Region Project. *Journal American Waterworks Association,* **72,** 391–400

JOSEPH, J. B. (1977) The effects of artificial recharge with a primary sewage effluent into alluvial deposits at Cilfynydd, South Wales. *WRC Report ILR 689.* Medmenham: Water Research Centre

LUCAS, J. L. and REEVES, G. M. (1980) An investigation into high nitrate in groundwater and land irrigation of sewage. *Progress in Water Technology,* **13,** 81–88

METCALF and EDDY, Inc. (1977) *Process Design Manual for Land Treatment of Municipal Wastewater.* US Department of Commerce, National Technical Information Service Report PB, pp. 299–655

MONTGOMERY, H. A. C., BEARD, M. J. and BAXTER, K. M. (1984a) Groundwater recharge of sewage effluents in the UK. *Proceedings of International Symposium on Reuse of Sewage Effluent,* Institute of Civil Engineers, London, October, pp. 219–226

MONTGOMERY, H. A. C., BEARD, M. J. and BAXTER, K. M. (1984b) Effects of the recharge of sewage effluents upon the quality of Chalk groundwater. *Water Pollution Control,* **83,** 349–364

TESTER, D. J. and HARKER, R. J. (1981) Groundwater pollution investigations in the Great Ouse Basin. *Water Pollution Control,* **80,** 614–628

WINFIELD, B. A. and BONE, D. A. (1981) The sewage farm reborn. *Public Health Engineer,* **9,** 185–186, 198

Chapter 12

Soil–aquifer treatment and reuse of sewage effluent: a new approach to sanitation

J. Bize, D. Fougeirol, V. Riou and N. Nivault*

Introduction

Reuse of sewage effluent for irrigation is already taken for granted in many countries where water is rare. However, operations are often carried out under deplorable sanitary conditions which have to be improved, especially as far as sewage disinfection is concerned. Seen from this angle, the classical approach to sanitation popular in temperate countries has to be reviewed, both in terms of the processes involved in the implementation of projects and from the point of view of the treatment techniques to be applied.

The techniques associated with soil–aquifer treatment are introduced here. A high level of purification (especially disinfection) can be seen in the results obtained from two pilot developments in France, one in the dunes around Port Leucate on the Mediterranean coast and the other in the chalky Parisian Basin at Flesselles. These results show the advantages of using the soil and aquifer to purify, store and transport water towards agricultural wells.

Reuse of sewage effluent: a new approach to sanitation

Advantages and disadvantages of reuse of sewage effluent

In countries where:

- Water is rare.
- The water regime is irregular.
- Regulating schemes are costly.

Sewage effluent represents:

- One of the components of the global 'water' assessment, in the same way as surface or subterranean water courses.
- A permanent flow resource.
- Controlled outflow which is immediately available.

In this context, the question 'Should we, or should we not, reuse sewage effluent for irrigation?' leads to a false debate. In fact, sewage, even in its raw state, is already being reused for market gardening in many countries, where millions of

* BURGEAP, 70 rue Mademoiselle, 75015 Paris

people depend on it for a living. However, this direct reuse is carried out under very poor sanitary conditions, can be a source of epidemics, and must be improved upon. The real alternative is either to pretend 'not to see' (that is, practise a policy of *laisser faire*), thus encouraging unofficial reuse, or improve and rationalize the existing system.

Thus, the question to ask is what degree of purification is necessary before reuse of effluent in irrigation. The problem is far from being solved, as is revealed in the variation in criteria from one country to another. These criteria concern both the health aspect (human consumption) and the agronomic aspect (protection of the flora and of the soil).

One point is, however, very clear: an enormous effort has to be made to eliminate bacteriological pollution before reuse of effluent for agricultural purposes (especially market gardening). Although the removal of other pollution factors (suspended solids, organic matter, etc.) seems important, it is of secondary importance in comparison with sewage disinfection. However, great attention must be paid to the elimination of toxic trace elements which can have harmful effects on both humans and plants. It is particularly advisable to separate and eliminate industrial toxic waste before reuse of effluent for irrigation.

Rigidity of classical sanitation schemes

Reuse of sewage effluent for irrigation can soon give rise to doubts about the quality of the effluent and the layout of the sewerage network, thus raising the question of the validity of classical sanitation schemes. These schemes have been questioned both in temperate and in arid countries every time increased protection for the receiving environment has become necessary. Classical sanitation schemes are designed for large towns in temperate countries, which – and this is not a coincidence – are generally situated near a water course. The layout of the sewerage system, with pumping stations allowing for a certain amount of play with the topography, is thus determined entirely by the location of the waterway (or of the sea).

The sewage treatment plant does not generally eliminate all suspended solids or organic pollution (BOD, COD). The self-purifying capacity of the river itself has to be relied on to eliminate residual pollution. However, if this residual pollution is eliminated through irrigation, without taking certain precautions to protect the soil, it may then prove to have an 'aggressive' effect on underground water. Moreover, the sewage treatment plant does nothing to solve the problem of bacteriological pollution.

The rigidity of this type of scheme became apparent when sanitation was extended to smaller towns, or towns situated at a distance from a waterway, and was implemented in several ways:

- *In calcareous areas:* the collection system is generally non-existent or not covering the whole area and underground aquifers become the inevitable receptacle for effluent, whether treated or not. And yet the law usually forbids the discharge of effluent into aquifers. Thus the installation of classical sanitation schemes consisted of building the sewerage, then the plant, and discharging effluent at any point without taking special precautions. This untreated waste creates two sorts of problems:

- If the soil is impermeable in places, stagnation of waste creates unhealthy conditions, accompanied by the usual environmental nuisances (smells, mosquitoes, etc.).
- If the soil is highly permeable (chasms, fissured limestone), the waste infiltrates and 'eliminates itself', thus polluting the aquifer.
- *In arid or semi-arid countries:* the absence of permanent waterways, or waterways with sufficient flow, is the rule, and one finds the same problems as in calcareous regions where there are no methods for dealing with waste. Moreover, these are the countries where there is a need to develop effluent resources under more hygienic conditions. This calls for pretreatment and an adjustment of the sewerage to enable it to serve irrigable agricultural land.
- *In coastal towns:* the protection of sensitive areas calls for the elimination of bacteriological pollution. Although the construction of a conventional sewage treatment plant is an economically viable solution for large towns, this is not the case in small and medium-sized towns; for these, the solution of overall development and treatment of waste is to be found within the earth.
- *In small rural districts or boroughs:* the investment costs and operational methods required for conventional sewage treatment plants are often not available, given the human and financial resource limitations of these communities. Furthermore, it has been noticed that there is often a low level of efficiency in small plants. This problem arises in numerous developing countries where, given an equivalent output, the preference is for simple techniques leading to a reduction in the number of operations and maintenance costs, and requiring a lower level of specialization on the part of the maintenance staff.

Given the imperfections of classical schemes, sanitation research has concentrated on:

- Developments in waste disposal to improve hydraulic conditions on the one hand and in the form of treatment systems on the other hand, in particular when the waste is eliminated on or in the soil (calcareous regions, arid countries).
- Disinfection of effluents before waste disposal in sensitive receiving-environments (coastal areas, aquifers, agricultural land).
- Simplification of treatment processes (developing countries, small districts in temperate zones) and research on greater reliability.

All these precautions have encouraged researchers in different countries to look into the purifying qualities of the natural elements, in particular soils, which are fundamental to the latest sewage treatment techniques.

The spectacular results that have been noted over the last few years – both in the domain of real technical developments and in experimental soil columns – have led to great enthusiasm for these techniques, which some consider to be completely new, but which are, in fact, as ancient as man's cultivation of the land. Thus we have witnessed the rehabilitation of techniques such as stabilization ponds or soil–aquifer treatment, both in the case of big urban developments, thanks to a rational implementation backed up by solid scientific basis, and reinforced by regular control of results.

Bacteriological pollution

Disinfection techniques applied to drinking water (chlorination, ozonization, ultraviolet) are much less reliable when applied to effluent, particularly in small treatment plants. All too often, the germs proliferate in the receiving environment (water course), and especially when the waste contains a high level of residual oxidizable matter.

Disinfection of effluent can be achieved, for example, by ensuring that the effluent is well dosed with chlorine. Unfortunately, these techniques are costly, difficult to implement and require the presence of a specialized technician. Because of this, they can only be envisaged in the case of large plants (serving tens of thousands of inhabitants). However, there remains the problem of the effects of the byproducts of chemical disinfection on the receiving environment. Taking these problems into consideration, the French administration recommends the implementation of these techniques only as a last resort, if it has been proved that no other method would be viable (dispersal in the natural environment, soil–aquifer treatment, oxidation ponds, etc.).

In fact, other techniques, using the purifying qualities of natural processes, are available. These techniques may be used either simply for the disinfection of purified effluent or as a complete treatment of raw sewage. They are much more flexible than conventional treatment plants; moreover, their function allows for the reception of storm waters (drainage sources), which gives them an important advantage over other methods.

The implementation of these 'extensive' methods essentially depends on the defects and advantages of the site, and on its physical characteristics as well as the land area available. Two methods are currently being used:

- *Oxidation ponds* constitute a satisfactory solution to the problem of disinfection. However, in hot countries, there is a contradiction between the resulting effluent and the objective, which is to reuse effluent. The contradiction is evident from the following calculation:
- Annual production of effluent/inhabitant: $0.1\,m^3 \times 365$ days $= 36.5\,m^3$/year.
- Evaporation by oxidation pond of $10\,m^2$/inhabitant: $2\,m/year \times 10\,m^2 = 20\,m^3$/year.

  Result: 55% loss of wastewater; production of a concentrated sludge, difficult to disinfect by final chlorination process.

  This technique requires a surface area of the order of $5\,m^2$/inhabitant (final oxidation pond) or of $10\,m^2$ per inhabitant (total surface pond receiving raw sewage). It is adapted to impermeable ground.
- *Soil–aquifer treatment* requires a much smaller surface area (approximately 0.5–1 m^2 per inhabitant). This technique is adapted to permeable and aquifer soils. Over and above its remarkable disinfecting power via the soil, it enables the aquifer to be used to store treated water, thus protecting it from evaporation, and permits natural transfer of this resource to agricultural wells.

These techniques can be used to treat any size of effluent flow, providing a sufficiently large surface area of land is available.

Need for a guide scheme for sanitation

The preceding considerations lead to a new approach to sanitation, aiming at non-polluting systems which are integrated into the environment. Prior to any sanitation development, a proper guide scheme has to be worked out; this is less a question of suggesting new techniques than of adopting a method based on the idea that it is the site that 'decides' and contains its own solutions.

As a result, the standard estimate-programme is now replaced by an unbiased preliminary exploration of the zone which is to be decontaminated, and of its perimeter:

- Urban shape.
- Topography of the site.
- Hydrographic context, water regime, climate.
- Hydrogeological context.
- Agricultural context.
- Protection: catchment of drinking water, sensitive areas.
- Land context.

The analysis must comprise a 'downstream–upstream' process. Sanitation consists of three phases: discharge of waste, treatment and collection. The first step is to decide on the location or locations and the waste disposal system which most effectively exploit the natural advantages of the site; the next is to take into account the local needs (agriculture, protection of the environment); then, following a precise logic, return to the 'pretreatment' link (to be organized strictly according to the requirements of waste disposal: infiltration capacity (for example) and finally to the collection system.

Transfer from a rigid to a flexible system makes it possible to:

- Break down the sanitation infrastructure into 'autonomous sanitation subunits' (for example, by urban drainage basin), which allow for substantial economies to be made on the collection system, pumping-stations and in energy requirements, especially in zones with a difficult topography.
- To split up investments, which is all the more feasible in that methods such as soil–aquifer treatment or oxidation ponds allow for variable devices.
- Diversify purification methods, soil–aquifer treatment being the method used in collective sanitation, individual sanitation and (an intermediary term) 'grouped' sanitation (as, for example, in the case of plots of land).

Soil–aquifer treatment: results obtained in two pilot development sites in France

Principles of soil–aquifer treatment

Soil–aquifer treatment (SAT) involves biological purification processes using fixed bacteria. It is enough to be present at a few conferences to be aware of the enthusiasm aroused by biological purification techniques using fixed bacteria in a mineral granular medium. All the firms specializing in the field have concentrated on research into these techniques, and now there exist innumerable processes for elimination of organic pollution and for nitrification or denitrification by fixed bacteria.

All these fixed bacterial processes are based on the same principles (*Figure 12.1*):

- Mineral granular media with large specific surface (activated carbon, biolite, glass beads etc.) with sufficiently large pores to encourage bacteria to settle.
- Need for permanent presence of a high concentration of bacteria.
- Continuous aeration to ensure sufficient oxygen for bacterial metabolism (with the exception of anaerobic processes such as denitrification).

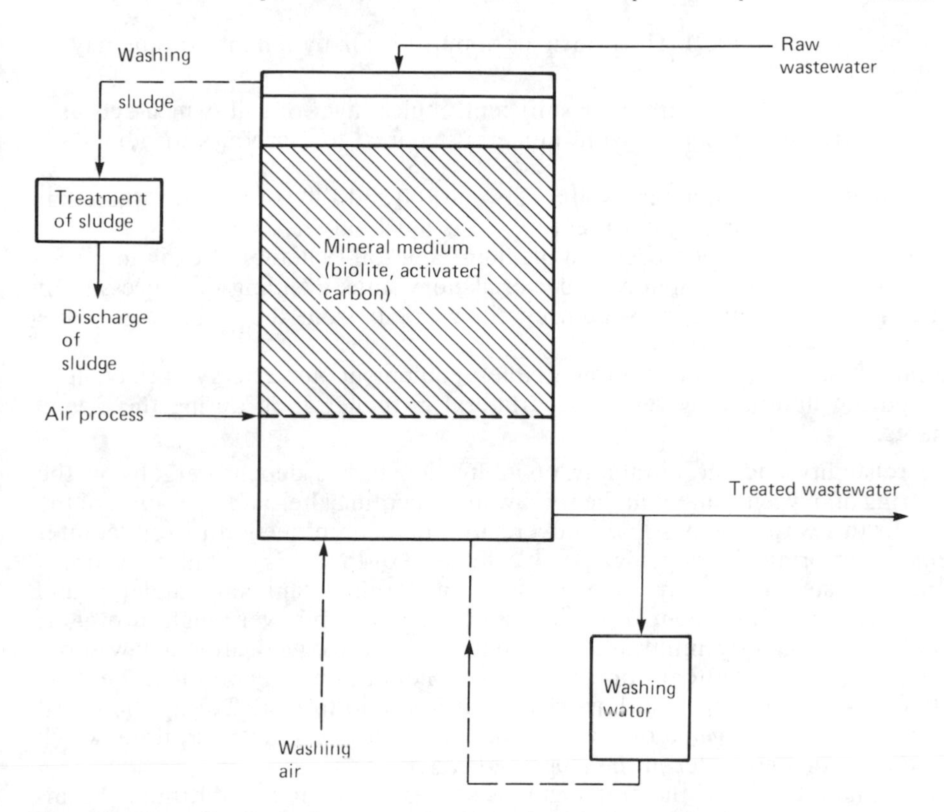

Figure 12.1 Example of basic diagram of a fixed bacteria biological purification process

These three points are fundamental to fixed bacterial techniques, and the following are the main difficulties involved in their implementation:

- Hydraulic supply: choice of direction for liquid and gas flow for running and washing processes.
- The washing process, whose aim is to remove excess sludge (suspended solids, excess biomass). The disadvantage of this washing process is the risk of eliminating too much sludge and creating low activity at the beginning of the next cycle;
- The increasing loss of load during the cycle, due to growing presence of mud in the system, occasionally requiring an increase in pressure at the base of the reactor in order to maintain regular aeration.

Despite these problems, fixed-bacteria purification processes are highly effective; their treatment efficiency is remarkable and the rates involved generally result in reasonable reactor dimensions. To sum up, these techniques provide an ideal environment for the formation of a highly concentrated biomass and for the physicochemical retention of suspended solids and micro-organisms (thus avoiding the need for a second phase of decantation).

It appears indispensible, and highly instructive, at this point, to establish a parallel between fixed bacterial systems, as just briefly described, and soil–aquifer treatment (*see Figure 12.3*). The principles of purification by soil infiltration may be described as follows:

- Percolation of effluent through a sufficiently thick layer of soil, which acts as a mineral granular medium with numerous pores to harbour an abundant microflora.
- The specific surface of natural soil is between 100 and 1000 times lower than that of a biolite, for example, but the 'reactor' is much larger.
- Permanent presence of a bacterial medium; any soil (whether already in place, or added) contains an abundant bacterial flora, thus avoiding the necessity of seeding and permitting maximum efficiency right from the beginning of the cycle.
- Aeration of the system requires neither equipment nor energy supply; it is carried out naturally by vertical and lateral air penetration during the drying phases.

The reliability and the durability of infiltration basins depend largely on the monitoring of oxygen content in the soil by implementing alternate operation of the basins. In this way, it is possible to overcome the main objection to soil–aquifer treatment performance, clogging. To be convinced of this, it is enough to examine an onsite waste disposal system consisting of a septic tank and underground leaching trenches. If effluent from the septic tank, which is very high in organic matter, is continuously infiltrated the trench will be clogged after a few days. However, this sort of system works for years without posing any problems because family use is intermittent and thus the trench has time to rest. During these rest periods, natural reoxygenation of the soil reactivates aerobic microflora which break down organic matter in the clogging layer.

Correct monitoring of the treatment of sewage effluent in infiltration basins relies on ensuring the equipment applies effluent on a rotational basis, so that during rest periods oxygenation can occur. During this period, cleaning of the basins can take place at the same time as washing of the fixed-bacteria biological reactors. This is simply a question of raking off the suspended solids on the surface*.

In summary, biological purification processes using fixed bacteria are simply the re-creation using compact equipment and accelerated phenomena produced during slow soil infiltration, where aeration and 'washing' occur naturally, without sophisticated structures or energy consumption.

Once fixed-bacterial processes have been mastered, there is no limit to the possibilities provided by using the soil for purification, and thence for sanitation in its broadest sense, possibilities which, though enormous, are as yet little known.

Two French developments for wastewater disposal and purification in infiltration basins will now be examined.

Infiltration in the dunes: the Port Leucate development site (Aude)

Port Leucate is a seaside resort situated on a narrow dune belt between a lagoon and the Mediterranean sea (north-east of Perpignan), which are connected by a

---

* Cleaning operations are facilitated if preceded by a drying period. It should be noted that one of the advantages, and not the least, of this process is to simplify the 'mud problem'. Moreover, it is advisable to choose a pretreatment process that follows the same logic; hence, for example, the advantages of sifting-type stations.

channel sheltering a pleasure-boat harbour. The influx of seasonal population was 25 000 in 1981, and is expected to reach 60 000 at the end of the century.

During the peak, or critical period, from 15 July to 15 September, the discharge of waste into the lagoon endangers the pursuit of shellfish growing, while discharge of wastewater into the sea via an outfall would not guarantee the safety of the bathing areas. Until 1979, all effluent – even during the tourist season – was discharged into the sea after biological treatment in a plant. When the treatment plant reached its capacity, in the summer of 1980, it became urgent to find a solution for the excess raw sewage flow, which reached 700 m³/day in 1980 and 1500 m³/day in 1981.

The solution involved taking advantage of the existence of the Corrège dune belt between the lagoon and the port channel, which was unsuitable for urbanization. The project is based on the acceptance of – and even a deliberate attempt to achieve – some degree of clogging in order to optimize sewage infiltration rate (of the order of 1 m/day) and to maintain an unsaturated medium between the infiltration basins and the water table. All these conditions taken together allow for the use of granular soils to their full capacity.

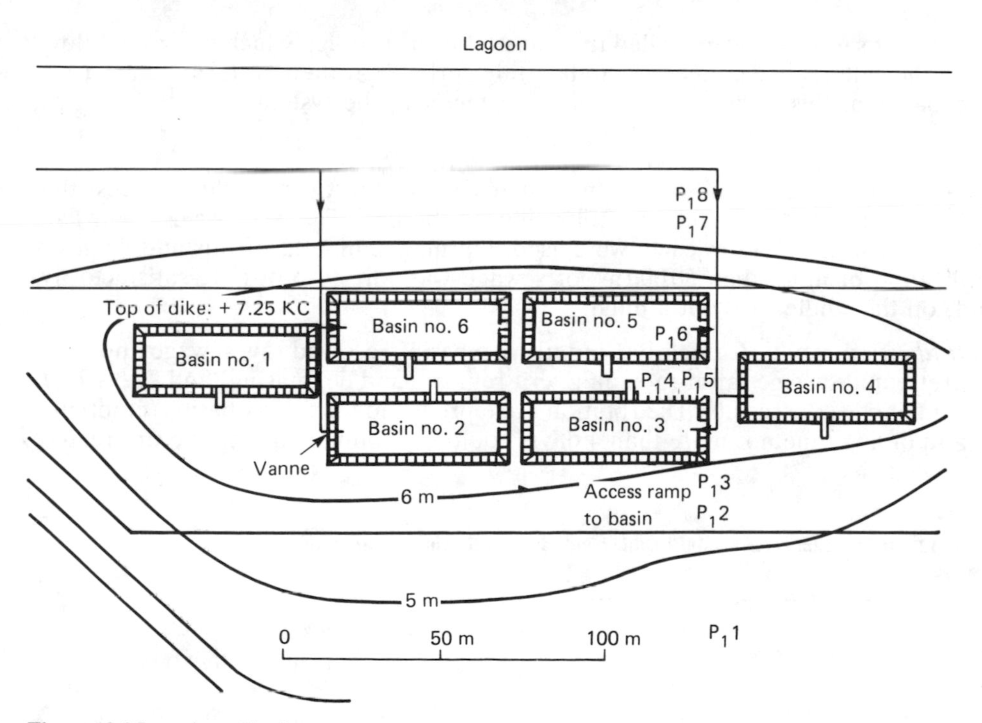

Figure 12.2 Location of basins

Six infiltration basins, each 50 m × 15 m totalling an infiltration area of 4500 m², were installed on the upper part of the dunes (*Figure 12.2*). Beneath the infiltration basins, the dune medium consists of two superimposed layers (*Figure 12.3*):

- An upper unsaturated layer, through which vertical flow occurs; this is called the infiltration layer and the main purification processes take place here.

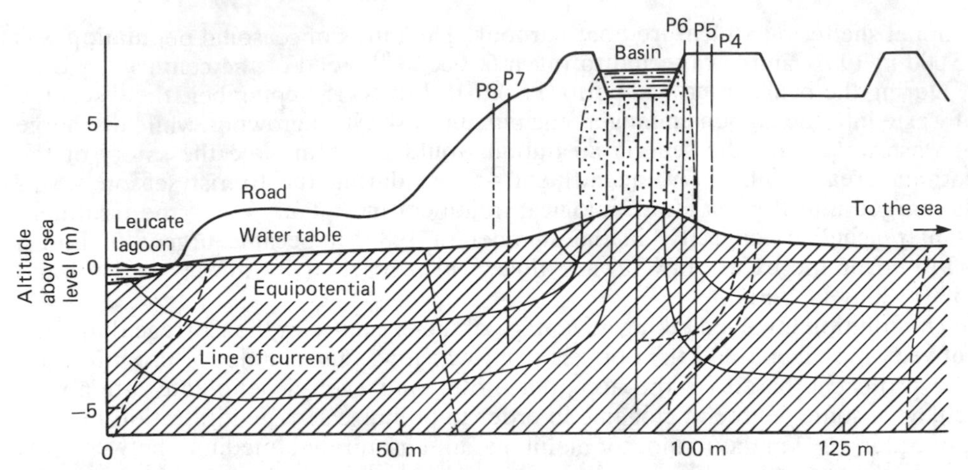

Figure 12.3 Flow network under an infiltration basin. ☐ Flow through unsaturated zone; ☐ flow through saturated zone. *Note:* distortion of flow network due to distortion of horizontal and vertical scales

- A lower saturated layer, called the transfer layer, through which horizontal flow in the aquifer occurs, used in the transport of purified water; the transfer capacity of this layer acts as the limiting factor in the system.

1981 Experiments

Experiments were carried out in the upper zone of the Corrège dune to test the spreading method to achieve infiltration in basins. The six basins, with an infiltration area of $750\,m^2$ each, were used. During the months of July and August, infiltration of more than $50\,000\,m^3$ of sewage was carried out quite satisfactorily and, on the whole, without a hitch.

- *Infiltration results:* for nearly a month, the basins received raw sewage and then pretreated sewage (after screening, grit removal and degreasing) and *Table 12.1* summarizes the results. The rotation cycle predicted for the six basins (emptying and drying time not more than 4 days) could be maintained with the pretreated sewage.

Table 12.1 Infiltration rates and infiltrable heads as a funtion of the nature of sewage

	Average infiltration rate	*Head of infiltrable sewage without declogging*
Raw sewage	0.6 m/day	1 m
Pretreated sewage	1 m/day	2 m

- *Clogging:* with raw effluent, the clogging deposit resembled black sludge under the water. As soon as it was dry, the deposit took on the texture of soft felt and it had only 10% water content. After this 'carpet' had been removed from the basins, the sand recovered its initial infiltration capacity. During drying, it was also noticed that a black colour appeared on the sand to a depth of 10–40 cm. In all cases, this coloration, which is the sign of a reducing medium, disappeared completely 24 h after removal of the clogging deposit, when the surface had dried.

- *Transfer capacity of the aquifer:* groundwater level measurements have shown that, under the 1981 experimental conditions (1000 m^3/day over a period of 2 months, rotation with six basins), the disposal capacity of the aquifer is still far from being saturated and the appearance of suspended aquifers was not recorded (maximum swelling 1.70 m, rapidly reabsorbed and leaving a non-saturated zone of 3.70 m.
- *Purification:* quality analyses were carried out on the sewage injected into the basins, the aquifer water and the Leucate lagoon water; the efficiency of the system is indicated in *Table 12.2*.

Table 12.2 Purifying efficiency of infiltration basins

	Concentrations measured in mg/l				*Purifying*
	Effluent		*Aquifer under basins*		*%*
	Average	*Maximum*	*Average*	*Maximum*	*Average*
SS	256	360	0	0	100
COD	600	768	50 (20–40)*	92	85–92
BOD_5	192	222	20		90†
N (all forms)	88	114	50		30–40
P	36	45	0.1	0.62	100

* Concentration in the aquifer before infiltration
† Not taking into account the aquifer concentration before infiltration (not measured)

Table 12.3 Decrease in bacterial pollution

Bacterial tests number/100 ml	*Intake points*				
	Raw sewage at basin inlet	*Dune aquifer before infiltration*	*Dune aquifer during infiltration (piezometers)*		*Lagoon water at 70 m*
			P6 (basin 5)	*Pr at 50 m*	
Faecal coliforms	1.8×10^4	90	4 000	30	10
Total coliforms	5.6×10^3	46	10 000	6	10
Faecal streptococci	1.4×10^3	10	280	6	10
Bacteria, *Escherichia coli*	540	0	0	0	0

Note: These results are the mean geometrical values of the bacteria counts carried out on: raw sewage: eight samples; preinfiltration aquifer: 12 samples; aquifer during infiltration: 16 samples; lagoon: eight samples

COD and BOD_5 were eliminated during the vertical flow and the original levels were, in fact, found in the aquifer. Organic phosphorus was held back by TSS, and the mineral phosphorus did not reach the aquifer. The total amount of nitrogen decreased from 88 to 50 mg/l (43% reduction).

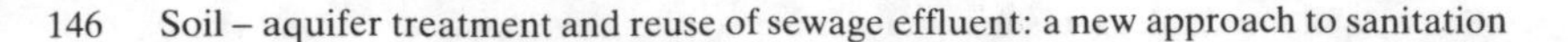

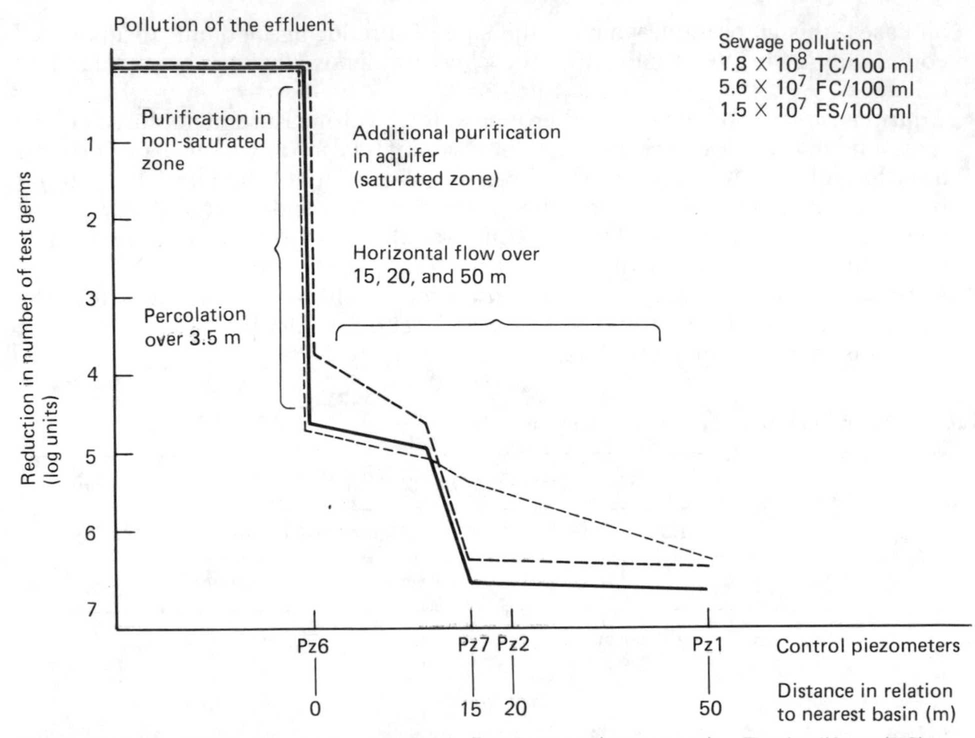

Figure 12.4 Graph of results obtained in 1981 in Port Leucate (raw sewage). – Total coliform (TC); -- faecal coliform (FC); --- faecal streptococci (FS). One log unit is the reduction in number of germs by a factor of 10

Table 12.4 Influence of the daily load and TSS on the performance of basins

	Level 1	*Level 2*	*Level 3*
Suspended solids load during 24 h (kg)	<300	300–400	>500
Unit load during 24 h (kg/m^2)	<0.4	0.4–0.55	>0.66
Suspended solids load			
4000 m^3/day	<75	75–100	>125
5000 m^3/day	<60	60–80	>100
Infiltration rate (m/day)	>10	5 to 10	<5
Drainage time (hour)	7	30	>48
Maintenance clogging	Very thin clogging layer: several floodings	Raking after each flooding	Excessive clogging

The main portion of the decrease in bacterial load occurred in the unsaturated zone, where it reached 10^4. These results show that, in the saturated zone close to the infiltration basin, the receiving aquifer plays a finishing role in the disinfection of wastewaters; at 50 m the reduction reached 5×10^6. At this stage, the initial concentration in the aquifer is achieved (*Tables 12.3* and *12.4; Figure 12.4*).

The 1981 tests, carried out on raw and semi-raw sewage, produced the most remarkable purification results and, moreover, were carried out at low cost. Only two workers, equipped with rakes, pitchforks and wheelbarrows, were employed for maintenance.

1982 Experiments

Once extension of the sewage treatment plant had been implemented, by the beginning of July, it was possible to treat all sewage from Port Leucate under optimum conditions. The infiltration project was then able to function in its definitive form: treatment in plant, then infiltration and tertiary soil purification of all the sewage produced by the resort (4000 m^3/day). The 1982 tests were aimed at studying the functioning of the basins receiving treated sewage which, naturally, had less of a clogging effect than the almost raw sewage treated in the preceding year.

The results obtained during this period of infiltration revealed that, with this effluent (TSS of at least 100 mg/l), infiltration occurred at average rates of 3–4 m/day for an infiltrated daily head of between 5 and 7 m, without interruption in the operation of basins. The basin floors had to be washed between each run but, after drying and raking, there was a 100% recovery in infiltration capacity.

1983 Experiments

What was remarkable in the 1983 studies was that extremely high (up to 50 m/day) infiltration rates were achieved, due to the combination of a relatively low load of raw sewage and the highly efficient performance of the physicochemical treatment works (*Figure 12.5*). However, the clogging element remained a reality.

Clogging of sewage In general, the decrease in infiltration rates during the flooding of the basin were the result of the accumulation of suspended solids on the basin floor and obstruction of the filter medium pores to a depth of a few centimetres. The first of these two phenomena was the most significant; indeed, the very short period of submersion of the basins did not allow for sufficient bacterial growth to have any significant effect on the pores of the rather coarse sand.

The clogging power can perhaps be connected with the load of solids, since there appears to be a connection between TSS concentrations and loads and the hydraulic parameters of infiltration. By referring to TSS loads recorded every 24 h, it is possible to identify three effluent quality levels. These latter can be seen to correspond with hydraulic infiltration conditions in the Port Leucate basins with flow rates of 4000 m^3/day and 5000 m^3/day (*see Table 12.4*). The qualitative relationship between the infiltration rate and organic matter purification shows that above 12–16 m/day the purification rate drops significantly. Inversely, below this threshold residual pollution underneath the basin diminishes when the infiltration rate increases. This unexpected trend implies the existence of a relationship wherein the polluting load could have an effect by superimposing itself on the infiltration rate factor in the case of the lowest rates (<12 m/day).

Microbiological purification The results are presented in *Table 12.5*. In its overall structure, the purifying system (infiltration basins, unsaturated zone, aquifer) varies in its effects according to the nature of the micro-organisms:

- Complete elimination of viruses (found in small quantities in sewage).
- Higher rate of elimination of faecal streptococci than of faecal coliforms.

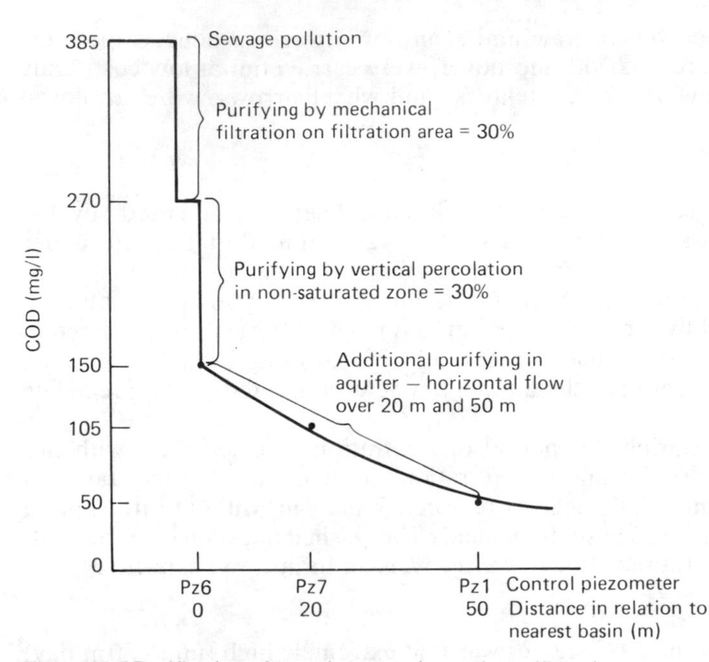

Figure 12.5 Purification of organic matter (secondary effluent).

Table 12.5 Global review of microbiological purification

	Reduction between sewage plant effluent and aquifer 50 m from basins	*Residual pollution in the aquifer*		
		Average	*Maximum*	*Minimum*
Coliforms/100 ml	3.8 log units	7×10^3	1.6×10^5	0
Faecal streptococci/100 ml	4.3 log units	30	1.4×10^2	0
Virus	Complete elimination	0	0	0

The 1983 results, using secondary effluent, show that despite rather rapid infiltration rates performance remained good.

Conclusions

In 1981, sewage which had simply been screened, degritted and degreased was found to create too much clogging to allow the basins to function efficiently. On the other hand, in 1983, the level of purification of the sewage was too high to obtain the rate of infiltration of 2 m/day, which was the aim. Optimization of effluent treatment in basins therefore depends on the maximum efficiency of pretreatment; conventional treatment processes consist of a discontinuous series of treatment levels and thus the finalizing of a series of intermediary steps seems to be necessary.

The Port Leucate development site has eliminated such nuisances as odours and mosquitoes. The investment cost of US$100 000 (or US$4/inhabitant) is 15 times lower than that of waste disposal via a sea outfall. Running costs are in the region of US$1 per year/inhabitant. The studies have shown that the pathogen removal in sand is definitely greater than for any other disinfection technique. Such developments can thus lead to the reuse of wastewaters, contributing to the optimal use of water resources.

Chalk infiltration: the Flesselles development site (Somme)

Flesselles has a population of 1900. A combined sewage collection system has been in existence for a long time. Until 1982, wastewater and stormwater flowed to a sedimentation basin and the overflow was directed towards two infiltration wells. These two wells rapidly became clogged and the flow downstream of the site caused some damage to crops. Then, a biological treatment plant (activated sludge), of capacity sufficient to provide for 2200 inhabitants, was built. Even after it had been installed (with some delay), the problem of disposal and tertiary purification remained.

Flesselles is situated on a calcareous plateau covered with silt; the nearest waterways are the Somme, at a distance of some 15 km and the Nievre, a small tributary of the Somme, 10 km away. At the request of the Departmental Agriculture Authority of the Somme and the Artois-Picardy Water Authority, research was initiated to investigate the possibilities for tertiary treatment processes and effluent disposal. The final report recommended the installation of the mechanisms described below, and the Departmental Amenities and Equipment Authority of the Somme was made responsible for their implementaion.

Implementation techniques

Figure 12.6 is a diagram of the infiltration mechanism. At the downstream end of the plant, the effluent is directed towards two infiltration basins which operate alternately. The infiltration rate aimed at is 0.20 m/day. The bottom of the basins reaches the chalk and is covered with a layer of 1 m of fine sand.

The excess water from the storm weir (mainly rainwater) is decanted into the first 2500 m^2 basin and then infiltrated into two basins, which can also function alternately, and whose bottoms are covered with a 0.50 m layer of fine sand. After a decennial rainfall, the volume of water arriving in the basins has been estimated at 10 000 m^3. The volume of the three basins taken together is sufficient to equalize this inflow. Outside these periods of exceptional rainfall, only one infiltration basin functions at one time.

The aquifer level lies at a depth of 40 m beneath the Flesselles site. The chalk is fissured in blocks of approximately 10 cm on the side and it is to be hoped that, over an area of 40 m of vertical flow in an unsaturated environment, additional purification would occur – not by absorption but because of the oxidizing medium provided by the fissured chalk (air circulation).

Results

By 1982, the plant was still not functioning, and all wastewater and rainwater went through the sedimentation basins and passed into the infiltration basins. Rainwater infiltration rates were initially 5 m/day but stabilized at 24 cm/day within a week. After a drying period of 1 month (without cleaning) infiltration rates were 2 m/day at the start and then rapidly stabilized at 20 cm/day. Infiltration rates remained at 0.20 m/day in dry weather and climbed to 2 m/day during heavy rainfall, the load in the basins increasing without causing an overflow.

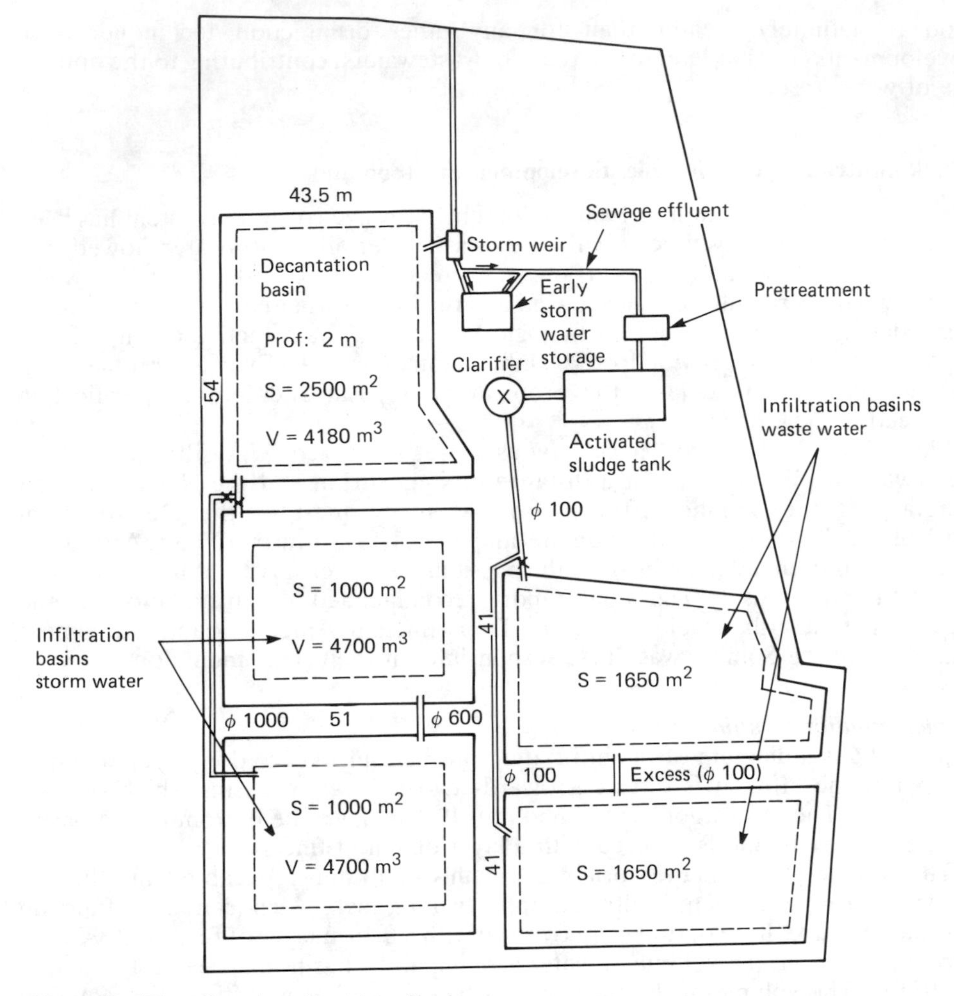

Figure 12.6 Flesselles: plan of location of basins

Treatment efficiency, such as provided by samples taken from the lysimetric bins situated beneath the layer of sand, was as follows: for percolation through 25 cm of sand, the BOD_5 reduced from 130 mg/l to 90 mg/l (31% reduction); and the COD dropped from 280 mg/l to 160 mg/l (43% reduction). The reduction observed in the case of bacteria is 2 logarithmic units for a thickness of 50 cm of sand.

The basins functioned for 1 year without being cleaned. At the end of that time, and after the basins had been dried, cleaning trials were carried out using simple methods (rakes) and proved to be rapid and efficient. Beneath the first 2 cm of slightly spoiled sand, the granulation level and the organic matter content had suffered very few modifications. Cleaning products were dry, easy to handle, and could be reused as an enriching agent for arable land.

In 1983, the activated sludge treatment plant was put into service. Treated wastewater was infiltrated alternately into the two basins, containing a 1 m layer of sand. On entry to the basin, the secondary effluent had 15 mg/l of BOD_5, 60 mg/l of

COD and 10 mg/l of TSS. Infiltration rates, however, remained at the same values (0.24 m/day) as in the case of effluent that had simply been decanted. After a 3 month rest period, without cleaning, the same rates were obtained.

Settled rainwater was infiltrated into the two other basins. Analyses of the quality of water in the layer of sand at the bottom of the basin were carried out. The reductions in faecal coliforms and faecal streptococci were of the order of 2 logarithmic units/50 cm of sand (with 10^7 and 10^6, respectively, bacteria/100 ml in rainwater at entry to the basin). The BOD_5 dropped by approximately 30% and the COD by 40%, for 50 cm of sand, and the beginnings of a denitrification process appeared. A drop varying between 70% and 85% was recorded in the total phosphorus content as the water flowed vertically through the sand layer; an additional drop also occurred in the chalk. To obtain the best results, cleaning should take place once a year for all the basins.

The total investment cost was US$73 000 (August 1982), or FF400 (US$40) per head. Maintenance costs are FF10 (US$1) per user per year.

The results obtained at Flesselles in 1982 and 1983 show that, in order to reach a very high level of purification and final disinfection of the effluent, the thickness of the layer of sand in the infiltration basins would have to be increased to approximately 1.50 m. In this case, the sewage treatment plant could be eliminated, leaving just the decantation basin, which would receive all water (whether waste or rain), and two alternating infiltration basins.

Chapter 13

Managing the sewage sludge composting process

George B. Willson*

Introduction

Sewage sludge composting operations increased from one to two plants in 1973 to 61 in production in 1983 in the United States. By spring 1985, the number had increased to 115. To be successful, these plants must consistently produce a good quality product to ensure the marketability of the compost and compliance with regulations on marketing sludge products. Their success will depend on their process control and the characteristics of the inputs, namely the sludge and the bulking materials. The specific quality required will also depend on the locally available markets. Typical uses of the sludge compost include soil conditioning for the establishment of vegetation for landscaping reclamation of disturbed lands and maintenance of park areas and, where permitted, farming, gardening and other private uses. Recommendations have been developed for these and other compost uses by the Agricultural Research Service (Hornick *et al.,* 1984).

Effects of sludge characteristics

Sewage sludges vary greatly in chemical and physical characteristics, depending on the methods of wastewater treatment used and the kind and amount of industrial wastes discharged into the sanitary sewers. Nearly all sewage sludges can be composted, as long as the biological wastewater treatment processes are viable. However, some sludges may require modest variations in operating procedure, while other sludges could yield a compost that was of little or no value for the usual uses of compost or was not marketable under regulations.

The stability, or conversely the digestibility, of the sludge determines the potential for self-heating and composting. The volatile solids content of the sludge is a rough indicator of its stability, although phosphorus-removal chemicals and conditioning chemicals for dewatering may obscure the interpretation. In most cases, it is possible to compensate for a 'too stable' sludge by reducing the aeration rate and/or sometimes increasing the thickness of the pile blanket. Although stability of the sludge does not limit the feasibility of the composting process, it does limit the potential for utilizing high aeration rates to maximize drying.

* Soil Microbial Systems Laboratory, Agricultural Research Service, USDA

It is well known that pH influences most microbial processes; the composting process is far less sensitive to pH than most. This is probably due to the diversity of microbial species involved. Experience at Beltsville indicates that there is no appreciable inhibitory effect on the composting process for sludges with pH as high as 11 (Willson, Epstein and Parr, 1976). At higher pHs there is sometimes a delay of several days in the onset of heating. Sludges with pH as high as 13 have been composted (Willson and Thompson, 1980). Less is known about the effect of low pH on sludge composting, but fruit wastes with pH as low as 4.5 have been successfully composted (Mercer *et al.*, 1962).

The ratio of carbon to nitrogen for sewage sludge (6:1–15:1) is below the recommended optimum of 25:1–35:1 for composting. Use of a carbonaceous bulking agent such as woodchips will increase the C/N ratio; however many sludges could be composted successfully without adjustment of the C/N ratio.

Sewage sludges contain varying amounts of soluble salts, depending on their source and the wastewater treatment processes employed. Potential problems can arise where alum or lime and ferric chloride are used as flocculating and conditioning agents during sewage treatment and sludge dewatering. The resultant increase in soluble salts in the sludge compost can be toxic to sensitive plant species. When prepared composts are used for agronomic and horticultural purposes, the danger of soluble salts is minimal because of dilution in soil after the compost is incorporated and mixed and due to the extent of leaching that occurs after application. For composts high in soluble salts, a preliminary leaching is especially important when the compost is used in potting mixes for house plants (Chaney, Munns and Cathey, 1980; Sterret *et al.*, 1983).

The amount of heavy metals present in the sludge is largely dependent on the type and amount of industrial waste effluents that are discharged into the sanitary sewers. The effects of excess heavy metals in sludges on agricultural crops have been extensively studied (Logan and Chaney, 1983). In general, acceptable concentrations of heavy metals in sludges for agricultural use are applicable to composted sludge. The composting process has little if any effect on the availability of metals for plant uptake, except for possible effects of increased pH and of higher organic matter per unit metals, both of which inhibit plant uptake of metals.

Heat treatment of sludge to improve dewatering may pose a unique problem. Unlike other sludges that contain a diverse population of micro-organisms adaptable to the composting process, these sludges are apt to be sterile. Thus, an initial inoculation would probably be necessary. Recycling used woodchips within the process would probably supply the needed inoculation thereafter.

Effects of bulking materials characteristics

Due to the large volume of bulking materials needed, local availability and cost are major factors in their selection. The bulking material, which is added to the sludge to condition it for composting, has major impacts on operating procedures, costs and the quality and quantity of compost produced (Willson *et al.*, 1983). The primary objective of adding the bulking material is to impart a suitable porosity, texture, structure and moisture content to the sludge so that aerobic microbial activity can be supported.

A significant volume of interconnected voids is needed to allow uniform passage of gases (oxygen entering the mass and carbon dioxide, water and other decomposition products departing) throughout the composting mass. As the thickness of the composting mass increases, there is a corresponding increase in the required air velocity: this necessitates larger and more uniform air passages. Therefore, the aerated pile requires a greater degree of porosity than is needed for windrows.

Structure refers to the ability of the bulking material–sludge mixture to retain its porosity by resisting settlement.

Texture, although closely related to porosity, has a separate significance. Considering the mechanics of oxygen transfer within the composting matrix, it is evident that the sites of aerobic activity are on the surface of material particles and that the particle interiors will be anaerobic. Since the ratio of surface to mass increases with decreasing particle size, the smaller the particle size, the larger the active aerobic area will be. However, a compromise is necessary to maintain the necessary porosity and to maximize aerobic activity. Ideally, a bulking material would provide a system of large voids for rapid movement of the gases and would absorb water from the sludge causing it to crumble into small particles that would only partially fill the voids. This is the effect of woodchips on sewage sludge when used at appropriate ratios for their moisture contents.

The maximum moisture content at which this consistently occurs is about 60–65% (or the mixture of sludge and woodchips) depending on the characteristics of both the bulking material and the sludge. The minimum acceptable moisture content for composting (about 40%) is rarely of concern with sewage sludges except for sandbed dried sludge. At lower moisture contents, the metabolic processes of the micro-organisms are increasingly inhibited.

There are several other characteristics of bulking materials that have important implications for the composting process or influence compost quality. Most of the materials that have been considered for use have a high carbon content and usually a low nitrogen content. It should be noted that the availability of these nutrients to microbes is more important to the process than the absolute C/N ratio. The desirable C/N ratio for soil conditioners (10:1–15:1) is substantially below the optimum for composting. Thus, a bulking material such as woodchips that increases the low C/N ratio of the sludge for composting, and which can be partially recovered, also lowers the C/N ratio for use of the compost.

Recovering the bulking material for reuse reduces costs of inputs to the composting process but does add a separation operation. The type of bulking material will largely determine the material handling characteristics and the required element.

Some prospective bulking materials may contain ingredients that are deleterious to the quality of the compost. An example of this would be the glass, metal and plastics in municipal refuse. Although these have little effect on the functional properties of the compost, they have a negative aesthetic effect. The broken glass also poses a hazard to people who handle the compost. The glass, metals and plastics may, however, provide structure for aeration during composting. With sufficient separation and grinding an aesthetically acceptable compost can be produced from municipal refuse.

Over several years, many potential bulking materials were studied at Beltsville. A typical analysis of the sludge from the Blue Plains wastewater treatment plant is given in *Table 13.1,* along with analyses of several composts made at various times from Blue Plains sludge and these bulking materials. The heavy metal contents of the woodchip compost were diluted about one-third by the wood fines that remained with the composted sludge after screening.

Table 13.1 Effect of bulking materials on composition of compost

Material	*TKN* %	*Zinc*	*Nitrogen*	*Copper* mg/kg wt	*Lead*	*Cadmium*
Blue Plains raw sludge	3.8	980	85	420	425	10
Compost with:						
Woodchips	1.6	770	55	300	290	8
Leaves	1.1	198	52	–	182	1.5
Peanut hulls	1.8	756	25	65	–	7.4
Municipal refuse	1.1	378	103	165	168	2.3
Air classified refuse	1.0	370	67	–	140	3.5

The dilution of heavy metals was about two-thirds for leaves, peanut hulls, municipal refuse and air-classified refuse. Leaves, peanut hulls and air-classified refuse do not require screening. In general, these bulking materials are very low in heavy metals, thus dilution is proportional to the added organic matter less the portion that is biologically oxidized. These materials do contain small amounts of nitrogen, except for the woodchips which have an insignificant amount. Thus, compost mixtures of this type make it possible to decrease the heavy metal concentration in sludge composts with little or no loss in fertility.

Leaves and air-classified refuse lack structure and were most effective when used in combination with a reduced amount of woodchips. Pelletized air-classified refuse (also called RDF, refuse derived fuel), retained its structure and provided adequate porosity through two composting cycles. The non-compostible fraction of the municipal refuse effectively provided the necessary structure and voids, but required separation from the compost and subsequent landfilling.

Other materials that have been identified as suitable for bulking sludge for composting were listed by Willson *et al.* (1980) and Epstein (1980). The use of shredded rubber tyres is described by Higgins (1982); zinc levels are increased in the compost due to the high zinc level in rubber products (usually 0.5–4% zinc).

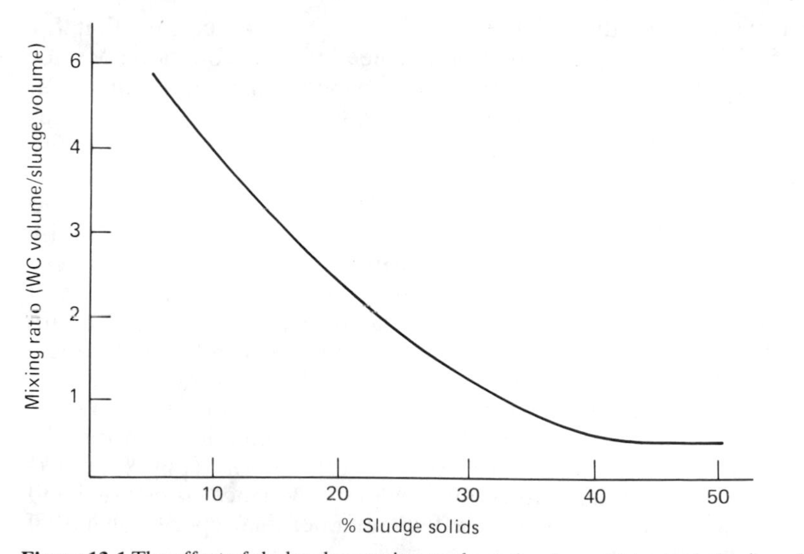

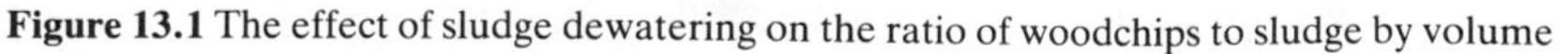
Figure 13.1 The effect of sludge dewatering on the ratio of woodchips to sludge by volume

The upper acceptable limit for moisture content of the bulking material/sludge mixture is usually about 60%. Thus, the acceptable ratio varies with the sludge solids content as shown in *Figure 13.1* for woodchips.

The possibility of combining two or more bulking materials, instead of finding a single material that meets all the necessary criteria, has been largely unexplored. The potential number of possibilities is so numerous that this option is best explored on a site-specific case-by-case basis. Opportunities exist for selecting or blending materials to reduce operational costs, reduce heavy metals concentrations and improve marketability of the compost.

Aeration of compost

Natural or forced aeration of a compost pile accomplishes several important functions:

- It supplies oxygen to the aerobic micro-organisms.
- It removes excess heat.
- It removes water vapour.
- It removes carbon dioxide, ammonia and other gaseous products of the decomposition process.

These functions must be considered in relation to their interactions and to the objectives of the process in determining a suitable aeration rate. The desired objectives that aeration can influence are elimination of pathogens or removal of excess moisture, thus facilitating subsequent handling and distribution.

Haug (1979) has suggested that the total oxygen requirement for composting can be determined from the chemical composition of the organic solids and the extent of degradation during composting. With an average composition for sludge organics of $C_{10}H_{19}O_3N$, the stoichiometric oxygen requirement can be approximated from the equation:

$$C_{10}H_{19}O_3N + 12.5\ O_2 \rightarrow 10\ CO_2 + 8\ H_2O + NH_3$$

He estimates that 33–82% of the volatile solids may be degraded and that for digested sludge 0.50 g of oxygen/g of sludge solids is needed. The equation does not accurately reflect the fate of the nitrogen, most of which will become microbial protein (Willson and Hummel, 1975). However, despite obvious oversimplification, the equation may still offer a reasonable estimate of the total oxygen requirement.

Several investigations (by Gray, Sherman and Biddlestone, 1971; Jeris and Regan, 1973; Schulze, 1962; Willson and Hummel, 1972) have experimentally measured oxygen consumption rates for composting materials. Most of these studies have related the consumption rate to temperature. Gray, Sherman and Biddlestone (1971) and Jeris and Regan (1973) indicated that the consumption rate increases with increasing temperature up to some optimum level and then declines. Maximum observed oxygen consumption rates corresponding to various materials and temperatures of operation are shown in *Table 13.2*. Oxygen consumption rates may be expected to vary with (a) microbial species and population density, (b) digestibility of the nutrients, (c) particle size, (d) moisture content, (e) pH, and (f) temperature. The optimum temperature may also vary since substrate and environmental conditions have a selective effect on microbial species, each of which will have its own optimum temperature.

Table 13.2 Oxygen consumption during the aerobic thermophilic composting of different organic wastes

Source	*Test material*	*Temperature*	*Oxygen consumption rate*
Schulze (1962)	Mixture of garbage and sludge	65°C	8 mg/h/gVs*
Gray, Sherman and Biddlestone (1971)	Ground-up garbage	48°C	13.7 mg/h/g (dry matter)
Gray, Sherman and Biddlestone (1971)	Straw and grass	55°C	2 mg/h/gVs
Willson (1972)	Dairy manure	60°C	4.8 mg/h/gVs
Jeris and Regan (1973)	Paper refuse	64°C	5 mg/h/gVs

* Grams of volatile solids (gVs)

Willson and Hummel (1972) developed some generalized relationships between the aeration rate, temperature, oxygen consumption rate and moisture content changes during composting, as shown in *Figure 13.2*.

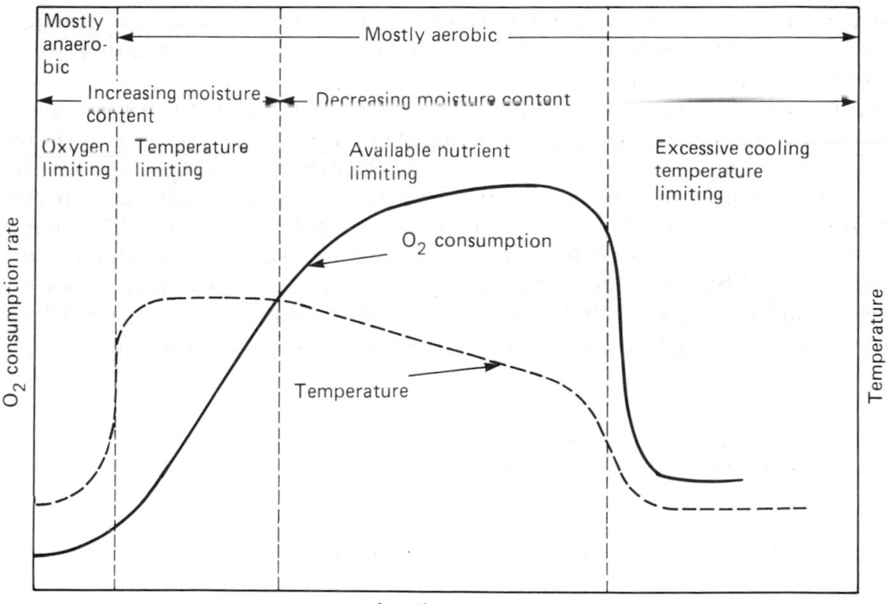

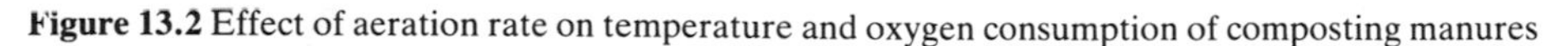
Figure 13.2 Effect of aeration rate on temperature and oxygen consumption of composting manures

At very low aeration rates, the oxygen supply is limiting and decomposition is mostly anaerobic, which is relatively slow and little heat is produced. The temperature of the compost will be near ambient in this range.

With increased aeration, the process becomes mostly aerobic. Aerobic dissimilation releases about 20 times as much heat per unit of waste decomposed as is released by anaerobic dissimilation (University of California, 1953). In this range, more heat may be produced than is removed by aeration and other losses.

The temperature rises to the level at which it restricts microbial activity. The capacity of the air to remove heat is the limiting factor; most of the heat removal is due to evaporation of water and removal of the water vapour by aeration.

In the next range of aeration rates, some cooling occurs and water is removed faster than it is produced. The factors limiting decomposition in this range are probably availability of nutrients and diffusion of the air throughout the mass.

Aeration rates above this range will result in heat being removed faster than it can be produced. This will lower temperatures below the thermophilic range where the rates of decomposition and heat production are slower. The temperature of the air supply will then be the limiting factor.

The micro-organism activity also varies with process time. At the start of the process, the temperature rises rapidly. This is accompanied by a build-up in the population of thermophilic organisms. If other conditions are favourable, the build-up will continue until the availability of nutrients limits the population. As the most readily decomposed nutrients are consumed, the process slows down. The effect of process time on *Figure 13.2* is to vary the scales.

Early research at Beltsville focused on maximizing temperature elevation to ensure pathogen destruction throughout the pile by manipulating the aeration rate, pile geometry and aeration pipe layout (Willson, Parr and Casey, 1978). Recently, research was conducted to determine survival of several pathogens and indicator organisms in the piles. This research indicated that an acceptable level of pathogen destruction resulted when temperatures were maintained at 55°C or above for 3 consecutive days (Burge *et al.*, 1981).

An important consideration for most sludge composting operations is removal of excess moisture during composting. An experiment was conducted to determine the effect of aeration rate on moisture removal (Willson, Parr and Casey, 1979). Compost piles, each containing 12 tonnes of sludge solids, were aerated at rates ranging from 3.1 to 37.5 m^3/h per tonne of sludge solids (*Table 13.3*). Rates were set by measuring the air velocity in the pipe and adjusting the timers for an appropriate time of blower operation during each 20 min cycle. The two piles with the lowest aeration rates were slightly slower than the others in warming up, while the two piles with the highest aeration rates had started to cool during the last few days.

Table 13.3 The effect of aeration rate on moisture removal during composting of a sludge–woodchip mixture by the aerated pile method

Air flow rate (m^3/h t)	*Initial moisture content of mix* (%)	*Final moisture content of mix* (%)	*Estimated tonnes of water removed* (15% Vs loss)
3.1	61.7	56.3	19.3
9.4	62.5	54.7	23.0
18.7	64.9	57.3	24.9
28.1	66.2	51.8	35.3
37.5	67.7	54.6	32.7

Table 13.3 shows the effect of aeration rate on moisture removal during the composting of the sludge–woodchip mixture. The piles were made up at the rate of one per day because of limited labour and materials. Rain during this week caused

significant increases in the moisture content of the sludge–woodchip mixtures, because of the increasing moisture content of the woodchips.

Table 13.3 includes estimates of the water removed from the sludge–woodchip mixture during the 3 weeks of composting. This calculation is based on the change in moisture content and on an estimated 15% loss of volatile solids as determined by Sikora *et al.* (1981) in a materials-balance study. The water removed from the mixture by aeration increases with an increase in aeration, but was not proportional to the aeration rate, as might have been expected since temperatures were similar.

If it is important to remove a maximum amount of moisture, the aeration rate can be increased to the optimum rate for decomposition and drying. As indicated in *Figure 13.2,* the process is unstable at this aeration rate; thus it is necessary to regulate the aeration rate on a temperature feedback system, as described by Miller and Finstein (1985). The maximum potential benefit of temperature-demand aeration is realized with an undigested primary sludge and decreases with increasingly stable sludges. The optimum temperature is about 55°C. A disadvantage of this system is that it is not as effective for pathogen destruction since substantial portions of the compost do not reach 55°C.

Design of an aeration system for composting also requires information on the amount of air pressure needed to overcome friction. The total air resistance is the sum of the resistance through each of the components of the system including the air distribution pipes. The pipe friction can be calculated from fluid mechanics. Resistance to air flow by the composting materials would be similar to that produced by granular materials, varying with particle size, shape and air velocity.

Airflow resistance of typical sludge–woodchip materials was reported by Willson (1983). The materials tested were:

- Woodchips (WC) which are used for the base of the aerated piles.
- Used woodchips (UWC) also used for the base.
- Unscreened compost (UC) which is used for the blanket.
- Sludge–woodchip mixture (MIX), and
- Screened compost (SC) which is used for the odour filter pile.

As is seen in *Figure 13.3,* the resistance varies not only with particle size and size distribution, but also with velocity. The screened compost was tested at two

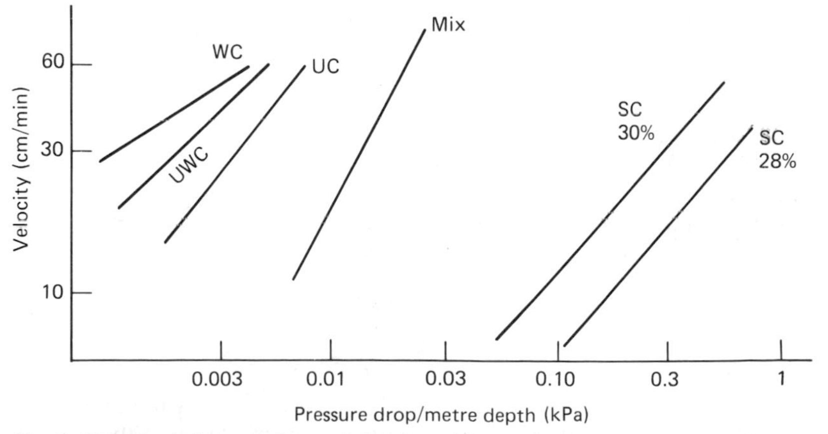

Figure 13.3 The resistance of composting materials to airflow as influenced by the rate of airflow. For abbreviations see text

moisture contents. Decreased resistance with increased moisture content was probably due to some aggregation of the particles, effectively increasing the particle size. Additional testing is needed to evaluate the effect of moisture content on all materials and to extend the data over a wider range of air velocities. However, moisture content is likely to have less effect on the resistance to air flow as the particle size increases.

Aerated pile method of composting

A variety of methods have been developed for managing the sludge composting process, ranging from simple methods requiring little equipment or facilities to highly mechanized and automated systems. The aerated pile method was developed (Willson *et al.*, 1980) to meet requirements of pathogen destruction and odour control with a minimum of equipment and facilities. It has been adapted to a variety of applications with varying levels of mechanization and automation.

Figure 13.4 is a schematic diagram of the aerated pile, which illustrates the basic concepts of the Beltsville composting method. The sewage sludge must first be mixed with a bulking material to impart a suitable porosity for air movement. Then the mixture is stacked over an air distribution system designed to provide uniform air movement with a minimum of resistance (in this case, air ducts above ground, covered with woodchips). A blanket of compost over the top of the pile provides insulation so that all the new sludge can self-heat to pasteurizing temperature for destruction of pathogens. The blanket prevents escape of odours from the sludge. A positive-pressure aeration system would cause cold areas adjacent to the air distribution pipes; thus a negative pressure system is used. The effluent air is then discharged through a small pile of compost for removal of odours. The pile is usually built incrementally as shown and is referred to as an extended pile.

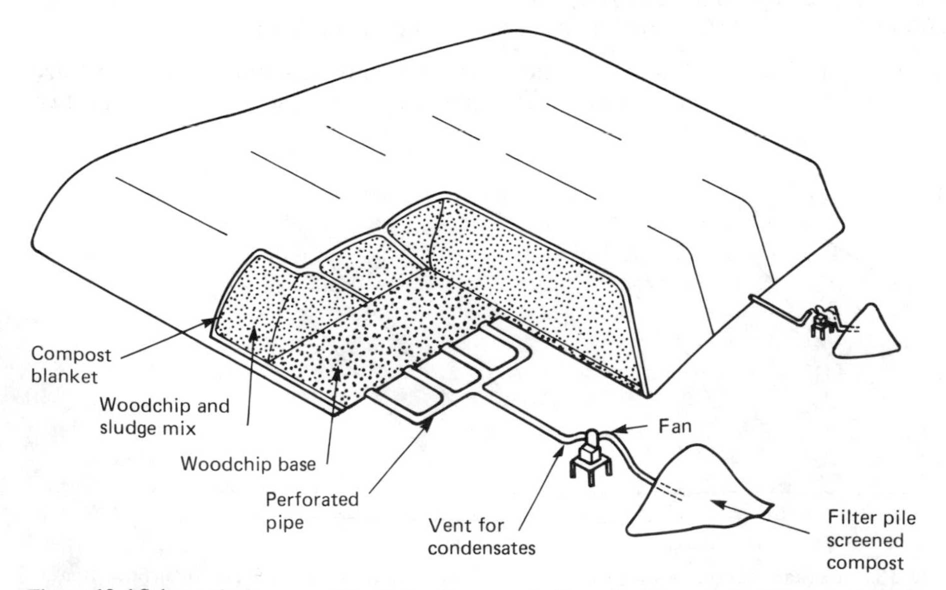

Figure 13.4 Schematic diagram of the Beltsville aerated pile method for composting sewage sludges

However, individual piles that are similar to the first segment on the left of the cutaway section can be constructed if sludge production in periodic.

Figure 13.5 is a materials-flow diagram indicating the unit operations for the process when woodchips are used as the bulking material. The sludge–woodchip mixture remains in the aerated pile for 21 days. Microbial activity causes temperatures to increase to a range of 60–75°C, where they remain for most of the composting period. The compost is then transferred to an unaerated curing pile for 30 days. Curing is an extension of the composting process during which time reheating occurs to the thermophilic temperatures. Curing ensures complete dissipation of phytotoxic gases and offensive odours.

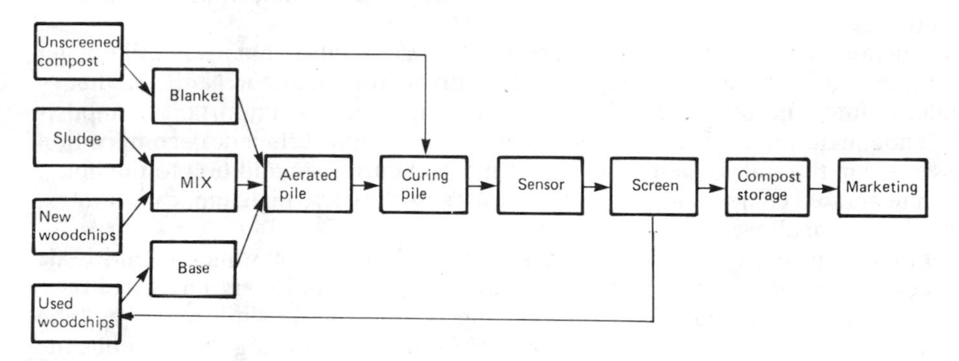

Figure 13.5 Materials flow for the Beltsville aerated pile method for sewage sludge composting

A portion of the cured compost is utilized to 'blanket' the aerated pile. The remainder, if necessary, is solar dried for screening. Woodchips, recovered in the screening operation, are reused as bulking material for new batches of sludge or to provide the porous base for the next pile. The screened compost can be marketed directly in bulk or bags, or can be stored.

Utilization of sludge composts

A good marketing programme for the compost is as important as the processing for a successful operation. This must be based on a knowledge of the characteristics of the compost and the soils, crops and climates with which it will be used. Hornick *et al.* (1984) summarized the research at Beltsville on compost uses and compiled recommendations for application rates and methods for numerous uses.

To determine appropriate application rates of compost as a fertilizer, both the nutrient content and its plant availability must be considered. Nearly all the nitrogen (N) in compost is in organic forms and must be mineralized in the soil before it is available to plants. Typically, about 10% of the N is available in the first crop season and smaller amounts become available in subsequent seasons. Phosphorus (P) availability is much more variable in sludge composts. This is probably due to the iron or aluminium used by many sewage treatment plants to condition sludge for dewatering. The P availability for one high-iron sludge compost was about 40% in the first year. Although the potassium (K) in sludge composts is considered to be mostly available, few sludges contain significant amounts. Thus, it is often advantageous to use mineral fertilizers in combination with sludge composts.

The optimum use of compost will usually be on problem soils that can benefit from the soil-conditioning effect as well as the fertilization. Compost can markedly improve the physical properties of soils that are deficient in organic matter. The benefits include increased water retention, increased soil aeration and permeability, increased water infiltration and decreased bulk density and surface crusting. Such changes in soil physical properties contribute significantly to reducing soil erosion and decreasing loss of plant nutrients by runoff. Compost has also been shown to be useful for controlling some soilborne plant diseases (Lumsden, Lewis and Millner, 1982). Information on this subject is limited and more research is needed.

Production of a consistent quality of compost is essential so that users will not get variable results. The compost should be analysed routinely for N, P, K, heavy metals, salinity and pH. The stability of the compost is also important; composts with inadequate processing can inhibit plant growth until sufficient decomposition takes place in the soil. Usually a month or more of curing, in addition to the initial high-rate aerated composting, is needed for uses that involve high application rates, such as in horticulture.

Composts made from sewage sludges that are low in heavy metals and toxic organics can be used safely and beneficially as soil conditioners and fertilizers. These benefits are largely dependent on the quality of the sludge, the bulking material used and the degree of stabilization achieved during the composting process. Research has shown that good quality sewage–sludge compost can be used as a valuable amendment in: the production of agricultural and horticultural crops; the establishment, maintenance and production of turf grasses; the formulation of potting media for nursery production; vegetable gardens; and the reclamation of marginal and disturbed soils.

References

BURGE, W. D., COLACICCO, D. and CRAMER, W. N. (1981) Criteria for achieving pathogen destruction during composting. *Journal of the Water Pollution Control Federation,* **53,** 1683–1690

CHANEY, R. L., MUNNS, J. B. and CATHEY, H. M. (1980) Effectiveness of digested sewage sludge compost in supplying nutrients for soilless potting media. *Journal of the American Society of Horticultural Science,* **105,** 485–492

EPSTEIN, E. (1980) Bulking materials. *Proceedings of the National Conference on Municipal and Industrial Sludge Composting.* Silver Spring, Maryland, Hazardous Materials Control Research Institute

GRAY, K. R., SHERMAN, K. and BIDDLESTONE, A. S. (1971) Review of composting, part 2 – the practical process. *Process Biochemistry,* **6(10),** 22–28

HAUG, R. T. (1979) Engineering principles of sludge composting. *Journal of the Water Pollution Control Federation,* **51,** 2189–2206

HIGGINS, A. J. (1982) Good marks for tire shredder. *BioCycle Journal of Waste Recycling,* **23,(3),** 37

HORNICK, S. B., SIKORA, L. J., STERRETT, S. B., MURRY, J. J., MILLNER, P. D., BURGE, W. D., COLACICCO, D., PARR, J. F., CHANEY, R. L. and WILLSON, G. B. (1984) *Utilization of sewage sludge compost as a soil conditioner and fertilizer for plant growth.* USDA, ARA, Agriculture Information Bulletin No. 464

JERIS, J. S. and REGAN, R. W. (1973) Controlling environmental parameters for optimum composting. *Compost Science Journal of Waste Recycling,* **14(1),** 10–15

LOGAN, T. J. and CHANEY, R. L. (1983) Utilization of municipal wastewater and sludges on land–metals. *Proceedings of the Colorado Sludge Meeting,* Denver, Colorado, p. 103

LUMSDEN, R. D., LEWIS, J. A. and MILLNER, P. D. (1982) Composted sludge as a soil amendment for control of soilborne plant diseases. *Research for Small Farms,* edited by H. W. Kerr Jr and L. Knutson. Special Symposium Proceedings, US Department of Agriculture Miscellaneous Publications, 1422, pp. 275–277

MERCER, W. A., ROSE, W. W., CHAPMAN, J. E., KAYSUYAMA, A. K. and DWINNELL, F. Jr (1962) Aerobic composting of vegetable and fruit wastes. *Compost Science,* **3(3),** 9–19

MILLER, F. C. and FINSTEIN, M. (1985) Materials balance in the composting of wastewater sludge as affected by process control strategy. *Journal of the Water Pollution Control Federation,* **57,** 122–127

SCHULZE, K. L. (1962) Continuous thermophilic composting. *Applied Microbiology,* **10(2),** 108–122

SIKORA, L. J., WILLSON, G. B., COLACICCO, D. and PARR, J. F. (1981) Materials balance in aerated static pile composting. *Journal of the Water Pollution Control Federation,* **53,** 1702–1707

STERRETT, S. B., REYNOLDS, C. W., SCHALES, F. D., CHANEY, R. L. and DOUGLASS, L. W. (1983) Transplant quality, yield and heavy metal accumulation of tomato, muskmelon and cabbage grown in media containing sewage sludge compost. *Journal of the American Society of Horticultural Science,* **108(1),** 36–41

UNIVERSITY OF CALIFORNIA (1953) *Reclamation of municipal refuse by composting.* Technical Bulletin, p, 9

WILLSON, G. B. (1983) Forced aeration composting. *Water Science and Technology,* **15,** 169–180

WILLSON, G. B. and THOMPSON, J. L. (1980) Dewatering of sludge compost piles. *Proceedings of the National Conference on Municipal and Industrial Sludge Composting,* pp. 46–54, Silver Spring, Maryland, Hazardous Materials Control Research Institute

WILLSON, G. B., EPSTEIN, E. and PARR, J. F. (1976) Recent advances in compost technology. *Proceedings of the 3rd National Conference on Sludge Management Disposal and Utilization,* pp. 167–172, Silver Spring, Maryland, Hazardous Materials Control Research Institute

WILLSON, G. B. and HUMMEL, J. W. (1972) Aeration rates for rapid composting of dairy manure. *Proceedings of the Cornell Agricultural Waste Management Conference,* Ithaca, New York, Cornell University

WILLSON, G. B. and HUMMEL, J. W. (1975) Conservation of nitrogen in dairy manure during composting. *Proceedings of the 3rd International Symposium on Livestock Wastes.* St Joseph, Michigan, ASAE

WILLSON, G. B., PARR, J. F. and CASEY, D. C. (1978) Criteria for effective composting of sewage sludge in aerated piles and maximum efficiency of site utilization. *Proceedings of the National Conference on Design of Municipal Sludge Compost Facilities,* Rockville, Maryland, Information Transfer Inc.

WILLSON, G. B., PARR, J. F. and CASEY, D. C. (1979) Basic design information on aeration requirements for pile composting. *Proceedings of the National Conference on Municipal and Industrial Sludge Composting.* Silver Spring, Maryland, Information Transfer Inc.

WILLSON, G. B., PARR, J. F., EPSTEIN, E. *et al.* (1980) *Manual for Composting Sewage Sludge by the Beltsville Aerated-Pile Method,* Beltsville, Maryland. Joint USDA-EPA Publication, USDA

WILLSON, G. B., CHANEY, R. L., CASEY, D., MORELLA, M. and COULOMBE, B. (1983) Effects of sludge and bulking agent characteristics on the composting process and product quality. *Proceedings of the 15th National Conference on Municipal and Industrial Sludge Utilization and Disposal,* pp. 140–144. Silver Spring, Maryland, Hazardous Materials Control Research Institute

Chapter 14

Discussion

Dr H. Bouwer, using slides, gave additional information not contained in Chapter 10. He talked about use of groundwater recharge or, rather, the use of soil–aquifer treatment because the unsaturated zone and the aquifer is used as a natural filter system. The Phoenix, Arizona system was used as an example; one of the treatment plants near Phoenix received waste from about 1½ million people, producing about 400 l of sewage effluent/day per person. The sewage treatment included primary sedimentation, activated sludge and secondary clarifier processes and the effluent was then discharged into a dry salt-river bed. This effluent stream was capable of irrigating about 10 000–20 000 ha but the quality of the water had to be adapted to existing farming practices. The farmer, who had been growing vegetables for years, could not be asked to stop growing vegetables and irrigate only fibre and seed crops for which the secondary effluent would be suitable. It was decided to treat the effluent so that it would meet the requirements for unrestricted irrigation.

In the United States, the public health requirements for unrestricted irrigation are formulated by state health departments and, without exception, they are very conservative; they usually require no viruses, no faecal coliforms and the absence of helminth eggs or parasitic worms. From the agronomic point of view the quality of the effluent water must conform with established standards for the chemical quality of irrigation water. Sewage effluent creates no particular problem here, except perhaps for the nitrogen concentration. If sewage effluent is the only source of water for irrigation in arid areas with high water demands, it can contain too much nitrogen and problems of excessive nitrogen arise. Not all crops react adversely but a number of crops do, so it is normal to reduce the nitrogen concentration to below 10 mg/l. Minor elements are not usually a concern in the case of domestic sewage but if there are industrial inputs to the system then the concentrations of certain metals and other elements can be too high. The solution to this is source control, enforcing city ordinances which prohibit industries from discharging their wastewater into the sewerage system without pretreatment or recycling.

All these requirements can be met by a soil–aquifer treatment system. However, this can only be achieved if hydrogeological conditions are favourable for groundwater recharge, and in this particular area near Phoenix a deep unsaturated zone overlain with mostly sand and gravel was ideal. The soil must be permeable enough to give high infiltration rates, yet fine enough to provide good filtration and good treatment. Then the underlying aquifer must be unconfined and sufficiently transmissive so that a groundwater mound is not built up which would drown out the whole system. The effluent is placed in infiltration basins at the top of the unsaturated zone, allowed to filter down to the groundwater (the soil treatment) then allowed to travel a distance through the aquifer (the aquifer treatment) before being pumped out and used for unrestricted irrigation.

These systems are not new. Bank filtration systems have been used for many years. Here, instead of taking water directly out of rivers, wells are placed along the bank a distance away from the stream and water is drawn through the aquifer, before being pumped out so that the benefit of aquifer treatment and soil filtration is gained. This is very good as a pretreatment, especially if the water in the stream is polluted or subject to accidental discharge of toxic materials.

Soil–aquifer wastewater treatment is the reverse of bank filtration, where partially treated sewage effluent, instead of being discharged directly into the stream where it can cause pollution, is infiltrated through the soil to the groundwater and then through the aquifer. Then, when it enters the stream, it is a lot better in quality, so it minimizes pollution and encourages reuse by downstream users.

A typical soil–aquifer treatment system has infiltration basins in two parallel rows with wells in between, from which the water is pumped out. These systems can be designed and managed so that all the sewage effluent which is infiltrated will indeed move to the wells and be pumped out as renovated water. No native groundwater will be pulled into the system and no renovated sewage water will go out into the aquifer outside the boundary of the system; so a portion of the underground environment is used as a treatment system. A variation on the system is where the infiltration basins are grouped together and surrounded by a ring of wells around the entire system. These wells pump a mixture of renovated sewage effluent and native groundwater from outside.

Where the water table is high, renovated sewage effluent from the aquifer can be intercepted by underground drains usually spaced about 100 m or so from the infiltration system, to give about 100 m of aquifer treatment. These systems are very site-specific, so in order to find out how they perform and how they should be designed and managed for optimum performance, it is usually best to install a small experimental project and make mistakes on that before proceeding with the large-scale project. Such an experimental project was installed in 1967 in the floodplain of an ephemeral stream in Arizona. Test basins were constructed and the amount of effluent infiltration into the ground was measured under operating conditions of alternate flooding and drying out. In addition, different forms of treatment, including vegetation in the basins, were evaluated. A gravel layer in the basin bottom was found to be unsatisfactory because it gets clogged up. Over time, sludge accumulates on the surface of the soil and occasionally the basins have to be cleaned. About once a year, or once every 2 years, deposited solids are removed, the basins are harrowed and then the original infiltration rates are restored.

During the period of inundation, the infiltration rate declines, then during drying, infiltration rates recover depending on the temperature and rainfall during drying. Usually it takes about 10 days in the summer and 20 days in the winter to get full infiltration recovery. So, maximum hydraulic capacity is obtained with flooding periods of about 2 weeks, alternated with drying periods of about 10 days in the summer, 20 days in the winter. Arizona, like Northern Egypt, is arid, with a hot, dry climate and not much rain.

Algae can give problems because they can clog the soil in these shallow basins, where the water depth is only about 15 cm. Filamentous algae grow on the bottom of the basins but eventually they float up and get carried away with the water, so they are not a problem. However, sometimes suspended algae proliferate and they can really clog up the bottom with a filter cake.

The soil profile at this site is a fine loamy sand on top with coarse sand and gravel layers down to the aquifer. This is very satisfactory because the fine material is on top, so anything that goes through that colloidal layer will also go through the coarser material below and will not clog in depth, a very difficult situation to overcome. Wells have been installed down to the groundwater, which is at a depth of about 10 m; the quality improvement of the sewage effluent can be monitored as it goes down to the groundwater and also at it moves through the aquifer. After the sewage effluent has passed through about 30 m of sand and gravel, suspended solids are completely removed, BOD is completely removed and heavy metals and other minor elements are below maximum allowable concentrations.

What happens to the nitrogen depends on how the basins are flooded and dried out. Aerobic conditions can be maintained in the soil by flooding and drying frequently, a few days wet a few days dry, then almost complete conversion of the nitrogen to nitrate and no nitrogen removal is achieved. On the other hand, if nitrogen removal is desirable, the basins must be flooded a little longer to allow anaerobic conditions to develop in the soil profile; this provides conditions that are favourable for denitrification and about 60–70% of the nitrogen is removed. In this way the final nitrogen content is below the 10 mg/l limit for unrestricted irrigation. Phosphate is removed in the unsaturated zone and also in the aquifer, but the approximately 90% removal of phosphate is primarily due to precipitation of calcium phosphate.

Faecal coliforms are greatly reduced in passage through the soil, but some faecal coliforms show up in the groundwater immediately below the basins. Further retention in passing about 100 m through the aquifer is necessary before complete removal of faecal coliforms is achieved. Viruses were detected in the effluent before it was chlorinated but at no time were viruses found in the water that was sampled from the underlying aquifer. Soil–aquifer treatment is an effective system for removing viruses. Laboratory studies have also been carried out to take a closer look at nitrogen transformations and virus removal and these have been very helpful in providing indications of how the field system should be managed for optimum results.

There is another test pond where there is tremendous instability between the algae and organisms that live off the algae, like *Daphnia* and rotifers (*Rotifera*). Under such conditions tremendous algal blooms occur. If fish are added to the basins a lot better ecological balance is maintained, because the fish eat the *Daphnia* and rotifers, and a much better quality of water is produced without fish kills caused by algal blooms.

The next generation was a larger system, started in 1975, to produce renovated water for an irrigation district. If the lower quality effluent could be used for irrigation and high quality ground and surface waters from the irrigation district could be used for municipal purposes the city of Phoenix would benefit. Here, two sewage effluent lagoons were used, with the lower one being subdivided into four infiltration basins. Severe algal problems occurred because the first lagoon acted as a facultative stabilization pond and the effluent passing to the four infiltration basis was very green. Suspended algae filtered out on the bottom of the basins and precipitation of calcium carbonate also occurred, giving very serious clogging problems. To solve this, the first pond was bypassed so that the sewage effluent from the treatment plant went directly to the infiltration basins, increasing the rate of turnover in these basins by a factor of 5.

In this system, the total dissolved solids increase a little because of evaporation in the basins, suspended solids are completely removed, total nitrogen is reduced by about 70% and the basins are operating in a denitrification mode. Phosphate–phosphorus removal is about 90%, which is good if the water is to be used in recreational lakes. Total organic carbon reduces from 10 to 1.9, with the residual being mostly synthetic organics, of real concern when it comes to use of this water for drinking. Faecal coliforms and viruses are essentially completely removed. This is very effective treatment at a very low cost. The main cost is in pumping the groundwater out of the aquifer.

Positively charged electrostatic filters have been used for trapping viruses and this has proved to be a very good technique. About 1000–2000 l of water are passed through these filters, which are then sent to the laboratory for virus assay. Water samples are also taken for organics analysis, the trace synthetic organics which are of concern if the water is to be used for potable purposes. The whole spectrum of organic compounds–trihalomethanes, like trichloroethylene and chloroform, and chlorinated aromatics, like chlorobenzenes etc.–are assessed. A significant removal of organics has been achieved in the infiltration basins themselves because of volatalization losses. Then there is removal in the unsaturated zone and in the aquifer but a residual still persists. Before use for drinking, activated carbon filtration and, possibly, reverse osmosis would be necessary. Residual organics at this level would not be a problem in agricultural reuse.

The plan is now to enlarge the system to about 36 ha (120 acres) with 12 wells. The main obstacles are the legal problems (who owns the sewage effluent, who owns the water after it has reached the aquifer), but they are being resolved. Of all the various alternatives for water resources development for the City of Phoenix, the soil–aquifer treatment of wastewater and exchange for high quality water with an irrigation district is the cheapest option. Now, secondary effluent from an activated sludge plant is used because the sewage treatment plant was already there. However, soil–aquifer treatment systems are very efficient in removing organic carbon so biological pretreatment is not really necessary. Primary sedimentation is sufficient before infiltration begins so soil–aquifer treatment can save a lot of investment in treatment plant. In Arizona, studies with primary effluent are proceeding. Because of the higher suspended solids load on the basins, a lower infiltration rate results and more basin area is required.

Mr A. T. Shammas, Saudi Arabia, asked Professor Mara if the addition of lime, and the subsequent increase in calcium activity in the effluent and in the soil solution, creates problems of nutrient availability in calcareous soils?

Professor Mara admitted that this was certainly a possibility. Clearly when one is investigating these, not necessarily new, techniques to remove pathogens one has to look at, obviously very carefully, the quality of the final effluent and how it is going to affect the soil structure in the area which is going to receive that treated effluent. The irrigation of calcareous soils with high-calcium effluents is unlikely to aggravate any existing problems of nutrient availability, depending on the quantity of calcium added. It would be sensible to undertake growth trials of the crops to be irrigated in order to be able to assess the problem under local conditions.

Mr D. Kypris, Water Development Department, Cyprus, first remarked about the boron content of sewage. He indicated that this had been a problem in Cyprus 10 years ago, when the content of boron in sewage was of the order of 2 mg/l. Realizing the danger, it has been arranged through the Ministry of Commerce and Industry that washing powders containing perborates are not allowed to be imported into Cyprus. About 2 years ago, the level of boron in Nicosia sewage was 0.7 mg/l and the domestic water supply in Nicosia had a level of boron of 0.3 mg/l, which means that it was quite a successful step. He suggested that the way to tackle this problem was to control boron at the source. He then went on to question Professor Mara. It was recognized, he said, that stabilization ponds remove helminth ova and protozoan cysts and also other biological material, apparently due to long storage in the sunshine. Since such systems may not be acceptable in Cyprus because marginal land is scarce and high in value, it might be more feasible to store treated effluent. Perhaps Professor Mara could comment on what would be the effects of storage of conventionally treated sewage effluent during the non-irrigation season, to be used in the irrigation season.

Professor Mara indicated that, in areas where suitable land for ponds is not available, it is certainly feasible to store conventional secondary biological effluents in deep reservoirs prior to their use for crop irrigation. Some considerable degree of treatment, especially as regards excreted pathogen removal is concerned, will occur (depending, of course, on storage time and temperature). A routine water quality monitoring programme would be recommended to provide operational guidelines to ensure that the irrigation water is of suitable microbiological quality.

Mr J. Al Shatti, Kuwait, questioned Professor Mara on the possibility of chemical removal in stabilization pond systems, when treating industrial effluents without any domestic sewage.

Professor Mara pointed out that ponds are rather more robust than conventional wastewater treatment systems in withstanding the effects of heavy metals; up to 30 mg/l can be well tolerated. Yet ponds are, like activated sludge and biofilters, a biological process and therefore cannot withstand toxic concentrations of heavy metals (or, indeed, of any organic compound). Pretreatment of wastewater from industrial areas may be necessary before it can be treated biologically; a decision can only be taken after a thorough analysis of the particular wastewater.

Dr A. Abdel-Ghaffar, Egypt, asked Professor Mara about the breeding of flies and the production of odours in ponds; also, if the growth of algae in ponds posed any problem when the effluent is used for irrigation.

Professor Mara replied to the effect that if properly designed and properly maintained, pond systems do not encourage fly or mosquito breeding, nor are they the cause of odour. The presence of microalgae in pond effluents used for irrigation is beneficial because the algae will act as slow-release fertilizers. However, research is needed to ensure the compatibility of algal-rich effluents and irrigation equipment in order to prevent clogging of the latter.

Dr G. B. Shende, India, mentioned that stabilization ponds in India had provided a very good quality of treatment but odour and mosquito breeding problems still continue, in spite of better maintenance, and people residing near ponds complain. Another point is that, during treatment in stabilization ponds, effluent pH increases and this has a certain effect in the soil, making some of the nutrients, particularly trace elements, unavailable to crops. He questioned Dr Bouwer on the life expectancy of soil–aquifer treatment systems under different soil (geological) conditions.

Professor Mara said that he had only seen a very few pond installations in India and some of them did smell very badly but it was quite clear they were overloaded. They were receiving a flow and strength of sewage far higher than they had been designed for, so problems were bound to arise, as would have occurred with trickling filters and activated sludge. The pH increase in ponds is desirable to accelerate the removal and increase the kill rate of excreted bacteria. A high pH also precipitates heavy metals, so trace elements may become unavailable in the soil. Remedial measures to provide the crops with those trace elements may have to be taken.

Dr H. Bouwer replied to the question on the useful life of rapid infiltration systems. He indicated that a lot of the processes occurring in the soil are renewable. Biodenitrification is a renewable process and so this can go on indefinitely without building up any residue in the soil. The removal of BOD, the mineralization of organic matter, is a bacteriological process and the end products are CO_2, sulphate, nitrate, phosphate etc; that can go on indefinitely. The concentrations of heavy metals that are immobilized are so low that even after, maybe, 50 years of use of these systems heavy metal accumulations will only occur in the top metre of the soil. Experience in Arizona has shown that the only thing that really accumulates to any quantity in the unsaturated zone and in the aquifer are phosphates which, under alkaline conditions, precipitate as calcium phosphate. Calculations have indicated that, with phosphate accumulation in the unsaturated zone and in the aquifer, maybe after 200 years the porosity of the materials is decreased by 1% or so. So, the answer to the question is: these systems have a very long useful life, maybe decades or centuries, but not an infinitely long life.

Dr H. A. C. Montgomery confirmed what Dr Bouwer said. The first of the sites mentioned in his presentation had been in operation for 53 years with crude sewage and was working as well at the end of the period as at the beginning. The second one, the city of Winchester, had been in operation for 60 years, working extremely well. He added that at a works where biologically treated effluent is infiltrated through subsurface drains, these are laid in gravel and the gravel becomes blocked after about 10 years. They then have to be dug up and new drains laid.

Mr D. Fougeirol added that in order to maintain good purification efficiency alternate operation of infiltration basins was necessary during the drying period; oxygen in the air penetrates into the sand layer and restores the infiltration rate.

Dr J. A. Aziz, Pakistan, asked Professor Mara if there were any examples of 2 m deep maturation ponds? Long-term experience with pilot-scale observation ponds in Pakistan showed that pond efficiency did not change significantly with temperature and no sound relationship between temperature, BOD removal and BOD loading was established. This suggests that it would be best to base the design of ponds on pilot-scale investigations. A third point raised was that, in Table 8.1 the retention time in an anaerobic pond was shown as 6.8 days. In Chapter 9 Professor Pescod suggested that anaerobic ponds could be used if the BOD is more than 250 mg/l, to achieve an optimal result, and a maximum retention time of 5 days should be sufficient. How was this 6.8 days chosen in Professor Mara's design?

Professor Mara replied that maturation ponds up to 3 m deep are used in some pond installations in the southern states of the United States and, recently, research into their performance has begun in north-east Brazil.

Mr M. Abu-Zeid, Egypt, commented on Professor Mara's paper and addressed a question to Dr Bouwer. The comment was related to the example on the designed Alexandria case. This was based on a winter population equal to the summer population, whereas Alexandria had double the population in summer, and he wondered how this would affect the area of ponds, estimated to be 68 km^2. The question to Dr Bouwer was related to the fact that the system is designed on the assumption of avoiding flow of groundwater from the system area outside the well system. He wondered if this was based on environmental control standards and, if not, what percentage of this water is really captured by the well system with practical spacing and how much is not. If it is not based on environmental control standards, is there not a well system that belongs to the salt-water project in this area? Would it not be easier to pump the water from these wells at a distance, with a better chance of obtaining better quality groundwater?

Dr H. Bouwer indicated that there were several reasons why they did not want this water to flow out into the aquifer. The main reason is that even though soil–aquifer treatment improves the quality of the wastewater, the quality of the resulting water is not usually as good as the native groundwater, so they tried to minimize the portion of the aquifer that will be affected by the renovated wastewater. This is, in a way, a groundwater protection argument. In certain areas, like southern California, sewage effluent is used for groundwater recharge and if the flow of water into the aquifer is not controlled it goes to the existing wells in the area and is used for drinking water. However, they have made a very intensive health-effects study which showed that the people who drank this water had no higher incidence of disease than people who did not drink the water; so, apparently, there are no immediate effects but there is still concern about the long-term effects. What will happen after drinking this water for 30 years or 50 years or 60 years? Consequently, there is a tendency to control water underground and, if it is used for drinking water, to post-treat after it is pumped out of the aquifer and before it goes into homes for drinking. In the system with two parallel sets of infiltration basins and wells in between, the movement of that water in the aquifer outside the system is controlled or restricted by putting in monitoring wells on the periphery of the system and then managing infiltration and pumping rates so that the groundwater levels in these monitoring wells never rise higher than the groundwater level in the aquifer outside the system. Then there is no grade intervention on the system and no movement of renovated sewage into the aquifer outside the system. Outside water is not allowed into the system because in many areas there are restrictions on groundwater pumping and already overdrawing of groundwater resources. Groundwater pumping rights have been adjudicated and legal problems would arise if native groundwater were drawn into the soil–aquifer treatment systems.

Professor Mara pointed out that the population of Alexandria seemed to vary according to whom one talked to but there had been a lot of debate as to what the population of Alexandria will be in the year 2005. The population at the moment may double in the summer, but the evidence suggests that the permanent population is going to increase at a far greater rate than the tourist population. Because of the temperature difference between winter and summer, ponds should be designed for winter conditions; therefore, in the summer when the ponds are operating at a higher temperature they are able to accept a much higher load. That should accommodate the additional load from the tourist population in the summer, so there should not be a great problem.

Dr S. Al-Salim, Jordan, questioned Professor Mara on the maximum depth for facultative and maturation ponds which would achieve the required efficiencies, especially in hot-climate countries. He also asked Dr Bouwer what was the percentage of evaporated water in the soil–aquifer treatment system. Referring to Chapter 12, he asked about the calculation of detention time for the oxidation pond, which was shown as about 100 days, taking into consideration that the depth was only 1 m. He remarked that this figure is not normally recommended for pond design, especially in hot-climate countries. In Jordan, evaporation had been limited to less than 15% of the design capacity of each oxidation pond, even in the desert.

Professor Mara replied that at present there is little reliable information on the performance of deep facultative and maturation ponds. Preliminary research in north-east Brazil suggests that such ponds with depths of 3 m are able to achieve very good removals of excreted bacteria and viruses; but more research is needed.

Dr Fougeirol accepted Dr Al-Salim's remark and confirmed that calculations were made with 1 m and the proportion of surface between the soil–aquifer treatment and the ponds was about 10/1 in France. If a deeper lagoon is used, there will be less evaporation.

Dr Arar asked Dr Bouwer what factors have to be taken into consideration in deciding on the distance between the recharge area and the pumping wells.

Dr H. Bouwer indicated that the factors which have to be considered are primarily the aquifer materials and the die-off of bacteria, especially pathogenic bacteria in the aquifer. In the aquifer the slower the rate of die-off, the greater the need to increase the distance of underground flow before the renovated wastewater is pumped out of the aquifer. As a rule of thumb, for most sand or sand gravel mixes at least about 100 m of underground travel distance and about 1 or 2 months of underground detention time is needed before the water is pumped out. A fine sandy aquifer will probably require less distance but a coarse aquifer material, such as a very coarse gravel, will probably require a longer travel. If the aquifer is fractured rock it should not be used at all because the bacteria can travel very long distances (kilometres) through fractured rock.

Ms M. H. Marecos do Monte, Portugal, asked a question in relation to a problem she had to solve. In one part of Portugal the soil is very limey and water is scarce. The water supply comes from rainfall and, of course, the soil is not only high in lime but there are caves and underground rivers. There is great concern about pollution of the groundwater because there are many industries, such as tanneries, in the area. Is it too risky to suggest infiltration as a sewage treatment process suitable for this area? Perhaps sand beds in the infiltration basins will be suitable, because in some cases the 10 or 20 m of soil depth to the aquifer is not available.

Dr H. Bouwer said that they would generally not recommend the use of soil–aquifer treatment in areas of fractured rock or cavernous limestone, unless there was a very thick soil mantle. It is true that, in soil–aquifer treatment systems, most of the purification takes place in the upper 1 m of the unsaturated zone, about 90% maybe, but to get the remaining 10% of the purification you need to go all the way down to the groundwater, about maybe 100–200 m through the aquifer. So, it depends on how thick the soil is above the limestone formations and also what can be done in the way of controlling the movement of the water in the limestone formation. How much treatment there is after the water leaves the aquifer, before it is used for drinking water, must also be taken into account. His recommendation was to avoid fractured rock and cavernous limestone formations for soil–aquifer systems.

Dr H. A. C. Montgomery agreed that normally one would avoid karstic rock formations for soil–aquifer treatment. However, it is just possible if the water is to be used for irrigating cotton, or some other crop which is not consumed by man or animals, that soil–aquifer treatment might still be adopted. Also, if there is dilution by natural water, this will help to reduce the concentration of pollutants in the effluent. So, although as a general principle one would not recommend it, there may be special circumstances which would allow use of such a system even in karstic conditions.

Mr A. T. Shammas asked Dr Willson if, in the greenhouse experiment, tomato plants grown in a potting material made up of composted 1:1 by volume mixture of peat moss and digested sludge suffered from nitrogen, potassium and phosphorus deficiencies. Would he advise enrichment of the composting material with these elements during the composting process or will this prove detrimental to microbiological activities?

Dr G. B. Willson suggested that the compost should probably be supplemented at the time of use because the nitrogen in particular will be mainly in the organic form during the composting process and will only become available for plant uptake when mineralized. One thing that makes compost useful as a soil conditioner is that so much of the nitrogen is in the organic form and thus the system is not being overloaded with readily available and leachable nitrogen.

Dr S. Al-Salim then asked if the sale of the product covered the cost of production?

Dr G. B. Willson replied that the marketing was not expected to cover the cost of production. It only offset a portion of the cost so that the composting process was less expensive than alternative methods of sewage sludge disposal.

Professor Pescod closed the discussion by mentioning that stabilization ponds should not be selected for sewage treatment where karstic conditions exist because lining, which is very expensive, will be necessary to prevent high infiltration losses.

Section 3

Use of sewage effluent

Chapter 15

The use of biologically treated wastewater together with excess sludge for irrigation

Rolf Kayser*

Introduction

Sewage farms where wastewater was irrigated by surface flooding were built in Germany as early as 1870. As their main purpose was to treat the wastewater, the availability of light sandy soils was a prerequisite for the installation of such farms. With irrigation and the fertilizer content of the wastewater, the crop yield from those light soils increased. Farming on sewage farms was rather laborious because only light machines could be used on the small plots surrounded by bunds. The main problem, however, was that most sewage farms were overloaded because of increased wastewater flows; as a result, they were abandoned and wastewater treatment plants were built.

Some cities, instead of building treatment plants, installed sprinkler irrigation systems and the first was in Dulmen (1913); Leipzig, Erfurt and Nordhausen followed in the 1930s. A very large system was planned north of Braunschweig for the wastewater of Braunschweig, Salzgitter and Wolfsburg but it was not constructed because of the war. Separate systems were built for Wolfsburg (1942) and for Braunschweig (1954–60).

Because of a shortage of fertilizers before, during and after the war, the purpose of sprinkling wastewater on farmland was not only to water the crops but also to make use of the nutrients in the wastewater in order to increase the productivity of light soils. Construction of wastewater irrigation systems was therefore subsidized by the government as an agricultural measure.

As the wastewater for irrigation was only mechanically treated by screens and sedimentation basins, this kind of rough treatment was not sufficient to allow its diversion to the rivers. Irrigation was therefore required all the year round. Light and permeable soils are necessary for such systems.

Today in the Federal Republic of Germany, apart from some smaller schemes, there exist the two larger irrigation systems of Braunschweig (40 000 m^3/d) and Wolfsburg (16 000 m^3/d). This chapter is based on experiences gained with these two systems.

* Professor, Technische Universitat Braunschweig, Federal Republic of Germany

Description of the systems

The design of the system was based on the following climatic conditions in the Braunschweig area:

8.5°C: mean annual temperature
13.9°C: mean temperature in the growing season (April to September)
625 mm: mean precipitation per year
366 mm: mean precipitation in the growing season
602 mm: mean evapotranspiration per year
488 mm: mean evapotranspiration in the growing season

These data indicate a deficit of 122 mm of water during the growing season. Due to the high permeability of the soil (pore size = 10 μm), irrigation is 150 mm during spring and summer and another 150 mm in autumn and winter, which adds up to 300 mm per year. It was calculated that the Braunschweig wastewater flow of 30 000 m^3/d would irrigate a net area of 3000 ha. During frosty weather, the wastewater would be diverted to rapid infiltration basins.

The Braunschweig system is shown in *Figure 15.1*. Rearrangement of the fields and roads was necessary in order to obtain optimal conditions, which are rectangular fields with a length of 300 or 600 m. The wastewater is distributed by a

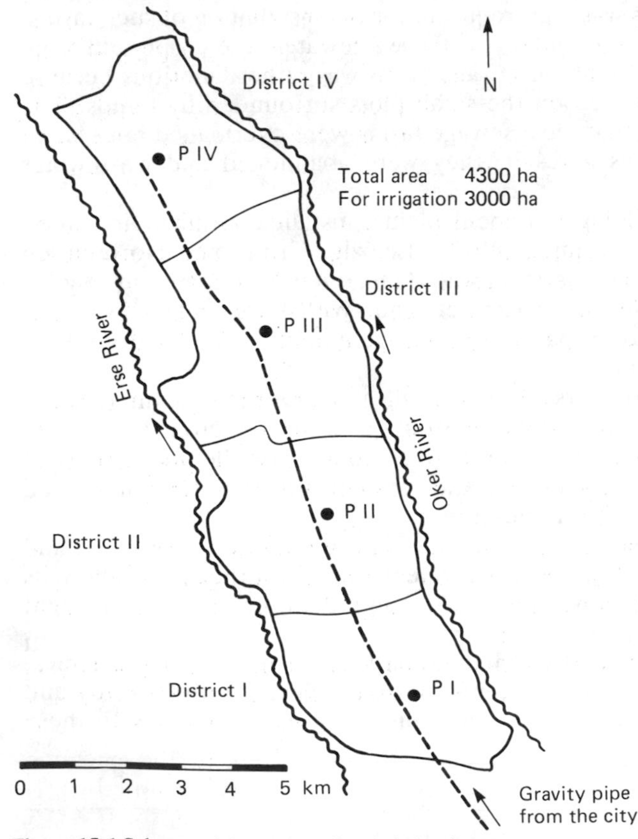

Figure 15.1 Scheme of the Braunschweig irrigation area

subsurface pressure pipe network (asbestos cement pipes, diameter 100–500 mm). Subsurface hydrants are arranged at distances of 90 m. A portable pipe system was chosen for sprinkling, with pipes 6.0 m long and 89 mm diameter. Circulating sprinklers at intervals of 30 m were mounted on the pipes. Conditions indicated an optimal total pipe length of 300 m and a dosage of 50 mm of water. Each sprinkler delivers about 7 m^3/h. An area of 300 m × 30 m served by ten sprinklers is watered for a period of 6–7 h.

Economic considerations led to the conclusion that, for each 500–1000 hectares to be irrigated, a pumping station should be installed. Because of the intermittent operation of sprinkling, balancing tanks are necessary. They were built in conjunction with the pumping stations and also serve as settling tanks. The settled sludge is diverted to the so-called sludge drying beds. The screened and degritted wastewater flows by gravity pipeline through the irrigation area. Four pumping stations with balancing tanks are located along the pipe.

The irrigation system is operated by the Braunschweig Wastewater Utilization Association (BWUA). Its members are 440 individual farmers, on whose land the wastewater is sprinkled, and the City of Braunschweig. The staff of BWUA works out an irrigation schedule every year in the winter according to the cultivation plan of the farmers. Depending on the actual climatic conditions, the general schedule may be improved from week to week, especially in summer. The staff of BWUA handle the whole irrigation operation, including the pumping stations, movement of pipes and maintenance and repair.

In order to avoid health risks, irrigation of vegetables and fruit for raw consumption is not allowed. Grain and sugar beets predominate in the irrigation area, although potatoes are also grown and there is some grassland. A minimum period of 3 weeks is specified between the last sprinkling and crop harvesting.

Operational problems and improvements

Irrigation

Although the displacement of the irrigation pipes was highly mechanized (about 18 000 m of pipes were moved every day by ten crews, each of two labourers), it was dirty work. Increasing wages made irrigation more and more costly. In 1972, therefore, BWUA started to test drum-coiled pipe-irrigation machines (*Figure 15.2*). The machine is moved by a tractor to the edge of a field. The tractor then pulls the pipe, with the sprinkler fixed to its end, across the field. When the machine is connected to the hydrant, the pipe drum is turned by a turbine driven by the water, thus pulling back the sprinkler. The pulling-in speed can be changed by a variable gear to between 5 and 20 m/h. At normal operation, the sprinkler takes 20 h to be pulled in over a 300 m long field. At a pressure of about 4 bars the sprinkler (nozzle diameter 20 mm) spreads the water over a 60 m diameter circle and delivers 40–45 m^3/h. In 20 h a strip of 300 × 60 m receives 50 mm of water. It was found that a heavy duty tractor (75 kW) was necessary, especially for pulling the heavy pipe (polyethylene, outside diameter 110 mm, walls 10 mm thick) over the field.

Sprinkling is stopped automatically when the sprinkler cart reaches the machine. On average, 20–30 min are required to move a machine and it can be done by just the tractor driver. Normally, about 50–60 machines are in operation at the BWUA but in the summer when, in addition to the wastewater, groundwater is used for

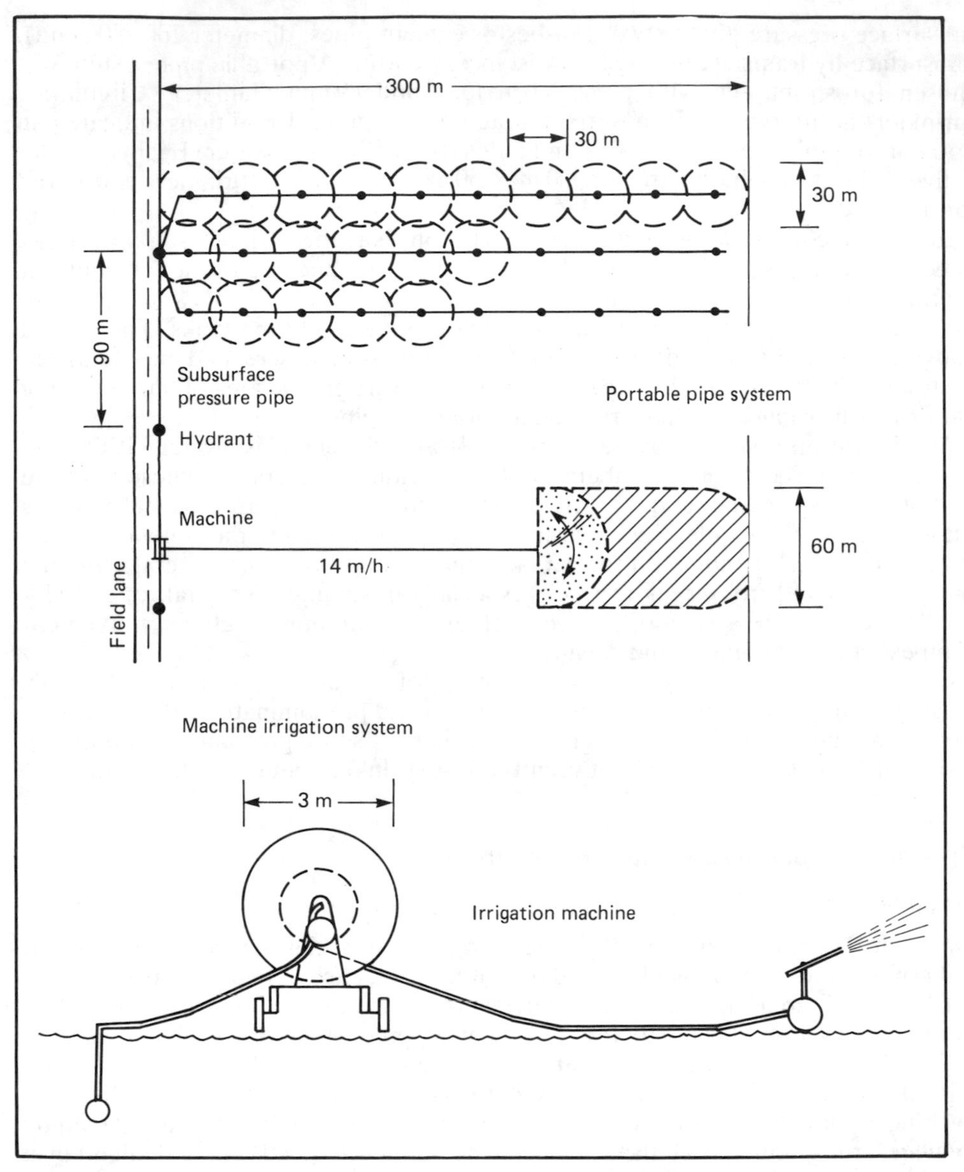

Figure 15.2 Schemes of irrigation systems

irrigation, as many as 100 machines may be in operation. A total of six tractors are required for the daily placing of all machines; in the summer, besides each tractor driver, an additional labourer helps to speed up the placement process. The change to machine irrigation reduced the number of labourers by about ten but, on the other hand, the investment costs for the machines were rather high.

Winter operation

After a few years of operation it was found that the infiltration basins, which were only used during very frosty periods, could not be loaded as planned. Irrigation with low pressure was therefore continued during the winter. The irrigators could only be moved by hand from time to time because the hydraulic switch-off mechanism of the machines froze at temperatures below zero. Depending on the duration of frosty weather, because most of the land is slightly sloping, the few level plots may receive a rather high load.

This kind of operation requires level land in order to avoid the water flowing via the next ditch to the river.

Sludge drying beds

The raw sludge, which was settled in the balancing tanks, used to be pumped to the sludge drying beds but they were too small for rapid dewatering. When they filled up to 1 m, the sludge became septic, unpleasant odours were emitted and the sludge dewatered too slowly. Since the irrigation machines with big nozzles came into operation, irrigating with more and more sludge together with the wastewater has been successful.

Environmental problems

In about 1970 the population living close to the irrigation districts began to complain about the odour and health risks. Intensive investigations were started.

The odour problem

Apart from the odour from the sludge drying beds, very strong odours were released from the sprinkling, because of its effective gas exchange capacity. Especially in the summer when the water temperature was high, the wastewater became septic on its long journey from the city to the fields (travelling time about 12–24 h).

At the BWUA, the addition of hydrogen peroxide was tested at one pumping station, with no success. In Wolfsburg, pure oxygen was injected into the main pressure pipe. Dissolved oxygen and hydrogen sulphide were measured at the sprinklers at the same time. It was concluded that, due to the hydraulic properties of the distribution network, some water took a short path while other water travelled for much longer.

To overcome odour problems it was felt necessary to remove at least most of the easily biodegradeable organics in the wastewater, in order to avoid septicity, and to abandon the sludge drying beds. The general idea for further planning was to irrigate the sludge together with the biologically treated wastewater, because land application is the best method of sludge disposal.

Bench-scale treatment experiments were carried out with aerated lagoons and activated sludge plants which were operated at different loading conditions. Besides the usual parameters such as BOD, COD, etc., the period after which the mixture of treated effluent and excess sludge became septic was measured. It was found that, in aerated lagoons with a detention period of 2–3 days and in the activated sludge process with a sludge age of 2–3 days, the soluble organics were sufficiently removed (filtered effluent BOD_5 approximately 10 mg/l) so that the mixture of effluent and excess sludge did not become septic within 2 days under anaerobic conditions.

The BWUA decided to build a single-stage activated sludge plant which should be extendable to a complete wastewater treatment plant. No facilities for sludge handling were built. The excess sludge only passes a fine bar screen (mesh 10 mm) before it is mixed with the final effluent. The plant (*Figure 15.3*) went into operation in 1979 and irrigation became almost odour-free within a few days.

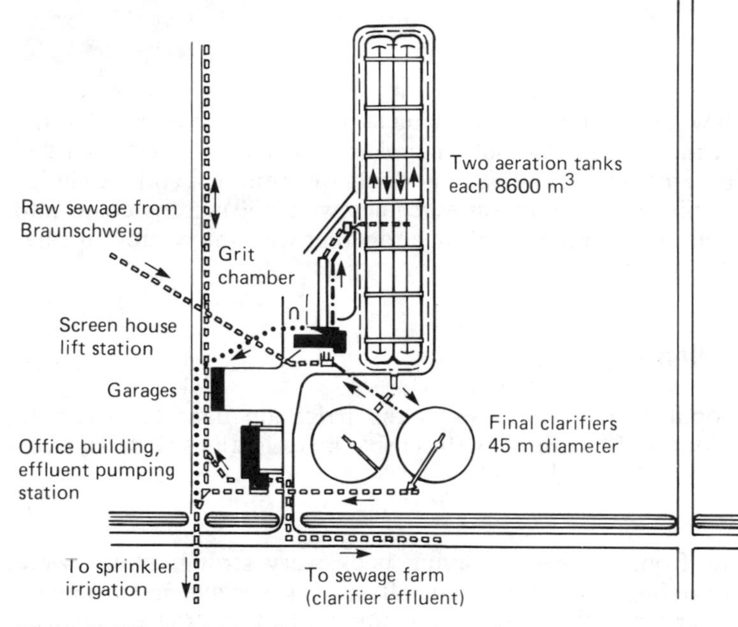

Figure 15.3 Plan of the pretreatment plant

The plant is designed for a flow of 72 000 m^3/d and 25 tonnes of BOD_5 per day. Annual average flow is now 55 000 m^3/d and 13 tonnes BOD_5/d; 15 500 m^3/d of treated effluent are still diverted to the old sewage farm. The remaining final effluent (34 400 m^3/d) is mixed with the excess sludge flow (5100 m^3/d), which is considerable because it is withdrawn directly from the aeration tanks and is not thickened. Some raw wastewater (4800 m^3/d) from villages near the irrigation area enters the gravity line. The annual average volume of wastewater, including excess sludge, which is used for irrigation is 44 500 m^3/d.

Heavy metals

The Federal Government of Germany issued standards for the disposal of sludge on farmland (*Table 15.1*), because heavy metals accumulate easily in the soil and subsequently in the crops. These standards may be used by analogy for wastewater irrigation too.

If even one element in the soil is found to have a higher concentration than permissible, no sludge (or wastewater) application is allowed. The most critical element is regarded as cadmium, especially in light soils where it is easily exchanged. It is expected that the standards for cadmium will be lowered in the future.

Table 15.1 German standards for heavy metals in soil and in sludge

Element	*Maximum permissible in the soil* (mg/kg dry soil)	*Maximum annual application rate* (kg/ha year)
Cd	3	0.033
Cr	100	2.0
Cu	100	2.0
Ni	50	0.33
Pb	100	2.0
Zn	300	5.0

An example may illustrate the cadmium problem. If the application rate of wastewater, containing the excess sludge, were 500 mm/year, the average concentration of cadmium in the wastewater must not exceed 0.0066 mg/l; if more water is applied, then the concentration must be decreased even more. In Germany, numerous cities have successfully demonstrated that by closely controlling industrial discharges to the municipal sewer systems, the cadmium concentration in the sludge can be reduced far below the critical concentration.

After more than 40 years of irrigation with untreated municipal wastewater from the city of Wolfsburg, the content of heavy metals in the soil is far below the limits and no higher than in adjacent fields. In Braunschweig, after more than 25 years of wastewater irrigation, the picture is not so good because of higher background concentrations (the Oker River was heavily loaded with mining wastes in former times and the farmers used the mud from the river banks as fertilizer).

Health risks

Even though it is forbidden to irrigate with wastewater vegetables and other fruits which may be consumed raw, there is still public concern that infections may be spread by wastewater being sprayed from sprinklers. Investigations have shown that at a distance of more than 150 m from the big machine sprinklers no bacterial infection from sprays is possible. If the wind is toward the sprinkler the distance of potential hazard is about 115 m.

If such a distance from public roads and housing areas were to have been established, it would have considerably reduced the irrigation area. Another investigation with special flat sprinklers was undertaken and it was found that the hazardous distance could be reduced to 100 m, or 30 m if the wind was toward the sprinkler. These investigations took place before the biological pretreatment plant went into operation. It should be added that from all the spray sampling no salmonellae were detected.

The board of control issued an irrigation decree which says that 10 m wide hedges have to be planted along housing areas and public roads and that the minimum distance of sprinklers from housing areas must be more than 100 m and from public roads 50 m. Irrigators closer than 115 m to streets and housing areas are only allowed with special flat sprinklers. Irrigation closer than 100 m to public roads is only allowed when the wind is from the road toward the field. BWUA has two employees working as internal controllers. They observe the wind direction and ensure that the irrigation decree is observed.

Investigations at the pretreatment plant have indicated that the mixture of treated effluent and excess sludge contains about 75% less salmonellae than the raw wastewater. With regard to public safety, the biological pretreatment plant is of some benefit. It should be mentioned that during all the years of raw wastewater sprinkling no case of infection has been reported.

Agricultural aspects

Nutrients

On average, the BWUA irrigates at present 2800 ha net irrigation area with about 44 500 m^3/d, giving an annual application rate of 578 mm. The annual nutrient loads are listed in *Table 15.2.* The light soils need some extra lime (CaO) to maintain the pH, and some additional potassium. Since wastewater irrigation was started, the content of phosphorus in the soil has increased continuously. The data indicate that about half the nitrogen is contained in the excess sludge and most of it is organic nitrogen. The treated effluent contains mainly ammonia, which is readily available for plants. Due to the fact that the need for water and the need for nitrogen are not coincidental, at certain stages the farmers may have to add some nitrogen fertilizer.

Table 15.2 Annual nutrient loads

Element	*kg/ha × year*	*In excess sludge (%)*
K	105	20
Ca	305	23
Mg	53	20
N	379	50
P	106	43
Suspended solids	2754	95

Sludge solids

Each hectare is loaded annually with 2.75 tonnes of suspended solids from the excess sludge, which sounds like a thick sludge layer but one has to keep in mind that it is applied in 10–12 doses. After a field is irrigated it is hard to detect any sludge. The farmers, however, found that there is more dust when they are threshing the grain; the reason seems to be that fine dried excess sludge particles remain in the ears.

Crop yields

It is difficult to obtain exact data about the increase in agricultural productivity, because such light soils are now irrigated with ground or river-water anyway and fertilizer is supplied in sufficient quantities. In *Table 15.3* the crop yields and percentages for the area covered before (1960) and after about 20 years of wastewater application (1979) are shown. It can be seen that there is a shift to produce the more valuable grains and sugar beets and less potatoes, but the same tendency is observable in areas not irrigated with wastewater.

Table 15.3 Crop yields and areal distribution before and 20 years after wastewater irrigation

	Before (1960)		*After (1979)*	
	Yield (100 kg/ha)	*Per cent of area*	*Yield* (100 kg/ha)	*Per cent of area*
Winter grain	2–6	23	6–16	29
Summer grain	6–16	16	6–20	22
Sugar beets	60–200	16	100–240	25
Potatoes	50–150	20	50–200	6
Grassland	–	9	–	8
Other	–	16	–	10

Treatment performance

Wastewater irrigation, which is a low-rate land treatment process, is comparable (if not superior) to most physicochemical advanced wastewater treatment processes. Average data for the Braunschweig system are shown in *Table 15.4*.

The biological pretreatment plant removes only organic substances (BOD, COD). Suspended solids formed by bacterial activity, as well as the apparently removed function of nitrogen and phosphorus, are contained in the excess sludge which is added to the treated effluent for irrigation. In the irrigation area only a small fraction of the total applied water is collected by tile drains and ditches and diverted to the river Oker. The remaining portion is evaporated or flows into the groundwater.

Table 15.4 Summary of treatment performance (concentrations, mg/l, except for pH)

	Raw	*Pretreated effluent*	*Effluent from irrigated area*
pH	7.4	7.6	7.0
Suspended solids	223	52	–
BOD_5	236	37	1
BOD_5-filt	85	13	–
COD	650	150	23
COD-filt	228	89	–
TKN	55	35	–
TKN-filt	39	31	–
NH_4-N	31.5	26.8	1.6
NO_3-N	0.2	1.1	25.2
Total P	15.7	11.7	0.15

The only parameter of concern is nitrate; its concentration in the effluent as well as in groundwater wells (~ 25 mg/l) is rather high. It is assumed that nitrate is washed out when the uptake by plants is low and/or when the application of water is higher than evaporation. In particular, the application of wastewater with its high nitrogen content in winter contributes a large portion of the total washout.

Costs of the BWUA system

The investment costs for the irrigation system (construction between 1956 and 1962) totalled about 30 million DM, of which 95% were grants from the State and Federal agricultural government agencies. It should be recalled that the irrigation system was visualized as an agricultural measure. Another investment of 5 million DM was made in 1974–76 when 100 irrigation machines were bought.

The investment costs for the pretreatment plant, for a design flow of 72 000 m^3/d from a population of 420 000, were 18 million DM (1978–79) of which 5.8 million DM were a grant (4.8 million DM from the Federal Ministry for Research and Technology). The annual expenditures shown in *Table 15.5* are almost totally covered by the City of Braunschweig. Farmers pay 120 DM/ha of irrigated area, which adds up to about 350 000 DM/year (5% of total annual expenditure or 7% of the expenditure for irrigation).

Table 15.5 Annual expenditures (1982)

	DM/annum	*DM/m³*
Pretreatment plant	2 900 000	0.15
Irrigation system	5 000 000	0.31

The comparatively high costs for irrigation are due to the considerable cost of labour, high costs for maintenance and repair and, last but not least, the high power consumption, which is about 0.30 kWh/m^3. Power requirements for the pretreatment plant are less, only 0.20 kWh/m^3. This indicates that sprinkler irrigation with its high pressure requirement of about 7 bars, which the pumps have to deliver, is rather power consuming. From the viewpoint of the City of Braunschweig, the high costs for irrigation are justified because the scheme serves as a sludge disposal system as well as an advanced wastewater treatment system.

Final remarks

Wastewater irrigation is a very effective means of wastewater treatment, because tertiary effluent quality is almost reached, regardless of the degree of pretreatment. Biological pretreatment is necessary in order to avoid odour problems in the irrigation area. Due to the high costs of irrigation it is desirable, and has been proved successful, that the irrigation system serves not only the agricultural needs and as an advanced wastewater treatment system but also for sludge disposal. Future problems may arise from the high washout of nitrates, especially during winter irrigation. The Wolfsburg Wastewater Utilization Association, therefore, is designing a pretreatment plant where, in winter, nitrogen will be removed by nitrification–denitrification.

It should be remembered, however, that the experience acquired under the climatic conditions of northern Germany may not be transferable to arid climatic regions without careful consideration of many of its components.

Chapter 16

Status of wastewater treatment and agricultural reuse with special reference to Indian experience and research and development needs

G. B. Shende*, C. Chakrabarti*, R. P. Rai†, V. J. Nashikkar*, D. G. Kshirsagar*, P. B. Deshbhratar* and A. S. Juwarkar*

Introduction

Developing or less-developed countries in the Asia and Pacific region, though lagging behind in technological development are by no means an exception to the recent trends in urbanization and industrialization. The increase in population and industrial growth, as in developed countries, has accelerated the generation of wastes of various kinds but, unlike affluent western countries, the developing countries of the region are short of resources for the systematic collection, treatment and hygienic disposal of such wastes. This creates environmental pollution problems and ultimately affects the quality of life and health of the community. The problems of community sanitation are, however, being solved in larger urban centres in most of the countries in the region by providing mechanisms for collection of household refuse and human excreta. While a few countries have opted for waterborne sanitation, as in India, others seem to have preferred waterless sanitation, probably for want of sufficient water and/or because of the costs involved in sewerage. Sewage or domestic wastewater is the product of waterborne sanitation wherein human excreta from waterseal latrines and wastewater from bathroom, kitchen and other domestic washings is collected through the network of underground pipes connected to every household. This is called a 'sewerage system', while the wastewater is called 'sewage'. The wastewater from kitchen, bathroom and washings (excluding human excreta) collected through surface drains is called 'sullage'.

Availability of sewage in Asia and the Pacific

Generally, the flow of domestic wastewater from a community is proportional to the water supply. This relationship operates only where water supply and sanitation (sewerage) are integrated. Unfortunately, this situation is not found in many

* National Environmental Engineering Research Institute, Nagpur, India
† Central Potato Research Institute, Simla, India

countries of the region where water supply has been provided for a large population on a priority basis but no parallel coverage is provided for sanitation through sewage or wastewater collection. *Figures 16.1* and *16.2* portray the water supply and sanitation situation in developing countries (Pickford, 1979). It can be seen from the figures that sanitation by sewerage is provided for about 25% of the urban population in the countries of South-east Asia, East Asia and the West Pacific region, while in the West Asia region there is very poor progress in sewerage.

The overall progress in extending sewerage facilities to the urban population in the region as a whole is very slow because of financial restraints, as a result of which the available flow of sewage is much less. The detailed data regarding coverage with sewerage in different countries are not available. However, the available information indicates that a large number of big townships in the region are having

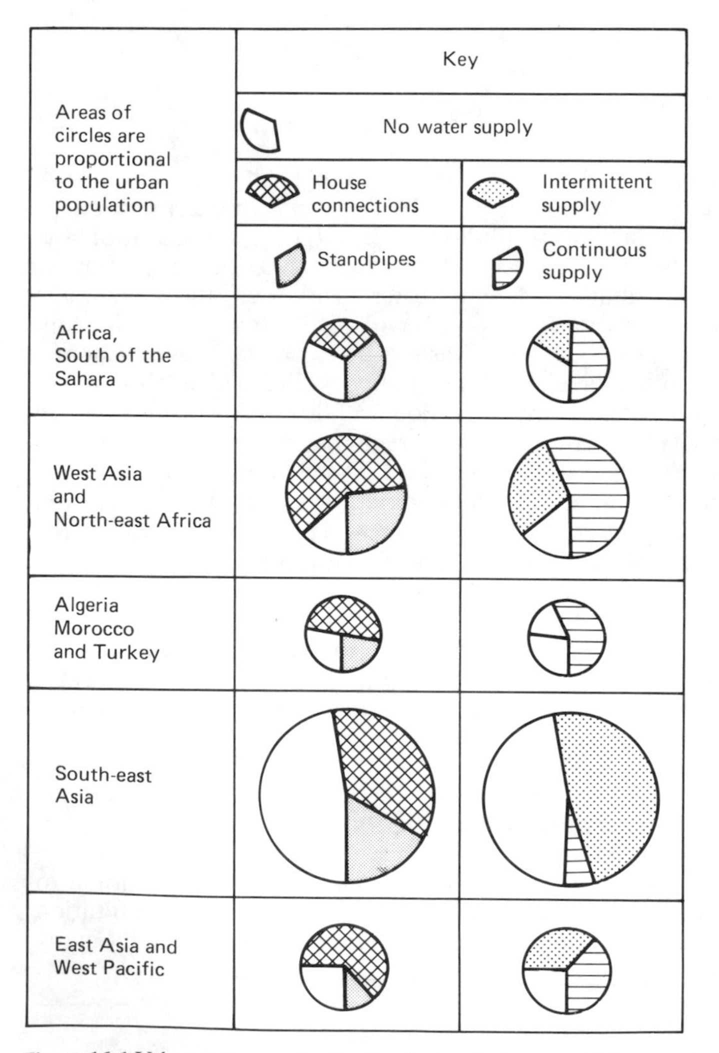

Figure 16.1 Urban water supply situation in developing countries

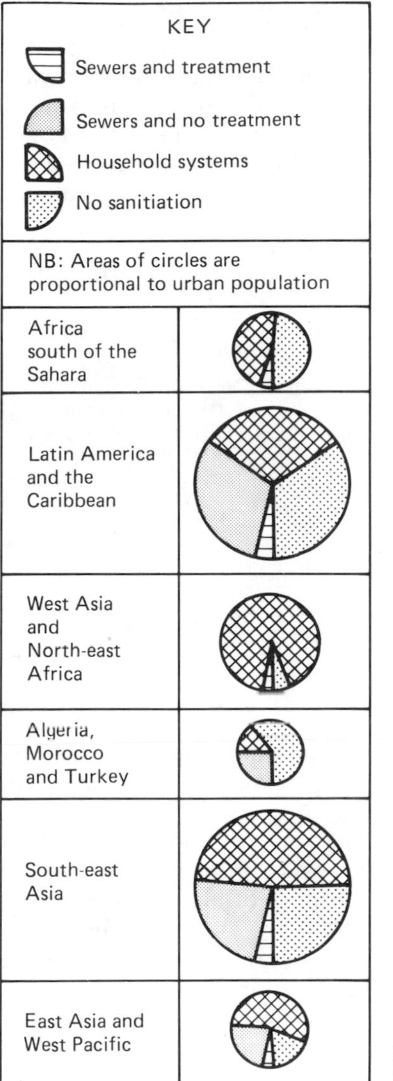

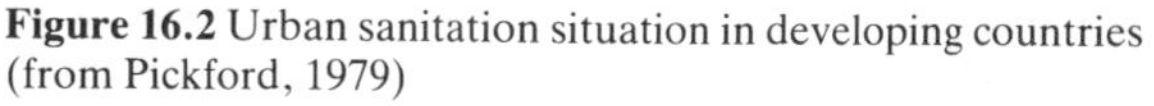

Figure 16.2 Urban sanitation situation in developing countries (from Pickford, 1979)

sewerage schemes planned and executed with the financial assistance or loans through different regional and global development agencies such as the World Bank and the Asian Development Bank under a number of aid and assistance programmes (Pescod and Okun, 1971). It is expected that waterborne sanitation through sewerage will be available for most of the population centres in the near future and slowly the flow of sewerage will go on increasing.

Information on sewerage and the flow of sewerage is available mainly from the two major countries of the region, India and China. In India, about 10 000 million litres of sewage are estimated to be available per day, while in China the flow is of the order of 100 million tonnes (100 000 million litres) per day (Wong, 1984).

Status of sewage treatment and reuse in agriculture

Overall progress with sewage treatment in the whole region is rather insignificant. The major bottleneck seems to be in the high cost of conventional methods which is prohibitive to the sustaining economy of developing countries. This point is well illustrated by the situation in two countries, India and China, as shown in *Table 16.1.* Besides tremendous capital costs, conventional watewater treatment also involves a high input of energy and a significant maintenance cost.

Table 16.1 Status of wastewater availability and treatment and estimated cost of treatment in India and China

Particulars	*India*	*China*
Estimate of available flow of wastewater (domestic)	10 000 million litres/day	100 million tonnes/day (100 000 million litres/day)
Percentage of wastewater treated		
(a) Primary treatment	37	–
(b) Secondary treatment	8	–
Estimated additional cost for secondary treatment		
(a) Capital cost	Rs 1865 million	Yuan 400–500 million
(b) Maintenance cost	–	Yuan 40–50 million

Source: Wong (1984) and Central Board for the Prevention and Control of Water Pollution (1979)

Approximately US$ 1 = Rs 12 = Yuan 2

Only a small fraction of the flow of wastewater in both countries receives treatment to a secondary level and mostly untreated or partially treated wastewater is available for utilization as irrigation water in agriculture or for land application (Central Board for the Prevention and Control of Water Pollution, 1979, 1980; Wong, 1984). Except for a few coastal towns or cities where sewage is disposed of into the ocean, most of the inland cities and towns dispose of about two-thirds of the available flow on the land. About 12 000 ha (30 000 acres) of land area were reported to be under sewage irrigation in 1971. This figure seems now to have reached about 73 000 ha (175 000 acres) due to the increased flow of sewage on account of commissioning numerous sewerage and sewage utilization schemes during the period of more than a decade (personal communication from Ministry of Agriculture, Government of India). In China also, sewage irrigation seems to have developed rapidly since 1958 and an area of over 1.3 million ha is reported to be under sewage irrigation (Wong, 1984).

Information on reuse of sewage for irrigation from many of the countries in the region is not available, probably because of a lack of specific interest by any agency in collecting such data. Through personal communications, it seems that an area of about 400–500 ha is irrigated with untreated sewage at Vientiane (Laos), and that in Manila (Philippines) recycling of domestic wastewater effluent from the Bliss Waste Treatment Plant (aerated facultative ponds) is under consideration.

Inefficacy of conventional treatment for resource conservation

Irrespective of high costs, the efficacy of treatment methods of the conventional type is limited only to the removal of the organic and microbial pollution load to almost a satisfactory level; while removal efficiency in respect of inorganic pollution due to nutrients, dissolved solids and heavy metals is very poor. Recently, the approach of in-plant wastewater treatment and its disposal in streams is being reviewed, particularly in the context of awareness of depleting water resources and the nutrient crisis all over the world. The importance of conserving water and nutrients is being increasingly realized and recycling and reuse of these constituents of wastewater is being emphasized. Wastes are now looked upon as a resource (Sundaresan, 1978; Mahida, 1981; Shende, 1984) since domestic wastewater often provides a fair to good quality irrigation water and has an appreciable concentration of nutrients, like nitrogen, phosphate, potassium and organic matter, which have irrigational, manurial and soil-conditioning values in agriculture. In economic terms, it is a big asset (*Table 16.2*).

Table 16.2 Nutritional and irrigational potential of domestic sewage in India

(1) Available flow: 10 000 million litres/day

(2) Nutrient and organic matter potential

Nutrients	*Average concentration* (mg/l)	*Contribution potential*		*Economic value** (Rupees† in millions)
		Tonnes/day	Tonnes/annum	
Nitrogen	60	600	219 000	876
Phosphate	20	200	73 000	292
Potash	40	400	146 000	584
Organic matter	400	4000	1 460 000	75
(3) Irrigational potential				
Perennial coverage			100 000 ha	80
Additional seasonal coverage			140 000 ha	56
Total annual value of available sewage				1963

Source: Shende (1984)

* Economic value has been computed by assuming the following rates:

(1) Nutrients Rs 4000/- tonne
(2) Organic matter Rs 50/- tonne
(3) Irrigation rates
  (a) Perennial Rs 800/- ha
  (b) Seasonal Rs 400/- ha

† Rupees 12 = US$1

Land application – a wastewater management alternative

Land application of wastewater is being given a higher preference and weightage over the conventional procedures of in-plant wastewater treatment and stream disposal, since the former provides scope for exploiting the reuse potential of wastewater for productive purposes. Land-applied wastewater also undergoes natural physicochemical and biological treatments in the soil matrix which is not only a highly effective low-cost alternative but also an ecologically balanced and environmentally compatible system of wastewater management. The efficacy of land treatment of wastewater in comparison with the conventional approach will be obvious from the data in *Tables 16.3* and *16.4*.

The efficiency of land treatment of wastewater, however, depends to a considerable extent on the type of soil, its effective depth and intake characteristics design application rate and the characteristics of vegetative cover with regard to the water and nutrient requirements and mode of its consumption (Environmental Protection Agency, 1981). A considerable amount of judiciousness with respect to

Table 16.3 Comparison of effluent quality for conventional land treatment and advanced wastewater treatment systems

Systems	*Effluent constituents (mg/l)*					
	BOD_5	*SS*	*NH_3N*	*NO_3N*	*Total N*	*P*
Conventional treatment						
Aerated lagoon	35	40	10	20	30	8
Activated sludge	20	25	20	10	30	8
Land treatment						
Slow rate (irrigation)	1	1	0.5	2.5	3	0.1
Overland flow	5	5	0.5	2.5	3	5
High rate (infiltration percolation)	5	1	–	10	10	2
Advanced wastewater treatment						
Biological nitrification	12	15	1	29	30	8
Biological nitrification, denitrification	15	16	–	–	3	8
Tertiary treatment, two-stage lime coagulation and filtration	5	5	20	10	30	0.5
Tertiary two-stage lime coagulation filtration and selective ion	5	5	–	–	3	0.5

Source: Environmental Protection Agency (1981)

Table 16.4 Comparison of effluents from activated sludge, trickling filters and sewage irrigation

Analysis	*Parts per millions except pH*				
	Effluent from Dadar Bombay activated sludge plant		*Effluent from Dadar Bombay trickling filters plant*		*Effluent from sewage irrigation Poona (India) from manhole no. 3*
	16.7.73	*19.7.73*	*28.7.72*	*30.7.73*	*20.4.65*
Free and saline ammonia	1.0	2.8	–	–	0.012
Albuminoid ammonia	–	1.0	–	–	0.056
Oxygen absorbed in 4 h at 37°C	10.8	9.2	18.8	29.0	0.756
Nitrates	0.48	0.88	0.10	–	4.753
Nitrites	0.20	0.10	0.20	–	Nil
Suspended solids	8.0	10.00	20.00	51.00	trace
pH	7.35	7.4	–	7.2	7.5
BOD at 20°C after 5 days	14.5	14.5	20.2	27.2	0.77*

Source: Mahida (1981)

* At 37°C

the possible interactive consequences in the wastewater–soil–plant system is required for successful operation of the system.

The region of Asia and the Pacific as a whole (with few exceptions) has a semi-arid and arid climate with a prolonged warmer period and seasonal rainfall from the monsoon. These conditions are conducive to a perpetual demand for irrigation water and congenial for land treatment-cum-utilization of wastewater for irrigation. An excellent degree of land treatment of wastewater, which enables conservation and exploitation of the resource potential (water, nutrients and organic matter) of wastewater for productive purposes, provides an appropriate technology for wastewater management for the socioeconomic and geoclimatic conditions in the developing countries of the region.

Preapplication treatment

Wastewater of domestic origin carries pathogenic and parasitic organisms which form a source of contamination of soil and crops and may cause infection to the farm labourers and consumers of sewage-irrigated crops (Doran, Ellis and McCalla, 1977). Although the soil–plant system does not provide favourable conditions for the growth and multiplication of these host-specific organisms, and this may actually add to the difficulty of their survival in the soil and plant ecosystem (Anon., 1973–83), the risk of contamination can still not be overruled. These risk factors can be guarded against to a considerable extent by preapplication treatment of sewage and by exercising a restriction on the choice of cropping system for sewage irrigation (Shende, 1984).

Recently, non-conventional low-cost wastewater treatment processes have been developed which are equally or even more effective in the removal of organic load and microbial pollution from wastewater. The performance of oxidation ponds or shallow stabilization ponds have been tested for a sufficiently long time at the National Environmental Engineering Research Institute (NEERI), Nagpur and other places in India and have been found to be the best treatment alternative under the climatic conditions of this country (Arceivala *et al.*, 1970). Since this treatment process utilizes algal-bacterial symbiosis in conjunction with the oxygenation of wastewater through the natural photosynthetic process energized

Table 16.5 Efficiency of oxidation ponds in sewage treatment

Parameters		*Raw sewage*	*Oxidation pond effluent*
pH		7.2	8.7
Total solids	(mg/l)	1052.0	749.0
Dissolved solids	(mg/l)	676.0	627.8
Total nitrogen	(mg/l)	60.50	25.90
Total phosphorus	(mg/l)	27.80	10.60
Total potash	(mg/l)	45.10	39.40
BOD_5, 20°C	(mg/l)	379.80	36.40
Per cent reduction in BOD_5	–	–	90.00
Algal cell count ($\times 10^4$/ml)	–	–	17.00
Per cent reduction in *E. coli*	–	–	99.99
Per cent reduction in faecal streptococci	–	–	99.99
Per cent reduction in bacterial count	–	–	82.5–99.9

through solar radiation, the operation and maintenance of this system is very simple. The data in *Table 16.5* demonstrate the efficiency of oxidation ponds in wastewater treatment.

In China also, an aerobic lagoon system has been described as being quite popular and it is being integrated with the land application system of wastewater management, wherein the self-purifying capacity of the water–soil–plant system is utilized. Like the stabilization ponds in this system also, plants such as reeds, algae, nepenethes, etc., can be grown in lagoons and harvested for their food value for livestock and poultry (Wong, 1984). In addition, duck and fish can also be raised in the lagoon. In India, stabilization pond effluent is found to be a congenial medium for fish culture where the algal biomass provides food for the fish.

Indian experience

Land disposal of wastewater has been practised in India from the second quarter of this century. It is commonly known as 'sewage farming'. Some of the sewage farms in India, such as Ahemadabad and Poona, are about 5–16 decades old.

Sewage farming practice

Sewage farming as a method of wastewater disposal is popular in India and is reported as being practised at more than 200 places (Control Board for the Prevention and Control of Water Pollution, 1979, 1980). Sewage farming is often done in a crude and irrational manner, although experience over decades has been gained. Practically no attempt seems to have been made at any level to correct the faults. A survey conducted by NEERI, Nagpur, covered a few representative sewage farms in India and revealed a few important aspects of sewage-farm practice: level of sewage treatment, dilution, hydraulic and nutrient loadings, crops grown, etc. The findings are presented in *Table 16.6*.

Rates of sewage application
The data in *Table 16.6* make it clear that the average application rate of sewage per unit area (available flow of sewage divided by area irrigated) is in excess of the normally used application rates in a properly managed irrigation system. The major factors responsible for the excessive rates are:

- systems are disposal orientated rather than for optimum resource utilization;
- increase in the design farm area proportional to the subsequently increasing volume of wastewater;
- overirrigation by farmers for tapping more nutrients on their land
- irrigation charges based on on area basis rather than the quantity of sewage supplied.

Irrespective of the causes of excessive application rates of wastewater, it is certain that this results in overburdening the capacity of soil to receive and purify sewage because, along with excessive hydraulic loading, the inorganic load on the soil also increases. As a consequence, problems such as sewage sickness, salinity and alkalinity of soil and secondary-environmental pollution effects arise. This also places a restraint on the extension of the reuse benefits of wastewater to the increased crop area. These data further show that excessive rates of wastewater application also result in excessive application of nutrients.

Table 16.6 Sewage farms in different cities in India, total area under command, nature of sewage used, application rates and different crops grown with sewage irrigation

Location	*Area* (ha)	*Volume of sewage used* (mld)*	*Treatment, if any*	*Dilution if any*	*Application rates* (m^3/ha day)	*Soil type*	*Crops grown*
Ahmedabad	890.3	299.9	Nil	Nil	336.8	Sandy loam	Pochia grass, paddy, maize, jowar, wheat, lucerne
Amritsar	1214.1	54.5	Nil	1:3	44.9	Sandy clay	Maize, berseem, sorghum, lucerne
Bikaner	40.4	13.6	Nil	Nil	336.8	Sandy	Bajra, wheat, grasses and vegetables
Bhilai	607	36.3	Secondary (stabilization pond effluent)	Nil	59.9	Sandy loam, clay loam	Paddy, maize, wheat, tuwar, vegetables
Delhi	1214.1	227.2	Primary and secondary	Nil	187.1	Sandy loam, loamy sand	Jowar, bajra, maize, barley wheat, pulses and vegetables
Gwalior	202.3	11.3	Nil	Nil	56.1	Silt loam, clay loam	Paddy, maize and guar, jowar, cowpea, wheat, potato, berseem and vegetables
Hyderabad	607	95.4	Primary	1:1.5	157.2	Loam	Para-grass and paddy
Jamshedpur	113.3	9.1	Secondary activated sludge	Nil	80.2	Clay loam	Napier grass, para-grass, guinea grass, berseem, jowar, maize
Kanpur	1416.5	31.8	Nil	1:1	22.4	Loam, silt loam	Wheat, paddy, maize, barley, potato, oats and vegetables
Madras	133.5	6.8	Nil	Nil	51.0	Sandy to silt loam	Para-grass
Madurai	76.9	136	Nil	Nil	177.3	Red sandy loam	Guinea grass
Trivendrum	37.2	8.6	Nil	1:1	231.9	Sand	Para-grass
Lucknow	150	300	Nil	1:3	–	Sandy loam	Maize, paddy, potato, vegetables and fruits, papaya, plantains, citrus

* million litres per day

Table 16.7 Application rates of sewage effluents and contribution of nitrogen at some typical sewage farms in India

Location of farms	*Form of sewage used*	*Average assumed nitrogen concentration* (mg/l)	*Average application rate** (m^3/ha day)	*Average daily contribution of nitrogen* (kg/ha day)	*Average annual (300 days contribution of nitrogen)* (kg/ha pa)
Ahmedabad	Raw	60	336.8	20.208	6052
Amritsar	Raw diluted	60	44.9	2.694	808
Bikaner	Raw	60	336.8	20.208	6062
Bhilai	Stabilization pond effluent	25	59.9	1.497	449
Delhi	(Treated, primary and secondary)	30	187.1	5.613	1684
Gwalior	Raw	60	56.1	3.366	1010
Hyderabad	Raw diluted (primary treatment)	40	157.2	6.288	1886
Jamshedpur	(Treated secondary)	25	80.2	2.005	601
Kanpur	Raw diluted	60	22.4	1.344	403
Madras	Raw	60	51.0	3.060	918
Madurai	Raw	60	177.3	10.638	3191
Trivendrum	Raw diluted	60	231.9	13.914	4174

Note: Original figures in gallons and acres converted to litres/m^3 and hectares, respectively.

* Taken from *Table 16.6*

Nitrogen is a very important plant nutrient and a widely occurring limiting factor in Indian agriculture; the application rates of nitrogen through prevailing overirrigation practices at typical sewage farms are calculated and presented in *Table 16.7*. This rate seems to vary from 403 to 6502 kg nitrogen/ha pa which seems to be several times the normal application rate, which may rise to 400 or 500 kg N/ha year. This excess nitrogen applied to the soil can neither be used by crops nor retained in the soil as it lost through various processes like ammonia volatilization, denitrification and leaching of nitrates to the subsoil and ultimately to the groundwater and water bodies (Environmental Protection Agency, 1981; Anon., 1973–83; Clapp *et al.*, 1977). Since a high nitrate concentration in water used for potable purposes is objectionable from the point of view of adverse health effects, the wasteful excess addition of nitrogen through sewage application is bound to be detrimental to environmental quality.

Health implications and crops grown

The potential of pathogenic contamination of soil and plants by untreated sewage and associated risks to community health due to sewage irrigation are well known. Data available (*see Table 16.6*) from sewage farms indicate that all types of crops are grown on sewage farms, irrespective of the extent of treatment received by the sewage and the mode of consumption of such crops. Edible crops like vegetables are not properly washed and/or disinfected before marketing. There seems to be no statutory control of the crops grown under sewage irrigation. Further, the farm workers are exposed to pathogenic and parasitic infections. Indications of infections of farm workers with various diseases and ailments are available from the data presented in *Table 16.8* from the few studies in India.

Table 16.8 Health status of sewage farm workers

A. Diseases observed at the time of study

B. Results of the stool samples tested for *Ancylostoma duodenalis* (hookworm), *Ascaris lumbricoides* (roundworm), *Tricuris trichura* (whipworm), *Enteriobius vermicularis* (pinworm), *Hymenolepis nana* (dwarf tapeworm), *Entamoeba histolytica, Ent. coli* and *Giardia intestinalis*

Studied by:	*Test group*		*Control group*	
	Number	*Per cent*	*Number*	*Per cent*
(I) *NEERI, Nagpur*				
(1) Total examination	360	–	306	–
(2) Total positive	303	84.0	179	59.0
(3) Positive for hookworm	231	64.1	123	40.1
(4) Positive for roundworm	245	68.0	159	52.0
(5) Infected with two or more parasites	156	43.3	44	14.4
(II) *Dr Kabir's study at Madurai sewage farm*				
(1) Total examination	663	–	2644	–
(2) Total positive	520	78.4	512	19.3
(III) *Dr Patil's study at two village farms near Baroda*				
(1) Total examination	152	–	479	–
(2) Total positive	114	75.0	287	60.0

Irrigation methods
The available information shows that at most of the sewage farms, surface methods of irrigation are used, depending on the nature of crops, their cultivation, type of soil and its topography and the level of scientific understanding and awareness of water economy of the management. Different types of surface irrigation procedures are employed, ranging from uncontrolled flood irrigation to the fairly well-managed ridge-and-furrow, border-strip or check basin irrigation. Subsurface or sprinkler irrigation is not practised on any sewage farm in India.

Environmental hazards
A scientifically planned and judiciously managed sewage farm, which requires skill and expertise, is rare in India. Many farms are known to have no proper levelling or adequate surface and subsurface drainage. A number of problems arise on such farms, particularly when application rates are excessive. Pockets of stagnant/standing sewage in canals and low-lying areas of the farm are frequently observed which create odour nuisance, provide favourable conditions for breeding flies and mosquitoes and spoil hygienic and environmental quality in the vicinity of the farms.

It is evident from the foregoing that the present state of sewage farming practice in India is characterized by many faults and defects due to lack of a rational and scientific approach in the judicious planning, designing and management of sewage farms. To exploit the optimum reuse of wastewater in agriculture, while subjecting the wastewater to land treatment, ways and means must be found to arrest the faults in the practice so as to improve it and tune up the reuse techniques to environmental compatibility and maximum productivity. This can be achieved by intensified research and development efforts in this field.

Research and development

Having identified the problems arising out of the sewage farming practice in vogue in the light of current trends in efficient, safe and environmentally compatible recycling of wastewater in agriculture through land treatment, NEERI has carried out long-term research and development activities in this field since 1969. Investigations planned and carried out at NEERI are aimed at developing guidelines for improvements in sewage farming practice with the object of maximizing the benefits of wastewater reuse, in terms of highest possible agricultural production per unit volume of wastewater, and minimizing direct and indirect hazards and risks to community health and the overall environment including soil, air and surface and groundwater.

Besides the results obtained from R and D activities carried out at NEERI, a few reports from sewage farm studies are also available on the crop and soil response to wastewater irrigation. Important findings from these studies are summarized in the following sections.

Yield response of crops to wastewater irrigation

Effect of untreated raw sewage with or without dilution and nutrient fortification
Results from long-term field studies carried out at NEERI (*Table 16.9* is for a wheat crop) have shown that the growth of crops under irrigation with undiluted raw sewage was as good as that under the standard practice of plain water irrigation

Table 16.9 Effect of irrigation with undiluted and variously diluted raw sewage with and without fertilizer application on the yield of crops (quintals/hectare; 8 years average)

Type of irrigation		*Dilution ratio, sewage: water*	*No fertilizer*	*Recommended dose of NPK*	*Supplemental dose of NPK*
Raw sewage	D-0	1:0	22.78	29.02	29.68
Raw sewage	D-1	1:0.5	21.60	29.18	30.09
Raw sewage	D-2	1:1	22.66	29.51	30.32
Raw sewage	D-3	1:2	18.70	30.17	29.29
Plain water	–	0:1	13.98	26.70	–

and application of a full dose of recommended nitrogen, phosphorus and potassium (NPK) through fertilizers, but the yield levels were relatively lower. The growth and yield of crops decreased further with increase in the proportion of dilution water. Application of recommended and supplemental doses (recommended NPK minus NPK applied through irrigation water) of NPK fertilizer brought about improvement in the yield levels of crops receiving irrigation with undiluted and variously diluted sewage and this level was either on par with or higher than that of the standard practice. Response of the crop to the application of the recommended and supplemental NPK was almost similar under irrigation with undiluted and diluted sewage in 1:0.5, 1:1 and 1:2 ratios. These results have indicated that application of a small quantity of fertilizer to supplement the nutrients carried by undiluted/diluted sewage was essential to ensure a balanced supply of nutrients for normal yield levels as obtained under the standard practice. The nutrient utilization efficiency (kg of grain/kg nutrient) also increased considerably due to dilution of sewage and application of supplemental NPK through fertilizers. Economic evaluation of these results have shown that the average reuse benefit, which was only Rs 0.54 for diluted raw sewage without nutrient fortification, increased to Rs 0.74 due to application of supplemental NPK through fertilizers.

The results (Mahida, 1981) from experiments conducted at Poona sewage farm with various vegetables and fruit crops also show that irrigation with undiluted and diluted sewage in 1:1 ratio resulted in considerably higher yield levels in comparison with the normal practice of canal irrigation and manurial treatment (*Table 16.10*).

It is clear, from the results, that wastewater treatment not only reduces potential health hazards but also reduces the organic load on the soil and the supply of nutrient to the crop under irrigation. This is desirable for reduction in the losses of soil nitrogen and also checks enrichment of nitrogen in the soil and the subsequent proliferation of these nutrients in surface and groundwater. In the case of primary treatment, about 30% of the nutrients are removed with the settleable matter and forms sludge. Dried sludge can be transported away from the treatment plant and used as organic manure at distant fields.

Effect of irrigation with untreated, primary-treated and secondary-treated sewage and rates of application The results obtained so far from these long-term field studies at NEERI have shown that the average yield of almost all the crops under test significantly increased due to irrigation with untreated, primary-treated and secondary-treated sewage (in conjunction with supplemental NPK) by comparison with the standard practice of plain water irrigation and application of the

Table 16.10 Yield levels of crops irrigated with undiluted and diluted sewage at Poona sewage farm (lb/acre)*

Crops	*Treatments*		
	Canal irrigation with manurial treatment	*Irrigation with diluted sewage in 1:1 ratio*	*Irrigation with undiluted sewage*
Beetroot	7 812	13 926	14 526
Carrot	8 670	7 782	10 494
Radish	6 486	5 480	7 440
Turmeric	–	18 432	19 280
Potato	5 465	6 257	8 331
Ginger	5 394	8 192	8 752
Papaya	23 856	24 924	33 036
KholKhol	8 664	10 504	14 796
Cabbage	8 280	10 110	10 830
Cauliflower	6 210	6 320	8 120
Lady's finger	2 514	3 210	5 256
French beans	5 920	7 320	7 200
Tomato	9 000	–	11 948
Tobacco	1 000	1 120	1 120
Groundnut	2 568	2 592	2 832
Sugarcane (tonnes/acre)			
(a) cane	–	47.10	48.60
(b) jaggery	–	5.06	5.16

* To convert lb/acre to kg/ha, multiply by 1.12

recommended dose of NPK (*Table 16.11*). Among the three types of wastewater representing various grades of treatment, an almost similar crop response was observed during the initial 3–4 year period. However, subsequently it was apparent that untreated sewage irrigation over a prolonged period tended to result in a relatively poorer yield level of some crops in comparison with primary-treated and secondary-treated sewage irrigation. The results have also shown that irrigation with this wastewater at high intensity resulted in yield levels which were almost on par or even slightly lower than that resulting from moderate-intensity irrigation with the corresponsing wastewater. The results have further indicated that irrigation with untreated sewage contributed to an excessive load of nutrients to the crop which can hardly be taken up by them, resulting in wastage of surplus nitrogen and very poor nutrient utilization efficiency. On the contrary, irrigation with primary-treated and secondary-treated sewage, supplying relatively less nutrients through the same number of irrigations to the crops, registered a considerably high nutrient utilization efficiency.

BOD of wastewater for irrigation The results from these studies carried out at NEERI with 5 levels of BOD_5 ranging from negligible to 1000 mg/l and 12 test crops, have shown that crop response to irrigation with moderate BOD_5 value of 150 mg/l was best in comparison with the other BOD_5 levels. Crop response under irrigation with wastewater BOD_5 levels of 50 mg/l and 400 mg/l was almost on par with that at the 150 mg/l level; whereas 1000 mg/l BOD_5 tended to result either in a relatively lower yield in the case of some crops or almost similar yield levels as resulting from irrigation with wastewater having 150 or 400 mg/l level of BOD_5 in

Table 16.11 Effect of irrigation with untreated, primary-treated and secondary-treated sewage at moderate and high intensities on the average yield of crops (quintals/hectare)

Type of wastewater	*Intensity of irrigation*	*Crops harvested**				
		Wheat (8)	*Moong* (5)	*Paddy* (7)	*Potato* (4)	*Cotton* (3)
Raw sewage	M	33.37	8.97	29.65	231.05	25.55
Raw sewage	H	34.05	8.83	27.50	196.52	21.74
Stabilization pond effluent	M	34.50	7.77	29.75	223.11	24.13
Stabilization pond effluent	H	33.76	8.17	27.11	171.56	21.03
Settled sewage	M	34.49	8.65	29.41	207.78	23.02
Settled sewage	M	31.97	8.46	26.65	178.79	17.68
Plain water	M	27.55	7.68	23.48	203.48	25.45
Plain water	H	26.96	7.22	20.28	171.56	17.00

M = moderate intensity; H = high intensity

* Figures in parentheses represent the years of harvest for calculating the yield

Note: The nutrients contributed through irrigation with different types of wastewater were supplemented to make up the recommended level by addition of chemical fertilizers; under plain water irrigation, the full dose of recommended NPK was applied through chemical fertilizers.

the case of other crops. This pattern of treatment effects will be evident from yield data of the few representative crops in *Table 16.12*. These results reveal that the BOD_5 level of 150 mg/l was generally suitable for all crops throughout the year in the temperature range prevailing in Central India. The results from the experiments already described in the foregoing also show that the adjustment of BOD_5 level, either by dilution or by primary treatment before irrigation, results in good yields of crops.

Table 16.12 Effect of varying levels of BOD_5 of wastewater irrigation on yield of crops (quintals/hectare)

BOD_5 level	*Cabbage*	*Brinjal (eggplant)*	*Wheat*	*Maize*
	Golden acre rabi 1970–80 1979–80	*Pusa purple cluster summer 1980 + rainy season*	*HDM – 1553 rabi 1980–81*	*Ganga 5 summer 1979*
Negligible	182.9	191.5	28.4	73.5
50 mg/l	202.3	203.6	35.3	78.1
150 mg/l	337.1	339.3	37.4	105.0
400 mg/l	327.4	410.6	36.2	98.3
1000 mg/l	266.6	413.6	34.3	103.0
SE (m)	18.5	19.5	1.0	3.6
CD 5%	40.3	42.5	2.2	7.8
CD 1%	56.6	59.6	3.2	11.3

Multiple reuse in aquaculture and agriculture A conceptual system of wastewater treatment through stabilization ponds, followed by first stage reuse of the pond effluent in fish ponds for pisciculture and second stage reuse of the fish pond effluent for irrigation has been tested (Shende, Juwarkar and Sundaresan, 1983) at NEERI for a period of 3 years. Fish pond effluent was also a good quality irrigation water but its nutrient concentration was considerably lower. For the normal level of crop yields, it required a supplementary application of fertilizers to the extent of about 20% of the recommended dose of nutrients through fertilizers.

Reuse of wastewater for forest irrigation Irrigation of forest species raised for fuel and timber with wastewater is an approach which helps overcome health hazards associated with sewage farming. Developing and enlivening the green belts around the cities with forest trees under wastewater irrigation also helps revive the ecological balance and improves environmental quality by self-treatment of wastewater through land application and forest irrigation (Shende, 1982). The results obtained in a two-year study (Baddesha and Chhabra, 1985) at the Central Soil Salinity Research Institute, Karnal, on eucalyptus indicated a good response of growth and wood production to sewage irrigation (*Table 16.13*). The results are also encouraging from the point of view of economic returns (*Table 16.14*).

Table 16.13 Effect of sewage water disposal on the height (m) of eucalyptus plants

Treatments	*Time* (months)						
	3	*6*	*9*	*12*	*15*	*18*	*24*
SW 15 cm daily	2.33	3.70	5.75	7.32	7.96	8.74	10.03
SW 15 cm fortnightly	2.26	3.75	6.10	7.56	8.17	9.05	10.09
SW 15 cm monthly	2.05	3.39	5.70	7.06	7.73	8.60	9.86
TW 15 cm monthly	2.00	3.23	5.45	6.90	7.50	8.33	9.80
CD at 5% treatments = 0.12; time 0.16							

SW = Sewage water. TW = Tube well water.

Table 16.14 Gross returns from 1 ha eucalyptus plantation under sewage water disposal system

Particulars	*First year*	*Second year*
Average height	7 m	10 m
Effective height for wood production	3.5 m	5 m
Average radius	3.2 cm	4.5 cm
Number of plants/ha	2178	2178
Total volume of the wood produced*	24.23 m^3	69.31 m^3
Weight of dry wood†	19.38 t	55.45 t
Gross returns (Rs)‡	9691.00	27 723.00

* r^2effective height
† Assuming the density of wood = 0.8 g/cm^3
‡ Assuming a selling price of Rs 500 per tonne of dry wood

Effects of wastewater irrigation on soil

The results obtained from the field experiments conducted at NEERI for a 5–8 year period are summarized in the following sections.

Effect on soil pH, EC and ESP Undiluted raw sewage did not markedly change pH and EC (electrical conductivity) of the soil but the diluted sewage tended to cause an increase. An increase in the proportion of dilution water increased the pH and EC of soil (*Table 16.15*), especially because of its poor quality (higher salinity and sodicity). The ESP (exchangeable sodium percentage) of the soil also slightly increased during the 8-year period of irrigation with undiluted and variously diluted sewage but the increase was more with the increase in the proportion of dilution water (*Table 16.15*), which had a higher SAR (sodium absorption ratio). Irrigation with untreated, primary treated and secondary treated sewage did not result in a significantly different effect on pH and EC of the soil. However, there was a slight but significant drop in pH and increase in EC and ESP of soil under irrigation with all the grades of wastewater during the 5-year period in relation to the base value (*Table 16.16*).

Effect on N, P_2O_5, K_2O and organic carbon content of soil Irrigation with undiluted raw sewage resulted in an increase in the total N, P_2O_5, K_2O and organic carbon content of the soil. An increase in the proportion of dilution water, however, brought down the level of this increase (*Table 16.16*).

Nitrogen enrichment of soil under irrigation with undiluted raw sewage is a matter of special significance. Total N, P_2O_5, K_2O and organic carbon content of the soil tended to increase under irrigation with untreated sewage in comparison with the corresponding level in the beginning (base value). However, the increases resulting from irrigation with primary treated and secondary treated sewage were insignificant (*see Table 16.16*).

Long-term effects on physical properties of the soil The long-ranging effects of sewage irrigation studied at Madurai sewage farm on various parameters attributing to the physical condition of soil (*Table 16.17*) indicate the following:

- There was a considerable build-up in organic carbon content in the surface layers of the soil, up to 45 cm depth, as a result of prolonged treatment with sewage. The longer the period of sewage irrigation, the higher was the build-up.
- The bulk density of soil up to 45 cm depth was considerably reduced as a result of 25 years of sewage irrigation, in comparison with the soil without sewage irrigation or with 15 years of sewage irrigation.
- Sewage irrigation also resulted in a considerable increase in the percentage of water-stable aggregates.
- Hydraulic conductivity of soil in the upper 30 cm layer tended to decrease as a result of sewage irrigation (this effect seems to be specific for the sandy loam soil of the farm).
- The water-holding capacity of the soil, to a depth of 45 cm, registered a considerable increase as a result of prolonged irrigation with sewage in comparison with the untreated soil.
- Clay content of the 15 cm top layer of soil increased due to sewage irrigation over 15–25 years but there was a decrease in the clay content in the next 15 cm layer (15–30 cm).
- Total capillary porosity of soil, even up to a depth of 105 cm, increased due to sewage irrigation, but the increase was remarkable up to 30 cm depth under 25 years of sewage irrigation.

Potential of heavy-metal pollution through wastewater irrigation
There is now considerable awareness about heavy metal pollution of soil and water resources and many reports have recently appeared in the literature. Land

Table 16.15 Effect of irrigation with differentially diluted raw sewage on important soil properties during an 8-year period

Treatment particulars	*pH (8.13 ± 0.26)*	*EC (0.22 ± 0.03 mmhos/cm)*	*ESP (6.66 ± 0.43)*	*Total N % (0.105 ± 0.025)*	*Total P_2O_5% (0.1103 ± 0.0097)*	*Total K_2O_5 % (0.66 ± 0.08)*	*Organic C % (0.986 ± 0.005)*
Undiluted raw sewage (100%)	8.50	0.23	7.68	0.143	0.106	0.60	1.00
Diluted sewage in 2:1 ratio (66%)	8.87	0.24	8.51	0.133	0.092	0.57	0.70
Diluted sewage in 1:1 ratio (50%)	8.93	0.26	9.58	0.120	0.089	0.56	0.65
Diluted sewage in 1:2 ratio (33%)	8.88	0.25	8.61	0.110	0.081	0.52	0.62
Well water	9.00	0.27	11.76	0.093	0.071	0.47	0.57
SE(m)	0.09	0.007	0.50	0.007	0.003	0.012	0.06
CD 5%	0.20	0.017	1.15	0.016	0.007	0.028	0.14
CD 1%	0.30	0.024	1.68	0.023	0.010	0.040	0.21

* Values in parentheses are the initial base values.

Table 16.16 Effect of irrigation with untreated, primary-treated and secondary-treated sewage at moderate and high intensities on some important soil characteristics during a 5-year period

Treatment particulars		*Soil characteristics*						
Type of sewage		*pH (8.5 + 0.12)**	*EC (mmhos/cm) (0.308 + 0.49)*	*ESP (6.66 + 1.01)*	*Total N % (0.107 + 0.013)*	*Total P_2O_5% (0.09 + 0.28)*	*Total K_2O % (0.531 + 0.061)*	*Organic carbon % (0.522 + 0.115)*
Raw sewage	M	8.2	0.50	7.66	0.116	0.117	0.504	0.561
Raw sewage	H	8.3	0.54	7.70	0.129	0.116	0.474	0.646
SP effluent	M	8,32	0.41	7.60	0.110	0.108	0.492	0.611
SP effluent	H	8.35	0.51	7.75	0.107	0.122	0.486	0.582
Settled sewage	H	8.37	0.43	7.40	0.106	0.117	0.495	0.568
Tap water	M	8.2	0.31	7.71	0.105	0.107	0.483	0.568
Tap water	H	8.32	0.30	7.59	0.098	0.106	0.453	0.546
SE(m)		0.09	0.02	0.38	0.009	0.006	0.037	0.067
CD 5%		–	–	–	–	–	–	–
CD 1%		–	–	–	–	–	–	–

M = Moderate intensity. H = High intensity.

* Values indicated in parentheses are the initial base values.

Table 16.17 Effect of sewage irrigation on physical properties of soil

Property characteristics	*Depth of soil in cm*	*Sewage irrigation*		*Without sewage irrigation*
		25 years	*15 years*	
Organic matter (%)	0–15	4.63	2.56	1.42
	15–30	2.02	1.57	0.97
	30–45	0.87	0.61	0.63
	60–75	0.55	0.30	0.45
	90–105	0.52	0.39	0.30
Bulk density (g/cm^3)	0–15	1.04	1.52	1.53
	15–30	1.22	1.54	1.57
	30–45	1.28	1.65	1.53
	60–75	1.36	1.64	1.44
	90–105	1.36	1.56	1.44
Per cent aggregate stability	0–15	83.48	84.38	72.43
	15–30	82.84	81.06	71.36
	30–45	82.58	72.79	66.70
	60–75	75.54	63.81	64.70
	90–105	67.45	32.34	54.13
Hydraulic conductivity (cm/h)	0–15	0.92	1.08	2.01
	15–30	0.91	1.03	0.98
	30–45	0.64	0.52	0.49
Water holding capacity (%)	0–15	59.78	49.70	33.30
	15–30	49.88	44.63	23.60
	30–45	38.27	34.16	24.47
	60–75	30.98	28.83	29.47
	90–105	33.95	26.29	29.30
Clay content (%)	0–15	26.63	23.56	19.10
	15–30	16.26	14.56	22.93
	30–45	25.33	29.90	31.06
	60–75	38.23	40.36	41.13
	90–105	50.10	47.16	49.63
Total porosity	0–15	60.40	40.30	36.20
	15–30	57.30	35.20	31.30
	30–45	50.20	39.20	33.90
	60–75	44.80	42.80	34.60
	90–105	46.00	37.60	35.20
Capillary porosity	0–15	54.30	35.30	33.00
	15–30	54.30	28.90	29.30
	30–45	46.70	33.80	31.20
	60–75	40.70	36.80	33.20
	90–105	40.80	27.80	32.20

application of wastewater is one of the promising answers to heavy-metal pollution (Leeper, 1978). In spite of the high capacity of land to lock up and arrest the movement of heavy metals, the problems of phytotoxicity are also to be looked into. These again are site-specific problems.

NEERI has lately started a few investigations on the fate of heavy metals applied to the land in wastewater irrigation. However, looking into the low level concentration of toxic heavy metals in the typical domestic wastewaters of India can still be considered a problem of tomorrow. Discharge of wastewaters from certain industries into municipal sewers, however, contributes heavy metals to the sewage and can be checked by proper treatment before the wastes from such industries are allowed into the sewers.

Conclusions

Countries in the Asia and Pacific region are in the process of development which, of course, is slow mainly due to financial restraints. Regarding health, the progress in water supply is relatively fast but sanitation is still considerably lagging behind. Some countries, like India and China, have made a noticeable start in providing sewerage facilities to at least a fraction of the urban population but the progress in sewage treatment is very very poor. Many sewerage schemes in these countries, as well as in other countries, have recently begun and the list is growing. Slowly, the volume of wastewater generated and collected will increase in many of the countries of the region.

Wastewater treatment by conventional means is too costly for the developing countries of the region and, in the light of recent developments, the problems of pollution are not satisfactorily solved with these techniques. Non-conventional treatment through oxidation/stabilization ponds and/or aerated lagoons followed by land application will not only take care of water pollution problems at low cost to these countries but will also help conserve the valuable resources from the wastewater (water and nutrients) which have a sizeable reuse potential in the promotion of agricultural production. This approach very much suits the geoclimatic and socioeconomic situations in most of the countries of the region.

Indian experience in wastewater treatment and reuse in agriculture, which has been backed by field tests over decades and research and development, can certainly providc at least interim guidelines for land application programmes in the other countries of the region. In the absence of facilities for wastewater treatment, environmental pollution through wastewater can be controlled to a considerable extent by land application-cum-treatment of wastewater with bountiful crop yields. The criticism of health hazards due to irrigation with untreated wastewater can be lessened by judicious and skilful management of the wastewater–soil–crop system and by restricting the choice of crops for wastewater irrigation. Tentative guidelines for selection of crops to suit various situations with regard to wastewater treatment are given in preferential order in *Table 16.18*.

In spite of the long experience of India and also China in wastewater utilization in agriculture, this field is still a new area for most of the countries of the region, including India, where land application of wastewater has been largely disposal orientated. This is especially true in the light of the concept of integration of wastewater recycling in agriculture with environmental pollution control. Therefore, the scope for research and development in this field is really vast in the context of the present status in the region. Considering the resource potential of wastewater and the tangible benefits that may accrue from its recycling and reuse for the development of agriculture, with special reference to the socioeconomic conditions of the developing countries of the region, a regional project should be conceived for the development and transfer of appropriate technology through research and extension efforts.

Interactive results from the soil–wastewater–crop system, being very much site-specific, cannot be made applicable directly to other sets of agroclimatic

Table 16.18 Suggested cropping patterns for irrigation with untreated, primary-treated and secondary-treated sewage effluent

Type of sewage effluent	*Suggested crops in order of preference*
Raw sewage (preferably diluted)	(1) Forest tree species, avenue trees, ornamental flowering shrubs etc. (2) Commercial crops, such as cotton, jute, milling type sugarcane, cigarette tobacco (3) Essential oil bearing crops, such as citronella, mentha, lemon grass (4) Any crop raised exclusively for seed production (5) Cereal and pulse crops with well-protected grains, such as wheat, paddy, greengram, pigeon-pea, blackgram (6) Oil seeds, such as linseed, til, castor, mustard, safflower, sunflower, soybean (7) Fruit crops (well protected), such as coconut, banana, citrus, etc.
Primary-treated sewage (preferably diluted)	(1) to (7) as above (8) Vegetables exclusively cooked before eating and borne on the plant away from the soil, such as brinjal (eggplant), lady's finger (okra), cucurbits, beans, etc. (9) Fruit crops borne on the plant sufficiently far from the soil, such as guava, chikoo, grape, papaya and mango
Secondary-treated sewage	(1) to (9) as above (10) All types of crops including vegetables borne near the soil surface but eaten after cooking, fodder grasses etc.
Secondary-treated and disinfected sewage	(1) to (10) as above (11) All crops without restrictions

conditions in another region. Nevertheless, the work at a particular location brings out a certain set of guiding principles which, with suitable modifications, assumes an applicability value after certain tests. A multilocational two-tier framework for a regional research and development project is therefore envisaged. The two-tier frame of the project is suggestive of the level of concentration of R and D effort and involves establishment of two or three well-equipped regional research centres (RRC) and the requisite number of technology transfer centres (TTC), which would be linked with the regional research-centres based on geoclimatic and socioeconomic similarities. The major role of the RRCs will be research and development of cost-effective and environmentally safe techniques for agricultural reuse of wastewaters to provide guidance and find solutions to the problems referred by the TTCs and to arrange training programmes at various levels. TTCs will have the responsibility of assisting the local authorities in design, planning and operation of land application systems and serving as a linkage between the users and the RRCs.

In the context of the present information on the status in the region, the following topics have been identified for the R and D efforts at the proposed Regional Research Centres:

- A detailed survey of existing land application systems, including evaluation of the performance of a few representative sites in the region, with regard to the direct and indirect benefits to agriculture, preservation of environment, improvements in socioeconomic conditions, health implications and environmental impact assessment.
- Preparation of a detailed state-of-the-art document on the existing and potential availability of wastewater, extent and degree of its treatment and mode of

disposal and feasibility of its reuse in agriculture. This will also include a preliminary feasibility report on possible locations where scientifically based land-application systems can be planned and operated with local cooperation.
- Characterization of wastewater with special reference to the constituents which may become limiting factors in the land-application systems.
- Soil survey of the possible areas in the neighbourhood of wastewater generating centres for selection of suitable sites for location of the system at the places identified under preparation above.
- Evaluation of the capacity of available soil types to receive and treat wastewater of varying qualities with regard to the specific pollutants under simulated or semisimulated conditions so as to ascertain the length of period over which the system can be operated without problems.
- Planning and conduct of field experiments on the interactive performance of the soil–wastewater–crop system on specific soils from the feasible sites, available wastewater quality and prevalent cropping patterns for determination of optimum application rates, needs and degree of treatment and/or dilution for maximum reuse efficiency and environmental and public health safety. This should also include monitoring the movement of hazardous constituents of wastewater in the soil, their migration to contiguous media and interlinked systems like ground and surface water, plant systems and animal (fish and mammals) life.
- Development of low-cost treatment processes using physicochemical and biological systems that will be effective in the conservation of reusable constituents, like nutrients, and elimination of specific constituents (which have been found to create problems or reduce the life span of the system in the light of the findings from evaluation above) for preapplication treatment.
- Establishment of soil–wastewater–crop systems on specific soils from the feasible sites, available wastewater quality and prevalent cropping patterns for determination of optimum application rates, needs and degree of treatment and/or dilution for maximum reuse efficiency and environmental and public health safety. This should also include monitoring the movement of hazardous constituents of wastewater in soil, their migration to contiguous media and interlinked systems like ground and surface water, plant systems and animal (fish and mammals) life.
- Microbiological investigations in respect of the rate of pathogenic and parasitic organisms in the soil–plant system and the interaction of microflora and fauna brought by wastewater with those inhabiting the soil to assess any changes in the microbiological properties of soils.
- Development of post-harvest techniques for premarket handling and treatment of wastewater-irrigated produce for control of health hazards to consumers.
- Development and evaluation of low-cost wastewater treatment and reuse and recycling systems, like forest irrigation, aquaculture, algal harvesting besides crop irrigation, to suit the needs and socioeconomic patterns in the different situations and countries in the region.
- Development of mathematical models from the data collected from various tests and experiments for planning and predictive purposes.

Acknowledgement

The author is grateful to the Director, NEERI, Nagpur, for providing all the facilities for the work reported upon as NEERI's work and to the Indian Council of

Agricultural Research, New Delhi for the financial support provided for most of the R and D activities at the National Environmental Engineering Research Institute, which are reported upon. The author also heartily acknowledges the help and encouragement extended by Mr F. J. Dent, Regional Soil Management and Fertilizer Use Officer, RAPA, Bangkok, for supplying relevant literature and personal communications received by him from different countries of the region, which have added to the dimensions of this chapter. The author also acknowledges the scientific and technical contributions and secretarial assistance of the staff of the Waste Water Agriculture Division of NEERI, Nagpur, in writing this chapter.

References

ANON (1973–83) Annual Progress Reports of the Research Centre at NEERI, Nagpur, under the *All India Coordinated Project on Use of Saline Water in Agriculture* (1973–74 to 1978–79) and *Microbiological Decomposition and Recycling of Farm and City Wastes* (1979–80 to 1982–83), sponsored by Indian Council of Agricultural Research, New Delhi

ARCEIVALA, S. J., LAXMINARAYANA, J. S. S., ALGARSWAMY, S. and SASTRY, C. A. (1970) Waste stabilization ponds – design, construction and operation in India, Nagpur, CPHERI

BADDESHA, H. S. and CHHABRA, R. (1985) Sewage utilization through forestry. *Paper presented at the National Seminar on Pollution Control and Environmental Management,* NEERI, Nagpur, 16–18 March

CENTRAL BOARD FOR THE PREVENTION AND CONTROL OF WATER POLLUTION (1979). Wastewater collection, treatment and disposal in Class I cities. *Status and Action Plan CUPS/4/1978–79.* New Delhi: CBP and CWP

CENTRAL BOARD FOR THE PREVENTION AND CONTROL OF WATER POLLUTION (1980). Status of water supply and wastewater collection, treatment and disposal in Class II towns of India. *CUPS/6/1979–80.* New Delhi: CBP and CWP

CLLAP, C. E., LINDEN, D. R., LARSON, W. E. and NYLUND, J. R. (1977) Nitrogen removal from wastewater effluent by a crop irrigation system. In *Land as a Waste Management Alternative.* Loehr, R. C. (ed.), Ann Arbor, Michigan: Ann Arbor Science

DORAN, J. W., ELLIS, J. R. and McCALLA, T. M. (1977) Microbial concerns when wastes are applied to land. In *Land as a Waste Management Alternative.* E. Loehr R. C. (ed.), Ann Arbor, Michigan: Ann Arbor Science

ENVIRONMENTAL PROTECTION AGENCY (1981) *Process Design Manual for Land Treatment of Municipal Wastewater.* (EPA-625-81-013). US EPA, Cincinnati, Ohio: Centre for Environmental Research Information

JAYARAMAN, C., RANI PERUMAL and SREE RAMULU, U. S. (1983) Influence of continuous municipal sewage effluent irrigation on soil physical properties. In *Proceedings of National Seminar on Utilization of Organic Wastes.* Agricultural College and Research Institute, Tamil Nadu Agricultural University, Madurai, 24–25 March

LEEPER, G. W. (1978) *Managing the Heavy Metals on Land.* New York and Basel: Marcel Dekker

MAHIDA, U. N. (1981) *Water Pollution and Disposal of Wastewater on Land.* New Delhi: Tata McGraw-Hill Publishing Company Limited

PESCOD, M. B. and OKUN, D. A. (1971) *Water Supply and Wastewater Disposal in Developing Countries.* Bangkok: The Asian Institute of Technology

PICKFORD, J. A. (1979) Control of water pollution and disease in developing countries. *Water Pollution Control,* **78**(2), 239–253

SHENDE, G. B. (1982) Wastewater reuse and self-treatment in the promotion of social and recreational forestry. *Proceedings of the Seminar on Social Forestry,* Nagpur, 20–21 February

SHENDE, G. B., JUWARKAR, A. S. and SUNDARESAN, B. B. (1983) Prevention of environmental pollution through reuse of wastewaters in agriculture. *Proceedings of National Seminar on Utilization of Organic Wastes,* held at the Agricultural College and Research Institute, Tamil Nadu Agricultural University, Madurai, 24–25 March

SHENDE, G. B. (1984) Status paper on agricultural use of sewage in Asia and the Pacific. *Paper presented at the annual meeting of Regional Organic Recycling Network Coordinators at RAPA, Food and Agriculture Organization,* Bangkok, 18–21 October

SUNDARESAN, B. B. (1978) Environmentally compatible water supply and waste utilization for developing countries. *Paper presented at Research Study Group Meeting on Appropriate Technology for Improvement of Environmental Health at Village Level, World Health Organization,* New Delhi, 16–20 October

WONG, K. K. (1984) Sewage irrigation in China. *International Journal for Development Technology,* **2(4),** 291–301

Chapter 17

Egyptian experience in the treatment and use of sewage and sludge in agriculture

A. S. Abdel-Ghaffar, H. A. El-Attar and I. H. Elsokkary*

Introduction

The disposal of wastes both in developed and developing countries has become problematical because of the increase in population or in *per capita* waste, or both. This problem in developing countries deserves serious attention owing to the generally lower standards of public sanitation.

Human habitation wastes are considered a public health hazard but they are a valuable source of organic matter and plant nutrients. The discharge of wastewater into a river, stream, lake or the sea results in water pollution and, at the same time, deprives agricultural land of two scarce materials, namely water and plant nutrients. However, the direct application of these substances on agricultural land is limited by the extent of contamination with heavy metals, toxic organic chemicals and pathogens. Also, continuous use of wastewater in irrigation may cause an increase in soluble salts and such increase may have a deleterious effect on certain crops (Abdel-Ghaffar, 1983).

At present, land application of wastewater is considered to be the best solution for disposal problems. The use of sewage for land irrigation is usually recommended for two main reasons: (a) it is a low-cost method for the disposal of wastewater; and (b) it permits the reclamation and reuse of valuable resources such as water and nutrients from sewage (Wang, 1984). Under arid and semiarid conditions, the two most important factors limiting soil productivity are water and organic matter (Abdel-Ghaffer, 1982). Both substances, as well as plant nutrients, are supplied if sewage effluent and sludge are used on agricultural land.

In Egypt, it is well known that in view of the climate and the characteristics of the soils, the importance of the organic matter comes directly next to that of water (Riad, 1982). Also, owing to increasing population and the consequent need for agricultural expansion, the utilization of every source of water and organic matter is necessary. Horizontal extension is one of the main targets for enlarging the cultivated area in Egypt. This could easily be achieved in the deserts existing around the Nile valley and delta (more than 96% of the area of Egypt are barren deserts). The soils of such areas are either sandy or calcareous. Cultivation of such areas requires the availability of a suitable source of water for irrigating a wide variety of plant crops. This could be achieved by the use of sewage produced from the surrounding cities. It has been reported by Abd El-Naim, Ibrahim and El Shal

* Alexandria University, Alexandria, Egypt

(1982) that there are about 4 million m^3 of sewage from Cario city and about 1 million m^3 from Alexandria city. These amounts could irrigate about 300 000 feddans (1 feddan = 0.42 ha; 1 ha = 2.4 feddans).

Data about the use of sewage for irrigating Egyptian soils are limited and only a few practical examples can be cited from El-Gabal El-Asfar and Abou Rawash Farms near Cairo. Also, pot experiments were carried out using sewage water from Alexandria city to irrigate horticultural crops (Shehata, 1983). Sewage sludge produced from the Eastern Treatment Plant of Alexandria was also tested to evaluate its potential as a source of plant nutrients and toxic heavy metals to different plant crops (El-Keiy, 1983).

El-Gabal El-Asfar Farm

Since 1911, the sewage effluent of Cairo city has been continuously used to irrigate El-Gabal El-Asfar citrus farm, after primary sedimentation in exposed basins. The area of this farm is about 3000 feddans (1260 ha) and it lies in the eastern desert, 25 km north-east of Cairo. The area has no availbale source of irrigation water and the soil is very poor in both organic matter and plant nutrients and is of loamy sand texture.

Table 17.1 The average composition of Cairo sewage used to irrigate El-Gabal El-Asfar farm

Constituent	*Unit*	*Sewage effluent*
EC	mmhos/cm	1.1
pH		7.2
SAR		3.5
B	mg/l	0.33
Cu	mg/l	0.07
Fe	mg/l	0.70
Mn	mg/l	0.20
Zn	mg/l	0.16
Cd	mg/l	0.01
Co	mg/l	0.034
Cr	mg/l	0.10
Ni	mg/l	0.11
Pb	mg/l	0.14

Source: El-Nennah *et al.* (1982)

The average composition of the sewage used is shown in *Table 17.1*. According to the US salinity laboratory diagram, this sewage effluent used to irrigate the soil of El-Gabal El-Asfar corresponds to class C–S1: water of high salinity and low sodium content. Boron concentration is less than 0.67–1.00 ppm, the permissible limits for sensitive plants (El-Nennah *et al.*, 1982). Since then, periodical soil analyses have been carried out to follow the changes taking place (Abd-El Naim, 1982).

As shown in *Table 17.2*, El-Shabassy *et al.* (1971) reported that after irrigation with sewage for 50 years, the colour of the surface soil changed from very pale yellow to dark brown while the colour of the subsurface layers remained almost

Table 17.2 Changes in soil properties due to sewage irrigation at El-Gabal El-Asfar farm

Parameters	*Profile 1 (control)*		*Profile 5 (50 years of sewage irrigation)*	
	0–28 cm	28–75 cm	0–40 cm	40–100 cm
Colour	Pale brown	Very pale brown	Dark brown	Very pale brown
pH	7.60	7.70	6.70	7.60
Total soluble salts me/100 g	0.13	0.13	0.07	0.04
Organic matter %	0.32	0.08	1.26	0.21
$CaCO_3$ %	5.10	5.20	2.80	5.80
Clay %	7.70	8.10	12.60	8.20
Silt %	2.20	1.10	1.10	1.30
Fine sand %	28.60	31.10	20.20	8.30
Coarse sand %	53.50	52.00	59.40	77.20
CEC me/100 g	8.90	5.70	11.30	6.60

Source: El-Chabassy *et al.* (11971)

unchanged. The differences in the colour of the top layer are mainly due to the increase in organic matter content. By sewage water irrigation, the organic matter content increased from 0.32 to 1.26% (*Table 17.2*) and in certain areas increased to 2.0% (El-Shabassy *et al.*, 1971) or 7.7% (Abdou and El-Nennah, 1980). On the other hand, there are no marked differences between the organic matter content of surface and subsurface layers.

Although there was no marked change in soil texture after irrigation with sewage, only the top soil of profile 5 showed some changes, indicating that the suspended matter of the sewage had affected the textural components. The clay was increased from about 8 to 13% (*see Table 17.2*).

It is clear, from *Table 17.2*, that irrigation with sewage water decreased the $CaCO_3$ content of the surface layer from 5.1% to about 2.8%. This is probably due to the fact that some $CaCO_3$ was dissolved by the organic acids present in sewage or biologically formed and leached downwards through the soil profile. The depth of maximum $CaCO_3$ accumulation became deeper due to irrigation. Moreover, the soil pH decreased from 7.6 in the surface layer to about 6.7 (*see Table 17.2*).

Application of sewage to this light soil did not show any salt accumulation. The initial salt content of the soil was lowered from 0.13 to about 0.07 me/100 g after

Table 17.3 Changes in the soil content of heavy metals after different periods of irrigation with sewage at El-Gabal El-Asfar farm

Parameters	*Unit*	*Length of sewage irrigation* (years)			
		0	*2*	*25*	*35*
Total Zn	ppm	90	150	190	230
Total Fe	ppm	6130	7362	21 470	36 810
Total Mn	ppm	340	370	510	520
NH_4OAC-Zn	ppm	0.4	0.7	1.4	6.3
NH_4OAC-Fe	ppm	3.1	6.1	7.6	15.3
NH_4OAC-Mn	ppm	1.4	1.7	8.5	10.8

Source: Abdou and El-Nennah (1980)

irrigation with sewage water for 50 years (*see Table 17.2*). It was also found that after 50 years of irrigation the CEC of the soil increased from 8.9 to 11.3 me/100 g soil (*see Table 17.2*).

Another study carried out by Abdou and El-Nennah (1980) on the El-Gabal Al-Asfar area showed that using sewage for irrigation markedly increased the total and soluble Fe, Mn and Zn in the soil (*Table 17.3*). These workers concluded that the use of sewage in irrigating sandy soils could be considered as a soil improver or as a plant nutrient source but attention should be given to environmental pollution. In addition, El-Nennah *et al.* (1982) showed that using sewage effluent in irrigation for 47 years increased available phosphorus and both total and soluble nitrogen in the soil (*Table 17.4*). Also, water-soluble B and the total-extractable and DTPA-extractable heavy metals Cd, Co, Cr, Cu and Pb were increased. These results also show that the prolonged irrigation with sewage effluent (47 years) could remove the organometal complexes to a deeper penetration than 50 cm. This may explain the lower amounts of organic matter and elements in the soil after 47 years than after 23 years of irrigation with liquid sewage (*see Table 17.4*).

Table 17.4 Changes in soil properties after different periods of irrigation with sewage at El-Gabal El-Asfar farm

Parameters	*Unit*	*Length of sewage irrigation* (years)					
		0		*23*		*47*	
		0–25 cm	*25–50 cm*	*0–25 cm*	*25–50 cm*	*0–25 cm*	*25–50 cm*
Organic matter	%	0.44	0.25	6.62	6.33	4.80	4.03
Total N	%	0.03	0.02	0.98	0.46	0.37	0.2
Available P	ppm	7.2	6.9	8.6	8.9	8.5	8.1
Total Cd	ppm	1.5	1.5	3.0	2.3	1.9	1.5
Total Co	ppm	37.4	30.4	38.2	31.2	41.4	34.3
Total Cr	ppm	89.0	75.0	403.0	344.0	221.0	189.0
Total Cu	ppm	29.0	74.0	716.0	484.0	228.0	169.0
Total Pb	ppm	67.0	67.0	382.0	267.0	186.0	170.0
DTPA Cd*	ppm	N.D.	N.D.	0.5	0.3	0.3	0.2
DTPA Cu	ppm	0.4	0.2	85.3	85.3	63.3	14.8
DTPA Pb*	ppm	N.D.	N.D.	22.8	15.1	11.8	7.5
Soluble B	ppm	0.1	0.1	1.0	0.3	0.3	0.6

Source: El-Nennah *et al.* (1982)

*N.D. Not detected

Abou Rawash farm

Abd El-Naim *et al.* (1982) evaluated the effect of using sewage from El-Giza city to irrigate sandy soil at Abou Rawash. The data in *Table 17.5* show the average chemical composition of sewage from El-Giza. The use of that wastewater for irrigation caused remarkable changes in the particle size distribution of the surface soil layer (*Table 17.6*). The total fine particles (silt + clay) and $CaCO_3$ contents showed variations between top soil (0–30 cm) and subsoil (30–160 cm). The fine particles increased from about 1.0 to 6.0% after 3 years of irrigation. The amount

Table 17.5 Average chemical composition of Giza sewage used to irrigate Abou-Rawash farm

Constituent	*Unit*	*Giza*
SAR		2.8
pH		7.1
EC	mmhos/cm	1.7
Cl^-	mg/l	320
SO_4^{2-}	mg/l	138
Ca^{2+}	mg/l	128
Mg^{2+}	mg/l	96
Na^+	mg/l	205
K^+	mg/l	35
Cu^+	ppm	0.4
Fe	ppm	15.3
Mn	ppm	0.7
Zn	ppm	1.4

Source: Abd El-Naim *et al.* (1982)

Table 17.6 Changes in soil properties after different periods of irrigation with sewage at Abou Rawash, Giza

Parameters	*Unit*	*Years of sewage irrigation*							
		Virgin		*1*		*2*		*3*	
		0–30 cm	*30–60 cm*	*0–30 cm*	*30–60 cm*	*0–30 cm*	*30–60 cm*	*0–30 cm*	*30–60 cm*
pH		7.70	7.60	7.20	6.80	6.70	6.50	6.00	6.40
TSS	%	0.10	0.05	0.05	0.05	0.11	0.05	0.09	0.07
Organic matter	%	0.19	0.10	0.32	0.10	0.46	0.23	0.56	0.20
$CaCO_3$	%	1.24	1.20	0.76	0.66	0.38	0.28	0.16	0.10
Coarse sand	%	67.00	75.00	69.00	77.00	73.00	75.00	67.00	66.00
Fine sand	%	30.00	23.00	28.00	21.00	21.00	22.00	27.00	30.00
Silt + clay	%	1.00	1.00	2.00	1.00	6.00	2.00	6.00	4.00
Bulk density		1.68	1.72	1.65	1.70	1.58	1.62	1.45	1.62
Field capacity	%	8.90	8.50	9.80	9.50	12.20	9.00	16.60	10.50
Wilting point	%	2.80	2.50	3.20	3.00	4.10	3.80	6.60	3.90
Average moisture	%	6.10	6.00	6.60	6.50	8.10	5.20	10.00	6.40
Water holding capacity	%	20.30	18.50	24.50	19.20	27.70	21.10	30.40	22.60
Zea mays maize grain yield (kg/ha)		–		1589		2643		3242	

Source: Abd El-Naim *et al.* (1982)

of $CaCO_3$ decreased from 1.24 to 0.16% while organic matter content increased about three times (from 0.19 to 0.56%) in the surface layer after three years of irrigation.

The results in *Table 17.6* show a remarkable decrease in soil pH (from 7.7 to 6.0) in the surface soil layer. However, the amounts of total soluble salts did not change by irrigation with sewage. Chloride and Ca^{2+} contents followed the same trend as TSS, while that of HCO_3^- was decreased. The amounts of Mg^{2+} and Na^+ were increased.

The changes in the physical properties of the studied soils, as influenced by sewage irrigation, are shown in *Table 17.6*. The values of the water-holding capacity of the soils followed the distribution of fine particles and increased from 20.3 to 30.4% for the surface soil layer and from 18.5 to 22.6% for the subsurface soil. The values of EC had also increased from 8.92 to 16.65% for the surface soil (0–30 cm) and from 8.51 to 10.5% for the subsurface soil. This was also noticed with the wilting point, available moisture and bulk density.

Maize was cultivated on these soils for three successive years (1980–82) and irrigated with wastewater from El-Giza city. The yield of maize showed an obvious increase as a result of using sewage for irrigation (*see Table 17.6*).

Use of sewage from Alexandria in irrigation

The estimated amount of wastewater from Alexandria city is expected to reach about 1 120 000 m^3/day by 1990 and by the year 2000 it will be about 1 470 000 m^3/day. The chemical composition of Alexandria sewage is given in *Table 17.7* as compared to El-Mahmoudia canal water near Alexandria.

Table 17.7 Chemical composition of El-Mahmoudia canal water and sewage at Alexandria

Constituent	*Unit*	*El-Mahmoudia canal*	*Wastewater*
EC	mmhos/cm	0.37	3.10
pH		7.60	7.80
SAR		1.10	9.30
Na^+	me/l	11.20	24.60
Ca^{2+}	me/l	0.80	1.50
Mg^{2+}	me/l	0.80	3.20
K^+	me/l	0.20	1.80
Cl^-	me/l	7.10	62.00
SO^{2-}	me/l	0.20	35.00
CO_3^-	me/l	0.00	1.10
HCO_3^-	me/l	6.70	6.60
NH_4^+	ppm	0.50	2.50
NO_3^-	ppm	1.80	10.10
P	ppm	8.50	8.50
Mn	ppm	0.50	0.20
Cu	ppm	0.30	1.10
Zn	ppm	0.40	0.80

Source: Shehata (1983)

An investigation was carried out by Shehata (1983) during the 1981 and 1982 seasons to evaluate the influence of applying sewage to different types of soils cultivated with grapes, oranges, olives and guavas. Their relative growth, relative tolerance and mineral composition of leaves and roots were compared when irrigated with water from a canal and with Alexandria sewage effluent. The studied soils included:

- a gravel soil from El-Khatatba, Behira;
- a sandy soil from 80 km south-east of Alexandria;
- a sandy loam soil (calcareous) from North Tahrir; and
- a clay soil from the Nile delta.

The main chemical and physical properties of these soils are given in *Table 17.8*. This study included the use of two water sources, namely Nile water obtained from El-Mahmoudia canal and raw sewage obtained from the upper layer of the canal leading to the East Wastewater Treatment Plant of Alexandria city. The chemical analyses of the two waters are given in *Table 17.7*. The results obtained in this study are summarized in the following sections.

Plant growth

The fresh weights of the whole plants were recorded at the termination of the experiment and used as indices of growth (*Table 17.9*). These results revealed, as an overall average and irrespective of soil type, that growth of the four plant species increased upon irrigation with sewage. Differences were found to be significant in the case of grapes and olives. When studying each soil separately it was noted, in general, that the above-mentioned trend was found with each soil; in other words, total fresh weight increased when sewage water was applied to irrigate the four fruit plant species. Differences, however, were only significant when sewage was applied to grapes grown on sandy soil.

Table 17.8 The general properties of four soils irrigated with Alexandria sewage or Nile irrigation water from El-Mahmoudia canal

Parameters	*Unit*	*Soil texture*			
		Gravelly	*Sandy*	*Sandy loam*	*Clayey*
EC	mmhos/cm	N.D.	5.0	6.6	5.7
SAR		N.D.	13.9	10.1	2.1
$CaCO_3$	%	N.D.	2.8	27.5	1.3
Saturation	%	N.D.	17.7	39.4	60.0
Field capacity	%	N.D.	8.9	19.7	30.0
PWP	%	N.D.	4.4	9.4	14.9
Sand	%	N.D.	98	68	28
Silt	%	N.D.	2	17	22
Clay	%	N.D.	0	15	50

Source: Shehata (1983)

Table 17.9 The relative total fresh weight (growth) of the four plant species with respect to the control

Treatments	*Grapes*	*Oranges*	*Olives*	*Guavas*
Canal water (control)	100	100	100	100
Sewage	138	105	146	116
Clayey soil (control)	100	100	100	100
Gravelly	44	111	119	77
Sandy	73	97	105	160
Sandy loam	68	101	95	58

Source: Shehata (1983)

The relative growth of plants subjected to sewage irrigation (*see Table 17.9*) indicated that the four fruit plant species differed in their response to sewage as the only source of irrigation. These species can be arranged according to their positive response to sewage application as follows: olives, grapes, guavas, oranges. The growth increments were 45.9%, 37.8%, 16.2% and 5.0% respectively.

Mineral contents of the plants

Generally, the N, P, K, Fe, Mn, Zn and Cu contents of the leaves and roots of the different plants species increased, while Mg decreased. Nutrient concentrations in the plants varied depending on plant species and soil properties (*Table 17.10*).

Table 17.10 The mature leaf mineral contents of the four plant species as affected by irrigation with El-Mahmoudia canal water or sewage

Water source	*Element*	*Grapes*	*Oranges*	*Olives*	*Guavas*
Canal	N %	2.81	2.80	2.18	1.99
Sewage		2.96	2.96	2.43	2.23
Canal	P %	0.21	0.09	0.09	0.14
Sewage		0.21	0.08	0.08	0.15
Canal	K %	2.20	0.95	0.81	0.90
Sewage		2.21	1.13	0.82	0.90
Canal	Fe ppm	385	105	103	99
Sewage		356	126	86	129
Canal	Mn ppm	89	23	37	45
Sewage		71	24	29	49
Canal	Zn ppm	287	52	46	73
Sewage		211	78	49	46
Canal	Cu ppm	124	40	40	51
Sewage		102	60	43	57

Source: Shehata (1983)

Results suggest that irrigating olives, grapes, guavas and, to some extent, oranges grown on sandy, or sandy loam soils, with sewage could be an effective means of reusing such wastewater safely for irrigation. The concentrations of N, P and K added in sewage, during the limited period, might be low in terms of part per million (ppm) but the continued use of such effluent for irrigation at high rates could add a significant amount of these nutrients. Thus, each ha cm of the sewage used for irrigation in the present investigation would add 13.2 kg N, 0.97 kg P and 7.7 kg K/ha.

Considering that approximately 114 ha cm of water per year are needed in the Nile delta, just to replenish the consumptive use of water by permanent green orchards, using this amount of sewage would add 1508 kg N, 110 kg P and 880 kg K/year per ha. In most instances, this would supply more P and K than presently being recommended and needed for horticultural use and almost all the N (Shehata, 1983). Therefore, the amount of fertilizer applied as part of the sewage will be a plus factor and must be accounted for when considering the fertilizer requirements for horticultural plants.

Sewage sludge use in agriculture

For a long time, a parallel was drawn between sewage sludge and agricultural wastes, particularly farmyard manure. The idea was that instead of dumping sludge at sea or in disposal pits, it could be utilized in agriculture. Unlike farmyard manure, sewage sludge contains, in addition to plant macro nutrients (N, P and K), components of industrial origin, for example heavy metals and other trace elements. Sewage sludge, because of its favourable soil-conditioning properties, due to a high organic matter content, is applied to agricultural lands. Unfortunately, many sewage sludges, particularly those derived from sewage having significant industrial inputs, contain considerable quantities of heavy metals. The metals in sewage sludge that are most likely to cause phytotoxicity problems when applied to agricultural lands are Ni, Cd, Cr and Hg (El-Keiy, 1983). Various proportions of the metals present in sewage sludges are available for plant uptake. This uptake depends on many variables such as soil texture, cation exchange capacity and pH, the amounts of sludge applied, the metal composition of the sludge and the plant species and varieties.

Sewage sludge produced in the East Wastewater Treatment Plant of Alexandria was used to evaluate its agricultural potential as a source of plant nutrients. Pot experiments in a glasshouse were carried out to study the effect of normal and high rates and residual effects of sludge on crop yield, elemental composition and heavy metal contents of the cultivated plants. The soil used in this study was taken from the newly cultivated land in the western desert at km 75 on the Alexandria–Cairo desert road. The main chemical and physical characteristics of the soil and the chemical composition of the sewage sludge are given in *Tables 17.11* and *17.12*, respectively.

The data in *Table 17.13* and *17.14* show the dry matter yield and the concentrations of metals in the leaves of the five plant crops as affected by sewage sludge at the rate of 50 g/kg soil. Due to the application of sewage sludge, the amounts of N, P, K, Fe, Mn, Zn, Cu, Co, Ni, Pb and Cd in the leaves of the five plant species were higher than those of the control (untreated).

Table 17.11 Some chemical and physical characteristics of the soil used

Soil variable	*Unit*	*Level*
EC	mmhos/cm	0.7
Soil pH (1:1 soil/water)		8.1
Total carbonate	%	8.7
Organic matter	%	0.13
Total nitrogen	%	0.02
$NaHCO_3$–P	ppm	4.0
NH_4OAC–K	mg/100 g	0.3
DTPA–Fe	ppm	0.4
DTPA–Mn	ppm	0.4
DTPA–Zn	ppm	0.4
DTPA–Cu	ppm	0.0
Sand	%	96
Silt	%	3
Clay	%	1

Source: El-Keiy (1983)

Table 17.12 The chemical characteristics and elemental composition of the sewage sludge

Variables	*Unit*	*Level*
EC (1:1 sludge-water)	mmhos/cm	2.8
pH (1:1 sludge-water)		7.1
Total carbonates	%	4.1
Organic matter	%	49.3
Total N	%	2.6
Total P	%	0.45
Total K	%	0.19
Total Fe	%	2.09
Total Mn	ppm	251
Total Zn	ppm	751
Total Cu	ppm	260
Total Co	ppm	10.6
Total Ni	ppm	35
Total Pb	ppm	340
Total Cd	ppm	2.6

Source: El-Keiy (1983)

Table 17.13 Effect of sewage sludge application on the dry matter yield of different plant crops in grams per pot

Plant species		*Control*	*Sludge* (50 g/kg soil)
Alfalfa (mean 8 cuts)		1.8	20.7
Wheat	Straw	3.6	31.6
	Grains	1.6	30.0
Fababean	Straw	6.2	33.2
	Grains	3.7	25.0
Soybean	Straw	3.1	22.9
	Grains	1.1	26.0
Sordan (mean 3 cuts)		2.1	56.2

Source: El-Keiy (1983)

Heavy metals removed by plants

The results in *Table 17.14* show that the amounts of heavy metals taken up by the top parts of the plants were extremely small. This could be attributed to their low availabilities due to both high soil pH and carbonate content. The five crops differed in their efficiencies to utilize the metals from the sludge and could be arranged in the order: sordan > alfalfa > wheat > soyabean > fababean. With respect to the phytotoxicity of the sludge used, Chaney and Harnick (1978) considered that sludges containing 2000 ppm Zn, 800 ppm Cu, 100 ppm Ni and 0.5 Cd/Zn ratio should not be applied to agricultural land. On that basis the sludge used in this study contained lower amounts of metals (*see Table 17.12*) than those suggested as being toxic. In addition, another factor has been proposed and defined as the 'zinc equivalent' which equals the concentrations of Zn + 2 Cu + 8 Ni in the

Table 17.14 Elemental composition of the different plant species due to sewage sludge application at the rate of 50 g/kg soil

Element		*Alfalfa*		*Wheat*		*Fababean*		*Soybean*		*Sordan*	
		C	*S*	*C*	*S*	*C*	*S*	*C*	*S*	*C*	*S*
N	%	2.9	4.0	1.7	3.0	3.5	4.1	1.6	3.1	0.7	1.4
P	%	0.1	0.2	0.2	0.3	0.1	0.2	0.2	0.3	0.2	0.3
K	%	1.9	2.1	2.6	4.1	1.4	2.0	1.4	2.6	21	2.1
Fe	ppm	90	173	242	277	247	251	93	101	83	88
Mn	ppm	35	51	24	29	84	101	65	52	28	33
Zn	ppm	101	133	310	579	165	246	45	151	24	29
Cu	ppm	9	16	9	16	15	20	8	9	4.7	5.9
Co	ppm	1.0	1.3	4	4	0.01	1.9	1	3	0.03	3.3
Ni	ppm	0.05	6.7	4	7	0.01	1.9	5	7	0.06	1.4
Pb	ppm	1.6	6.0	5	24	7.7	12.1	2.5	4.8	4.40	5.9
Cd	ppm	0.08	1.3	0.3	1.3	1.0	1.9	1.1	1.5	0.04	0.14

Source: El-Keiy (1983)

C = control; S = sludge

sludge. This value should be lower than 250 ppm in soils of pH > 6.5 (Webber, 1972). In the present work the 'zinc equivalent' value at the rate of 50 g sludge per kg soil equals 77 ppm which is much lower than the critical value. Due to the toxic effect of Cd, a 'metal equivalent' value was proposed by Bingham *et al.* (1979). This value equals the concentrations of 4.55 Cd + Cu and should not exceed 600 ppm in lime or alkaline soil. In this study, the metal equivalent of the sludge used at the rate of 50 g/kg soil was 13.6 ppm, which is very much lower than the critical value. Also, according to the US Environmental Protection Agency (1977) criteria for sewage sludge application, the metal levels used in this study are below the critical limit and there is no hazard from using the sludge produced by the Alexandria East Treatment Plant on agricultural lands.

Effect of sludge on metal contents of soils

The data in *Table 17.15* show that appreciable amounts of metals were added to the soil after the application of 50 g sludge/kg soil. However, the results in *Table 17.16.* show that only a very small portion of the added metals was available to the plants, as indicated by the amounts extracted by a DTPA reagent. Many heavy metals are less available under neutral to alkaline conditions than in acidic conditions (Wang, 1984).

Table 17.15 Total amounts of metals added to the soil from using 50 g sewage sludge/kg soil

Metal	*mg/kg Soil*	*Metal*	*mg/kg Soil*
Fe	1045	Co	0.5
Mn	12.6	Ni	1.8
Zn	37.6	Pb	17.0
Cu	13.0	Cd	0.1

Source: El-Keiy (1983)

Table 17.16 Mean amounts in ppm of DTPA-extractable metals from soils after sludge application and cultivation with alfalfa, wheat, fababean, soybean and sordan

Metal	*Soil treatment*	
	Sludge added	*g/kg Soil*
	0	*50*
Fe	2.8	64.3
Mn	2.9	6.9
Zn	3.5	5.2
Cu	N.D.	7.9
Pb	N.D.	0.5

Source: El-Keiy (1983)

Future reuse of sewage in Alexandria

The sewage effluent of Alexandria is considered as an excellent resource to develop the agricultural production in an area estimated to be about 40 400 ha in the western Nubaria region. The soils of this region are of marine sediments, highly calcareous, very poor in organic matter content and plant nutrients. These soils are of coarse-to-medium textures with good permeability. Almost all the land reclamation projects in the region suffer from shortage of irrigation water.

Table 17.17 Water quality criteria for agricultural irrigation in China

Item	*Criteria Not more than, mg/l*
pH	5.5–8.5
Total salt	1500
Chlorides (as Cl)	300
Sulphide (as S)	1.0
Mercury and its compounds (as Hg)	0.001
Cadmium and its compounds (as Cd)	0.005
Arsenic and its compounds (as As)	0.05
Chromium (VI) and its compounds (as Cr)	0.1
Lead and its compounds (as Pb)	0.1
Copper and its compounds (as Cu)	1.0
Zinc and its compounds (as Zn)	3
Selenium and its compounds (as Se)	0.01
Fluoride (as F)	3
Cyanide (as CN)	0.05
Petroleum	10
Volatile phenol	1
Benzene	2.5
Trichloroacetaldehyde	0.5
Acrolein	0.5

Source: Wang (1984)

The water quality criteria for agricultural irrigation in China (*Table 17.17*) are applicable to sewage irrigation (Wang, 1984). Accordingly, the quality of Alexandria sewage is suitable to irrigate the soils and crops for long periods, except for its salinity hazard. This could be solved through mixing with water from the Nubaria canal. The problem of industrial waste, to be expected in Alexandria sewage, should be considered. However, pollution problems could be alleviated by selecting crops with a low rate of uptake or by selecting non-edible crops. The high soil pH (> 8), plus its high contents of calcium carbonate and gypsum, would make the availability of heavy metals to plants very limited. The sewage treatment method must have a low energy consumption and cost to make the maximum net return. All possible renewable sources of energy must be utilized, including biogas production from the treatment plants.

References

ABDEL-GHAFFAR, A. S. (1982) The significance of organic materials to Egyptian agriculture and maintenance of soil productivity. *FAO Soils Bulletin,* **45,** 15–21, FAO, Rome

ABDEL-GHAFFAR, A. S. (1983) Use of organic amendments and some microbiological aspects of Egyptian desert soils. *Workshop on Uses of Micro-biological Processes in Arid Lands for Desertification Control and Increased Productivity.* Albuquerque and Santa Fe, New Mexico, USA, 6–12 October, 1–46

ABD EL-NAIM, E. M. (1982) Effect of different organic materials on certain properties of the calcareous, sandy and alluvial soils of Egypt and on crop yield. *FAO Soils Bulletin,* **45,** 211–220, FAO, Rome

ABD EL-NAIM, E. M., IBRAHIM, A. E. and EL SHAL, M. E. (1982) A preliminary study on the effect of using sewage water in sandy soils. *Ain Shams University, Faculty of Agricultural Research Bulletin,* **1965,** 1–12

ABDOU, F. and EL-NENNAH, M. (1980) Effect of irrigating loamy sandy soil by liquid sewage sludge on its content of some micronutrients. *Plant and Soil,* **56,** 53–57

BINGHAM, F. T., PAGE, A. L., MITCHELL, G. A. and STRONG, J. E. (1979) Effects of liming an acid soil amended with sewage sludge enriched with Cd, Cu, Ni and Zn on yield and Cd content of wheat grain. *Journal of Environmental Quality,* **8,** 202–207

CHANEY, R. L. and HARNICK, S. B. (1978) Accumulation and effects of cadmium. Conference, San Francisco, California, January–February, 1977. London: *Metal Bulletin Ltd.*

EL-KEIY, O. M. Z. (1983) *Effect of sewage sludge application on soil properties and plant growth.* PhD thesis, Faculty of Agriculture, University of Alexandria, Alexandria, Egypt

EL-NENNAH, M., EL-KOBBIA, T., SHEHATA, A. and EL-GAMAL, I. (1982) Effect of irrigation of loamy sandy soil by sewage effluents on its content of some nutrients and heavy metals. *Plant and Soil,* **65,** 289–292

EL-SHABASSY, A. I., ABDEL-MALIK, S. H., ZIKRIE, B. S., MITKEES, A. I., NAIROOZ, F. I., HASAN, H. M. and ABD EL-NAIM, E. M. (1971) Effect of sewage water on the properties of sandy soils (El-Gabal El-Asfar Farm). *Agricultural Research Review, Cairo,* **49,** 97–116

RIAD, A. (1982) Potential sources of organic matter in Egypt. *FAO Soils Bulletin,* **45,** 22–25, FAO, Rome

SHEHATA, M. M. E. (1983) *Effects of irrigation with sewage water on growth and mineral composition of young grapes, oranges, olives and guavas grown in different soils.* PhD thesis, Faculty of Agriculture, Alexandria University, Alexandria, Egypt

WANG, H. K. (1984) Sewage irrigation in China. *International Journal of Development Technology,* **2,** 291–301

WEBBER, J. (1972) Effects of toxic metals in sewage on crops. *Journal of the Water Pollution Control Federation,* **71,** 404–413

US ENVIRONMENT PROTECTION AGENCY (1977) Municipal sludge management. *February Register,* **42** (211) 57420–57427 US Environmental Protection Agency

Chapter 18

Treatment and use of sewage effluent for agriculture in Cyprus: immediate prospects

V. D. Krentos*

Introduction

During the last 25 years, Cyprus has witnessed a rapid advancement of its urban and rural domestic water supplies to the extent that today the need for further expansion and improvement in *per capita* consumption is met as it arises. However, central sewerage schemes have only recently been implemented and, at present, Stage I of the Nicosia system has been in operation since 1983. Before this date, only communal sewage treatment plants were in operation at Akrotiri, Episkopi and Dhekelia. The oldest sewage treatment plant was established in 1955 at the Nicosia Central Prison to treat sewage from the Nicosia General Hospital. It was of the percolating filter type and the treated effluent was used to irrigate alfalfa, ornamentals, even vegetables and potatoes.

With the implementation of a number of refugee housing schemes, small-scale communal sewage treatment plants have been constructed and some are in operation. In addition, a number of self-contained units serving large hotels and tourist establishments are functioning, some since 1970. The effluent is treated to a high degree of purity and is used for grass-lawn and landscape irrigation.

In recent years, municipal authorities and major communities have been undertaking feasibility studies, and some are at the stage of implementation, for the introduction of central sewerage systems and effluent treatment plants. From the very inception and planning stages of such schemes, strong public awareness underlines the necessity to use treated effluent for irrigation because it serves both as efficient reuse of scarce water resources and easy and useful disposal of wastewater.

In a semiarid country like Cyprus, water shortage can become endemic, while the future demand for water above the present consumption levels will be growing. Under these conditions, the utilization of treated effluent for irrigation or groundwater recharge constitutes an integral part of any sewerage scheme contemplated. Apart from the direct benefits from the use off treated effluent for crop production, its disposal either in dry river beds or into the sea raises strong public objections because of possible environmental pollution hazards. The potential use of treated effluent for agriculture formed the subject of a study by government bodies in the early 1970s. Subsequently, two reports commissioned by

* FAO Consultant (Ex-Director, Agricultural Research Institute, Ministry of Agriculture) Nicosia, Cyprus

UNDP, with FAO acting as the executing agency, dealt specifically with the assessment of water quality, soil and crop suitability for sewage effluent irrigation (Ayers, 1979) and optimum utilization of sewage effluent (Florin, 1980) from the Nicosia treatment plant.

Under the UNDP project GLO/80/004, research and development in integrated resource recovery in Cyprus, two consultants' reports were commissioned. One report dealt with the use of effluent from the Zenon-Kamares II treatment plant as a demonstration project for crop irrigation and/or aquaculture (Hydroconsult, 1983). The other report (Haydar, 1984) dealt with the use of effluent from the Nicosia sewage treatment plant for irrigation.

At present, government agencies are in the final stages of formulating acceptable effluent quality standards, chemical and biological, and establishing a code of practice.

Experimental work on the use of treated sewage effluent for irrigation was initiated in 1984 at the Agricultural Research Institute in Nicosia. The objective is to evaluate the potential for agricultural use of treated sewage effluent based on chemical and biological criteria, nutrient availability, as well as on considerations of soil and plant contamination with heavy metals.

In this paper it is proposed to focus on the need to recycle treated sewage effluent, evaluate the present situation and assess the immediate prospects for extending irrigated agriculture within the overall development of the country's water resources. Since effluent quality depends on the method of treatment, a brief description will be made of the principles involved and the processes employed in existing sewage effluent treatment plants.

Effluent treatment processes

General

The degree of treatment before application is an important consideration in the planning, design and management of irrigation systems for effluent use. Preapplication treatment is essential to protect public health, prevent nuisance conditions during treatment, storage and application, and to prevent damage to crops and soils. Moreover, the physical properties and the chemical and biological constituents of effluent are just as important parameters in the design of collection, treatment and disposal and in the protection of the environment. Some constituents of concern are suspended solids, biodegradeable organic matter measured as BOD, pathogens indicated by total and faecal coliform bacteria, plant nutrients (nitrogen, phosphorus, potassium), heavy metals, dissolved inorganic salts, residual chlorine, etc.

Wastewater treatment in itself consists of a combination of physical, chemical and biological processes and operations to remove or to reduce to an acceptable level solids, organic matter, pathogenic organisms and sometimes nutrients in excess. According to the degree of treatment, a process is described as preliminary, primary, secondary and tertiary or advanced. Disinfection to remove pathogens usually follows the last step in treatment for reuse in irrigation.

Preliminary treatment

This treatment includes coarse screening, comminution of large pieces and grit sedimentation.

Primary treatment

The purpose of primary treatment is to remove settleable organic and inorganic solids by sedimentation. About 50–70% of suspended solids (SS) and 35–50% of the influent biochemical oxygen demand (BOD_5) are removed.

Secondary treatment

In most cases, secondary treatment follows primary as a matter of course and involves the removal of biodegradeable dissolved and colloidal organic matter by aerobic microorganisms in the presence of oxygen in the form of inorganic metabolites such as CO_2, NH_3 and H_2O. During secondary treatment, one of the several aerobic biological processes may be used, depending on the manner in which oxygen is supplied to the microorganisms and on the rate at which organic matter is metabolized. Accordingly, biological wastewater treatment processes may be divided into high-rate and low-rate processes. Examples of high-rate processes include activated sludge, trickling filters (biofilters) and rotating biological contactors. High-rate biological treatment processes in combination with primary sedimentation remove 85–95% of BOD_5 and SS originally present in the influent and a high proportion of the heavy metals. Activated sludge usually produces an effluent of slightly better quality in terms of BOD_5 and SS than biofilters and rotating biological contactors. Followed by disinfection, these processes provide substantial but not complete removal of bacteria and viruses. Low-rate biological processes involve the suspension of low concentrations of microorganisms in large earthen ponds or lagoons. Commonly used low-rate biological processes include aerated lagoons and stabilization ponds. Aerated lagoons have detention times of 7–20 days and water depths of about 2.5m. Oxygen is supplied by mechanical aerators agitating the upper layer. Near the bottom of the lagoon, anaerobic conditions develop which provide for the decomposition of organic settlement by anaerobic bacteria. Stabilization ponds (oxidation ponds) use algae to supply oxygen. Hydraulic detention times range from 20 to 30 days and basin depths from 1.8 to 2.5m. The zone near the bottom is also anaerobic.

Tertiary or advanced treatment

This treatment is employed when a higher degree of purity is required: $BOD_5 < 10$ mg/l, SS < 5 mg/l. Special processes, including coagulation and sand filtration, are needed for removing suspended solids, nitrogen, phosphorus, heavy metals, etc. to satisfy the treatment criteria and minimize the probability of human exposure to enteric viruses when the treated effluent is used for landscape irrigation in parks, school yards and playgrounds.

Disinfection

The disinfection process involves the injection of a chlorine solution in a contact basin. Doses of 5–10 mg/l are normally used at contact times of at least 15 min. However, in instances of specific irrigation use of effluent, a contact time of up to 120 min may be required.

The effectiveness of disinfection is measured in terms of concentration of total coliform or faecal coliform bacteria remaining in the effluent after chlorination. *Table 18.1* provides an indication of the criteria of disinfection required for the various types of irrigation in California.

Table 18.1 Wastewater treatment and quality criteria for irrigation

Treatment level	*Coliform limits*	*Types of use*
Primary		Surface irrigation of orchards and vineyards: fodder, fibre and seed crops
Oxidation and disinfection	≤23/100 ml	Pasture for dairy animals; landscape impoundments, landscape irrigation (golf courses, cemeteries, etc.)
	≤2.2/100 ml	Surface irrigation of food crops (no contact between water and edible portion of crop)
Oxidation, coagulation, clarification, filtration and disinfection	≤2.2/100 ml max. = 23/100 ml	Spray irrigation of food crops; landscape irrigation (parks, playgrounds, etc.)

Source: California Department of Health Services (1978)

Storage

In most cases, although storage is not considered a step in the treatment process, it is nevertheless an essential link between the treatment plant and the irrigation system. Storage facilities balance variations in flow from the treatment plant, meet peak demand, minimize disruptions and further improve quality.

Existing sewage treatment systems

With the exception of small sewage treatment plants, it is only recently that urban sewage schemes have been implemented, or are being contemplated, which embody the use of effluent for irrigation as an integral part of the projects. In addition, hotel establishments mainly in the coastal areas have introduced individual sewage plants, the treated effluent of which is used for landscape and grass lawn irrigation. Four of the most important schemes which trace the development of sewage treatment schemes in Cyprus and incorporate effluent irrigation are briefly described below.

Nicosia Central Prisons sewage treatment plant

This is the oldest plant in operation, having started in 1955. It is of the conventional percolating (trickling) filter type consisting of two primary settlement tanks, two percolating filters and two humus tanks. The effluent, after passing through the humus tanks, is used for irrigation of alfalfa and other forage crops. In earlier times, irrigation of tuber crops such as potatoes was practised. This plant was used to treat both the sewage from the prisons and that of the Nicosia General Hospital. However, after the recent connection of both sewerage systems to the Nicosia central sewerage, the treatment plant ceased to operate.

Akrotiri sewage treatment plant

This plant is also of the percolating filter type, consisting of four primary settlement tanks, three percolating filters, two humus tanks, six sludge drying beds and one final settlement balancing tank. It has been in operation since 1962, treating sewage from the Akrotiri Sovereign Base Area and its performance has been very satisfactory. It produces effluent meeting the Royal Commission Standards. *Table 18.2* summarizes the results of analyses of effluent sampled in June 1983 from different source points of the plant.

Table 18.2 Effluent analysis of samples taken from Akrotiri treatment plant

Source point	*BOD_5,* mg/l	*SS,* mg/l	*Dissolved oxygen,* mg/l	*pH*
Humus tank	55	14	3.3	7.4
Storage tank before chlorination	45	28	1.8	6.9
Storage tank after chlorination	40	18	5.2	7.2
Reservoir mixed with dam water	30	42	10.8	9.4

Source: Department of Fisheries, Nicosia

Treated effluent reuse at Akrotiri

After disinfection with chlorine, treated effluent is used to irrigate the grass lawn of the sports fields at Akrotiri. The total area irrigated by sprinklers is about 10 ha of grass lawn and 2 ha of roundabouts and road hedges irrigated by stand pipes. The available daily flows of treated effluent are of the order of 1500 m^3, sufficient to irrigate about 15 ha. The surplus is discharged into a nearby large reservoir where it mixes with dam water and is subsequently used to irrigate table grapes (furrow) and citrus (minisprinkler).

Analyses for boron carried out in 1973 indicated a boron content of 0.84 ppm. At that time, a few vegetables such as pepper, eggplant, marrow and tomato as well as flowers and ornamentals were grown satisfactorily, with the exception of some toxic symptoms on marrows and tomatoes, then ascribed to boron toxicity; no ill-effects were noticed on other vegetables or flowers, nor on citrus and vines. However, for the first year of establishment of landscape ornamentals, fresh water was used for irrigation.

Zenon-Kamares II sewage treatment plant

This is typical of a number of treatment plants established to serve government housing estates for refugees. It was completed in 1984 and it is now in operation but not at its design capacity. Eventually, it will serve an estimated population of 2600 at 250 litres per caput day (lpcd) and a BOD_5 load of 68 g per caput day. The treatment is the activated sludge process with contact stabilization producing an effluent within the Royal Commission Standards of 20 mg/l BOD_5 and 30 mg/l SS. In addition, there is tertiary treatment consisting of an upward flow gravel clarifier to achieve 15 mg/l BOD_5 and 15 mg/l SS. After chlorination, the effluent is pumped to a balancing reservoir for sprinkler irrigation of the grass lawn of a football stadium. This treatment plant was recommended as a demonstration project under UNDP GLO/80/004.

In another refugee housing estate, a similar process of activated sludge with extended aeration and incorporating tertiary treatment, chlorination and a balancing storage reservoir is used to treat effluent used for irrigation of alfalfa for a nearby dairy farm.

Nicosia sewerage system

This is the largest scheme and is being implemented in three stages. Stage I is already completed and Stage II is in the process of implementation. The treatment plant consists of two lined aerated lagoons with a total area of 0.8 ha, followed by four unlined facultative lagoons in series, of 1.4 ha each, and a chlorination facility at the river outflow. The plant is now in operation and the daily flow is about 5000 m^3 with an outflow into the river of about 3500 m^3.

In Stage II, the working capacity of the treatment plant will increase to 10 000 m^3/day and, at Stage III, to 15 000 m^3/day.

The utilization of effluent for irrigation in an area adjacent to the Nicosia sewage treatment plant formed the subject of a comprehensive study in 1973 which, on the basis of the anticipated effluent quality EC_w and B content, recommended the use of reclaimed water for irrigation of forage crops such as alfalfa, sudan grass, rhodes grass and forage barley on a pilot area of about 100 ha, including properly designed field experiments for research and demonstration purposes.

More recently, the use of treated effluent for irrigation formed the subject of a report by Ayers (1979), commissioned by FAO. The emphasis was placed on two main quality parameters, namely EC and boron content of the effluent. Further reference to the findings and recommendations of this report will be made in later paragraphs.

Another study by Florin (1980) financed by UNDP dealt with the very subject of optimum utilization of treated effluent from the Nicosia sewage treatment plant. The recommendations in this report substantially corroborated those proposed in 1973 by the *ad hoc* committee of the Ministry of Agriculture and Natural Resources, namely that salinity and boron-tolerant forage crops could be irrigated. At present, the treated effluent, after chlorination, is discharged into the Pedieos river bed.

Limassol and Larnaca central sewerage systems

Similar urban sewerage schemes, but employing faster rate treatment processes, are being implemented by the municipalities of Limassol and Larnaca. Treated effluent of acceptable biological and chemical quality will be used for crop irrigation.

Treated effluent quality standards

To a large extent, the quality of sewage effluent depends on the quality of the domestic water supply, the nature of the waste added during use and the degree of treatment of the influent. Although groundwater from different sources may vary, in general the quality for potable purposes is acceptable for irrigation. However, another factor affecting the quality of the effluent from sewage treatment plants, such as that of Nicosia, is the intermittent supply (every 48 h) in the summer months. The direct effect of this water economy measure is to increase the salinity of the final effluent.

Chemical quality

The higher the concentration of dissolved salts in effluent used for irrigation, the greater the chance of creating problems to the physical properties of soils and crops. These problems may arise either from the growth of salinity or specific ion concentration or excessive concentration of one or more trace elements. These problems, however, are not specific to treated effluent but they also, as usual, apply to fresh water irrigation.

Guidelines for evaluating water quality for irrigation have been put forward by the University of California Committee of Consultants (1974) and by Ayers and Westcot (1985). A reference to their contribution to water quality criteria in Pettygrove and Asano (1984) will suffice for the purposes of this chapter.

Salinity affects the uptake of water by crops, while toxicity from a specific ion such as boron affects sensitive crops. Nitrogen, bicarbonate, pH and residual chlorine have other effects on susceptible crops. The degree of restriction on use depends on the magnitude of these parameters taken singly or in combination with the other factors.

In the absence of analytical data of the effluent from the Nicosia sewage treatment plant, Ayers (1979) evaluated the inflow water as having an average salinity of EC_w 1.40 mmhos/cm, and adjusted SAR of 14, boron 0.3–0.5 mg/l and chloride 8.3 mg/l. Under these conditions, moderately tolerant crops should be chosen for irrigation. However, the effluent water from the facultative lagoons under Nicosia conditions could be expected to be in the range of EC_w of 2.0–2.5 mmhos/cm. Such a salinity in the effluent used for irrigation would probably create greater management problems. Even today no representative data on effluent analyses are available, but EC_w values of 2.5 mmhos/cm or even higher would not be unrealistic, at least during intermittent domestic water supplies.

Data on relative salt and boron tolerance in supplies have been published by Maas (1984) and would apply to a great extent to Cypriot conditions.

From experience so far, the choice of crops to be irrigated with treated effluent should be restricted to moderately tolerant to tolerant to both salinity and boron. The choice of more sensitive crops would require a very careful consideration of all factors, including quality of influent supplies, land suitability and, above all, irrigation management. Leaching, for example, would seem to be an appropriate

Table 18.3 Plugging potential of irrigation water used in drip systems

Type of problem	*Restriction on use*		
	little	*slight/moderate*	*severe*
Physical			
Suspended solids (mg/l)	50	50–100	>100
Chemical			
pH	7.0	7.0–8.0	>8.0
Dissolved solids (mg/l)	500	500–2 000	>2 000
Manganese (mg/l)	0.1	0.1–1.5	>1.5
Iron (mg/l)	0.1	0.1–1.5	>1.5
Hydrogen sulphide (mg/l)	0.5	0.5–2.0	>2.0
Biological			
Bacterial population maximum number/ml	10 000	10 000–50 000	>50 000

Source: Nakayama (1982)

practice in most instances. The choice of the irrigation method would also play an important role in overcoming problems. For example, some constituents in the treated effluent could make drip irrigation problematical. *Table 18.3* lists the types of problems likely to arise from the use of an effluent through a drip irrigation system.

Treated sewage effluent contains high concentrations of plant nutrient elements, such as nitrogen, phosphorus and potassium. These are largely from sewage, while natural contents in fresh water are increased and range between 20 and 60 mg/l. The form of nitrogen, however, whether ammoniacal, nitrate or organic, depends on the degree and type of treatment. Although nitrogen is essential and beneficial to plant growth, particularly during the early stages of growth, excessive and, in the case of treated effluent irrigation, uncontrollable dosages promote profusive vegetable growth, lodging, delayed maturity and inferior produce quality. According to the guidelines for interpretation of water quality, for irrigation, it is anticipated that irrigation water containing 5–30 mg/l total nitrogen would present slight to moderate restrictions on its use, while concentrations above 30 mg/l of total nitrogen would certainly be problematic.

Phosphorus in the effluent from secondary treatment may range between 5 and 15 mg/l, a concentration hardly adequate for Cypriot calcareous soils, which are naturally poor in available phosphorus and total phosphorus. Supplemental fertilization is necessary in irrigated soils, should the available phosphorus be below the adequacy level of the particular crop.

Cypriot soils are generally rich in potassium supply. Fresh water supplies also contain potassium, while effluent would be expected to contain higher concentrations.

Treated effluent contains sufficient boron to satisfy crop nutrient needs. Of concern, however, for Cyprus is excess of boron, emanating from household detergents containing perborates, causing toxicity and consequently reduced crop yields. Treated effluent in most cases in Cyprus contains around 1.0 mg/l boron.

No data are available on the trace element contents of treated effluents. However, there are initial indications that irrigation with effluent may correct zinc deficiencies. Toxicities from heavy metals are not aniticipated because in Cyprus there are very few cases of industries releasing effluents containing appreciable amounts of these metals.

Biological standards

The use of effluent for crop irrigation requires that the water be treated effectively to remove all pathogens and viruses which would cause serious health hazards. As discussed earlier, the sewage treatment processes and/or operations should produce an effluent of high quality, low BOD, low SS and be free from coliform bacteria and viruses. Although the limited experience of using treated effluent for irrigation over the years has not created serious problems, yet there is a strong awareness from the public, health authorities, and environmentalists of the direct and indirect hazards involved. Experience elsewhere indicates that such fears are exaggerated and that properly designed, managed and maintained sewage treatment plants produce effluent of good quality. Continuous monitoring of the quality and strict compliance to a sound code of practice for the safe use of effluent for crop irrigation is essential.

After long experience and experimentation with the use of effluent for irrigation in countries with similar climatic and agricultural conditions, effluent quality criteria have been established for four groups of crops (*Table 18.4*). In addition, the World Health Organization assembled the standards for the use of treated effluent for irrigation in force in four countries (*Table 18.5*).

In Cyprus, an interdepartmental Technical Working Committee was set up in 1984 to study the issue of sewage effluent use in all its aspects and to propose appropriate standards. So far, the Committee has tentatively proposed the following quality standards for effluent used for amenities of limited access:

$BOD_5 < 20$ mg/l
SS < 30 mg/l
Total coliforms MPN $< 1000/100$ ml
Free residual chlorine 0.2 mg/l minimum
Combined residual chlorine 1.0 mg/l minimum

For amenities of unlimited access, such as public parks and playgrounds, effluent quality standards will be more severe. Furthermore, the committee proposes the following code of practice which will ensure the safest possible use of treated effluent for irrigation purposes:

- effective operation and maintenance at all times by a skilled operator to ensure smooth and uninterrupted operation;
- daily recording of all operations performed;
- suitable apparatus for daily measurement of EC_w, pH and residual chlorine to be available;
- the choice of irrigation systems be restricted to subsurface, drip and surface methods; for spray irrigation a 300 m buffer zone be allowed;
- outlets, taps and valves in the irrigation system be of the type that can be secured and clearly marked to prevent use by the public; and
- cross-connection between pipelines conveying sewage effluent and potable water be avoided.

Research and development in sewage effluent utilization

During recent years, the utilization of treated sewage effluent for irrigation has attracted increased research attention resulting from the awareness of its usefulness for crop production and as a convenient and environmentally acceptable means of disposal. The beneficial effects of sewage effluent from additions of nitrogen, phosphorus, micronutrients and organic matter had already been recognized. More recently, it has been demonstrated (McCaslin and Lee-Rodriquez, 1977) that sewage effluent provides nutrients more efficiently than commercial fertilizers. Striking results have been obtained from soils deficient in micronutrients and particularly in iron.

In arid and semiarid regions, water remains the limiting factor in crop production and treated sewage effluent is a potential source of irrigation water. Although the use of treated effluent for agriculture is widely practised in a number of developed countries, central sewerage schemes and treatment plants are only beginning to emerge in developing countries.

The use of treated effluent and sewage sludge raises environmental hazards such as groundwater pollution and the build-up of toxic materials in the environment (CAST, 1976; Sanks and Asano, 1976; Jacobs, 1977; Elliot and Stevenson, 1977). Another serious problem, to which much research effort has been devoted, is the

Table 18.4 Criteria for effluent use for crop irrigation[1]

Constituents (mg/l)	*Group A* *Cotton, sugar beet plant for seed production*	*Group B* *Olives, groundnuts, fruits where the skin is not eaten, grass for animals*	*Croup C*[2] *Gardens, vegetables for cooking, fruit for the conservation industry*	*Group D*[3] *Other plants (unlimited irrigation)*
Effluent quality:[4]				
BOD_5 – total	60	45	35	15
BOD_5 – filtered	–	–	20	10
Suspended solids	50	40	30	15
Dissolved oxygen	0.5	0.5	0.5	0.5
Coliform count/100 ml	–	–	250	12 (80%) and 2.2 (50%)
Residual chlorine	–	–	0.15	0.5
Mandatory treatment:				
Sand filtration	–	–	–	required
Chlorination – minimal contact time, minutes	–	–	60	120
Distances (m):				
Residual areas	300 (sprinkling)	250	–	–
Paved roads	30	25	–	–

[1] The criteria are not limited to the treatment method, however for oxidation ponds they should be adapted to a retention time higher than 15 days
[2] Irrigation should be terminated two weeks before the pick-up season and no fruit which falls on the ground should be taken
[3] Observation and examination frequency in accordance with the irrigation season
[4] Sample test should be satisfactory at least in 80% of the cases

Table 18.5 Existing standards governing the use of renovated water in agriculture

Agricultural use	*California*	*Israel*	*South Africa*	*Federal Republic of Germany*
Orchards and vineyards	Primary effluent, no spray irrigation, no use of dropped fruit	Secondary effluent	Tertiary effluent heavily chlorinated where possible. No spray irrigation	No spray irrigation in the vicinity
Crops for human consumption that will be processed to kill pathogens	For surface irrigation, primary effluent. For spray irrigation, disinfected secondary effluent (no more than 23 coliform organisms per 100 ml)	Vegetables for human consumption not to be irrigated with renovated wastewater unless it has been properly disinfected (1000 coliform organisms per 100 ml in 80% of samples)	Tertiary effluent	Irrigation up to 4 weeks before harvesting only
Crops for human consumption in a raw state	For surface irrigation, no more than 2.2 coliform organisms per 100 ml. For spray irrigation disinfected, filtered wastewater with turbidity of 10 units permitted provided it has been treated by coagulation	Not to be irrigated with renovated wastewater unless they consist of fruit that is peeled before eating		Potatoes and cereals, irrigation through flowering state only

Source: World Health Organization (1973)

survival of pathogens and viruses in sewage effluent after irrigation. In reviewing the subject, Shuval (1974) concluded that the most prudent policy to achieve maximum social benefits from effluent use is to combine low-cost treatment processes, which reduce pathogens to reasonable levels, with restriction of crops to those presenting the lowest possible public health risk. Among the irrigation methods, drip irrigation in spite of the high initial cost presents the least possible contamination hazards.

A significant development in the proper use of sewage effluent has been the publication in 1984 of the guidance manual on *Irrigation with Reclaimed Municipal Wastewater*, published by the California State Water Resources Control Board (Pettygrove and Asano, 1984). This publication compiles a wealth of information on practically all aspects of sewage effluent utilization emphasizing the beneficial use of reclaimed wastewater. Apart from providing a wide literature coverage, it focuses on water quality, crop water use, crop selection and management, public health aspects, irrigation system design, etc. This manual would be of particular interest to those involved in the planning, design and operation of agricultural and landscape irrigation systems using treated sewage effluent.

In 1978, UNDP financed a study for the optimum utilization of sewage effluent from the Nicosia treatment plant for irrigation purposes. Subsequently, based on the quality and availability of effluent for irrigation, it was recommended that salt-tolerant and boron-tolerant forage crops such as barley, alfalfa, maize and sudan grass be grown. Eucalyptus was another new alternative proposed (Florin, 1980) which, on financial feasibility grounds, could wholly replace forage production which involved high investment costs.

With the UNDP global project GLO/80/004 titled 'Research and Development in Integrated Resource Recovery in Cyprus', two concurrent feasibility studies were commissioned by the World Bank. One dealt with the alternatives for using effluent from the Zenon-Kemares II treatment plant as a demonstration project for irrigation of wheat-berseem, olives and eucalyptus and sportsground grass lawns. Another attractive use of effluent was for fish culture. It was feasible to combine the crop alternatives and fish culture into an integrated demonstration project (Hydroconsult, 1983). The other feasibility study (Haydar, 1984) suggested the implementation of the project recommended in 1980 by Florin.

The experimental work on the use of treated sewage effluent initiated in 1984 at the Agricultural Research Institute is of specific relevance to the demand for a sound basis of local knowledge and experience in utilizing wastewater for crop production. Preliminary results, indicating a clear superiority of treated sewage effluent over fresh water, were obtained in 1984 using cotton as the test crop. During the current irrigation season the test crops will include alfalfa, sudax sunflower and cotton (Papadopoulos, 1985, personal communication). Apart from the beneficial effects on yield, the effects of the effluent on soils, crops and the environment will be evaluated by continuous monitoring through analyses of the effluent, chemical and biological, and soil and plant analyses.

Sewage effluent reuse potential

The need for using treated sewage effluent for irrigation must be viewed as an integral part of the planning and development of the water resources of the island and as an efficient means of recycling scarce water resources on economic, social and environmental grounds.

Water resources development

In an average year, it is estimated that rainfall accounts for about 4600 Mm^3 (million m^3) of water falling over the whole of the island on a total area of 9251 km^2. The maximum surface runoff in an average year is estimated to be about 600 Mm^3, of which 400 Mm^3 is lost to the sea. Groundwater from wells and boreholes is extracted at the rate of about 400 Mm^3 for crop irrigation, domestic, industrial and tourist purposes. However, the safe yield of the aquifers is only about 350 Mm^3, leaving a deficit of 50 Mm^3 in any one year. Of the total rainfall, 350 Mm^3 are being developed through groundwater and springs, leaving 600 Mm^3 of surface runoff available for potential development. The development of surface water resources involves the impounding of runoff water in dams, and conveyance of water to areas to be irrigated or to water treatment works for domestic supplies. By the end of 1984 the total dam storage capacity reached 151 Mm^3, and by 1987 this capacity will be increased by an additional 140 Mm^3 (Department of Water Development, 1984).

By the end of this decade, roughly one-third of the total surface runoff of 600 Mm^3 will be developed. This will allow an increase in the area of irrigated land of 16 000–60 000 ha from the present 44 000 ha. The additional water will also relieve aquifer deficits and supplement domestic water supplies.

Irrigation policy

The scarcity of the water resources has been the major constraint to the impetus of agricultural development and increased crop production. Groundwater resources, on which both domestic and agricultural supplies depended, were being exhausted by exploitation beyond the rate of natural recharge. In devising policies for development of the water resources of the island, the main strategy is to satisfy present and future demands by the various competing sectors by:

- the development of the surface water resources in an integrated and coordinated manner;
- the control of groundwater extraction to a safe limit for the aquifer;
- the achievement of economies in domestic water supplies and, in particular, high irrigation efficiencies in the agricultural sector, which is by far the largest water consumer; and
- the reuse, to the maximum possible, of treated effluent for irrigation purposes.

Recognizing the key role of water as the means for increasing agricultural production, although acknowledging the necessity to augment domestic water supplies, the government allocated over 60% of the development budget for the agricultural sector as a whole, for water development and increased efficiency of use. Any increase in the present irrigated area is wholly dependent on the extent to which surface water resources are developed.

With the parallel impounding of runoff water in dams, the increase in irrigated land, past, present and future, is shown in *Table 18.6*.

Table 18.6 Trends in irrigated agricultural land

	1978	*1980*	*1985*	*1990*	*1995*
Area under irrigation (ha)	43 610	48 000	52 900	56 060	60 000

Source: Department of Water Development (1981)

Domestic water supply

The major source of water for domestic supplies is groundwater pumped from boreholes or free-flowing from springs, mainly in mountainous areas. At present, the only surface water used for domestic purposes is the water stored in the Lefkara dam and treated at the Khirokitia waterworks, at a present annual output of about 6 Mm^3 supplementing the water supplies of Nicosia, Larnaca, Famagusta, as well as a number of rural communities within reach of the main conveyor pipeline. With the completion of the Vassilikos-Pendaskinos Project, the water impounded in the Kalavassos and Dhypotomos dam will be partly used for domestic supplies after treatment at the Khirokitia water works, and the newly constructed Kornos treatment works. In this way, the water supply of Nicosia, Larnaca and Famagusta and a number of villages will be further augmented and will more or less satisfy the projected demand in the early 2000s.

Urban water supplies
With the exception of the town of Paphos, the urban distribution of water for Nicosia, Larnaca and Limassol, is the responsibility of water boards.

Nicosia For the Nicosia supply in particular, the operation and maintenance of the pumping units on boreholes and of the Khirokitia treatment plant is the responsibility of the Department of Water Development which sells water to the Nicosia water board at a price. In turn, the Board fixes the rates chargeable to the consumers. The total annual consumption in 1984 for the whole of Nicosia was of the order of 9 500 000 m^3 at an average daily consumption of 26 000 m^3. Because of shortages of water, particularly during the summer months, the supply is cut off and resumed after 40 h. This has had an economizing effect; the annual consumption for the last 5 years remaining, to all intents and purposes, unchanged in spite of increases in the number of household consumers. Based on these figures, the average daily consumption is 130 lpcd (litres per caput day).

Limassol The town of Limassol provides for its own water supply from boreholes in neighbouring river aquifers and also from natural springs. Total annual production from these sources amounted to 7 831 000 m^3 in 1984. Limassol has a number of manufacturing industries as well as a thriving tourist industry. At an estimated total population of 140 000 including tourists, the average daily consumption is of the order of 160 lpcd.

Larnaca The water supplies of Larnaca come from boreholes in neighbouring river aquifers owned by the Water Board, as well as from the main conveyor pipeline from Khirokitia water treatment works. Total supply from these two sources was 2 900 000 m^3 in 1984. The total local population is estimated to be 40 000 but increases considerably during the summer months from the influx of tourists. Average daily supply in 1982 was a little less than 8000 m^3 and the average daily consumption somewhat less than 200 lpcd.

Paphos The local population of Paphos is around 13 000 but increases to over 16 000 during the tourist season. The main sources of water supply are boreholes owned by the Municipality of Paphos and located in the river aquifer upstream of the Asprokremnos dam. These boreholes can produce 8000 m^3/day. In 1984, the supply was 1 465 000 m^3; on average, daily supply was 4000 m^3 and the average daily consumption was 250 lpcd.

Rural water supply
The water supply in rural areas comes either from springs, for villages in mountainous areas, or from groundwater sources. In 1982, the total rural population was estimated at 186 700. At an estimated average daily consumption of 130 lpcd, the total annual consumption was 8 858 000 m^3. With the migration to urban areas, the rural population remains constant in spite of increases from births. There are, however, rural areas where tourist development increases consumption to higher levels.

Industry
The average water consumed by industrial enterprises is estimated at the equivalent of 12 lpcd or 7200 m^3/day for a population of 600 000, with an estimated total annual consumption of 2 620 000 m^3.

Tourism
Water consumption in the tourist sector varies according to the category of tourist dewelling from 250 to 600 lpbd (litres/bed day). Assuming an occupancy of 75%, the average daily consumption is about 400 lpbd. The total number of beds in approved tourist establishments in 1984 was 25 700, thus the total daily consumption is 10 280 m^3 or about 3 750 000 m^3 annually.

Present and future prospects for irrigation with sewage effluent

Summarizing the data in the previous paragraphs concerning the supply of water to urban and rural areas and the industrial and tourist sectors, it emerges that the total water consumption adds up as follows:

Urban	
Nicosia	9.50 Mm^3
Limassol	7.83
Larnaca	2.90
Paphos	1.46
Rural	8.86
Industry	2.62
Tourism	3.75
Total consumption 1984	36.92 Mm^3

Future projections based on population increases, but mainly on an increase in the average daily consumption, and the increase in tourism, place the annual consumption for Nicosia at 16.3 Mm^3, Limassol 12.1 Mm^3, Larnaca 4.0 Mm^3, and Paphos 3.0 Mm^3.

For the purpose of assessing the water reuse potential through effluent recovery for irrigation, and assuming only the present day level of consumption, it may be estimated that the regional consumption will be as follows:

Nicosia district	15.00 Mm^3
Limassol district	12.00
Larnaca, Famagusta districts	7.00
Paphos district	3.00
	37.00 Mm^3

Assuming a maximum possible water recovery of 90% and taking into consideration the limitations in agricultural use, proximity of effluent treatment plants to suitable soils, etc., an overall useful water recovery of 70% could be achieved, namely:

Nicosia district	11.00 Mm^3
Limassol district	8.00
Larnaca district	5.00
Paphos district	2.00
Total	26.00 Mm^3

With the recovery of 26.00 Mm^3 annually of effluent suitable for irrigation of a cropping pattern consisting of 50% fruit trees, 20% ornamentals and grass lawns, 15% table vines and 15% forage crops with an irrigation efficiency of 70%, the average annual irrigation requirements would be in the region of 10 000 m^3/ha.

Thus the extent of land which could be irrigated from treated sewage effluent, at modest estimates, would be:

Nicosia district	1100 ha
Limassol district	800
Larnaca district	500
Paphos district	200
Total	2600 ha

This area represents about 6% of the total land presently under irrigation and would contribute, to a significant degree, to irrigated crop production. The economic impact would be considerable when one takes into account the very high cost of developing surface water resources.

Acknowledgements

The author wishes to express his sincere thanks to the professional staff of the Department of Water Development, the Department of Agriculture, the Agricultural Research Institute, the Department of Town Planning and Housing, the Environmental Conservation Service and the Nicosia Sewage Board, who provided most of the technical background information.

References

AYERS, R. S. (1979) Use of sewage effluent for irrigation in Cyprus. *Report on a Consultancy in 1979. FAO Report AGDP/CYP/78/003* Rome

AYERS, R. S. and WESTCOT, D. W. (1976) Water quality for agriculture. *Irrigation and Drainage Paper No. 29.* FAO, Rome

AYERS, R. S. and WESTCOT, D. W. (1985) Water quality for agriculture. *Irrigation and Drainage Paper No. 29. Rev. 1,* FAO, Rome

CALIFORNIA DEPARTMENT OF HEALTH SERVICES (1978) Wastewater reclamation criteria. *California Code. Title 22, Div. 4, Environmental Health.*

CAST (1976) *Application of Sewage Sludge to Crop Plant: Appraisal of Potential Hazard of Heavy Metals to Plants and Animals. Report No. 64.* Ames, Iowa: Council for Agricultural Science and Technology, Iowa State University

DEPARTMENT OF WATER DEVELOPMENT (1981) *Irrigation in Cyprus.* No. 1/31, Nicosia

DEPARTMENT OF WATER DEVELOPMENT (1984) *Major Water Development Works.* Nicosia.

ELLIOT, L. F. and STEVENSON, F. J. (1977) *Soils for Management and Utilization of Organic Wastes and Wastewaters.* Madison, Wisconsin: American Society of Agronomy

FLORIN, R. (1980) Optimum utilization study of sewage water for irrigation in Nicosia. *Report on a Consultancy.* FAO, Rome,

HAYDAR, A. (1984) *Effluent Irrigation from Nicosia Sewage Treatment Plant.* UNDP GLO/80/004

HYDROCONSULT (1983) *Alternative Use Systems: Kamares Sewage Effluent.* UNDP GLO/80/004

JACOBS, L. W. (1977) *Utilizing municipal sewage wastewaters and sludges on land for agriculture production.* East Lansing, Michigan: Michigan State University Publication No. 52

MAAS, E. V. (1984) Salt tolerance of plants. In: *The Handbook of Plant Science in Agriculture,* edited by B. R. Christie. Boca Raton, Florida: CRC Press Inc. (Cross-reference from Westcot and Ayers)

McCASLIN, B. D. and LEE-RODRIQUEZ (1977) Effect of using sewage effluent on calcareous soils, pp. 598–614. *Scientific paper No. 20,* New Mexico State University

NAKAYAMA, F. S. (1982) Water analysis and treatment techniques to control emitter plugging, pp. 97–112. *Proceedings of the Irrigation Association Exposition Conference,* Portland, Oregon

PETTYGROVE, G. S. and ASANO, T. (1984) (eds). *Irrigation with Reclaimed Municipal Wastewater – A Guidance Manual.* Report No. 84–1 wr. Sacramento: California State Water Resources Control Board

SANKS, R. L. and ASANO, T. (1976) *Land Treatment and Disposal of Municipal and Industrial Wastewater.* Ann Arbor, Michigan: Ann Arbor Science Publications

SHUVAL, H. I. (1974) Disinfection of wastewater for agricultural utilization. *7th International Conference on Water Pollution Research,* Paper No. 7c(ii) 9–13

WESTCOT, D. W. and AYERS, R. S. (1984) Irrigation water quality criteria. In: *Irrigation with Reclaimed Municipal Wastewater – A Guidance Manual,* edited by Pettygrove, G. S. and Asano, T. Report No. 84–1 wr. Sacramento: California State Water Resources Control Board

WHO (1973) *Reuse of Effluents: Methods of Wastewater Treatment and Health Safeguards. Report on WHO Meeting of Experts.* Geneva: WHO Technical Report Series No. 517

Chapter 19

Ongoing research on the use of treated sewage effluent for irrigating industrial and fodder crops

I. Papadopoulos and Y. Stylianou*

Introduction

Agriculture in Cyprus depends on irrigation and, because conventional water resources are seriously depleted, it faces the serious challenge of improving upon or at least maintaining crop productivity while coping with less and/or lower quality water available for irrigation. The problem is an ancient one but, recently, due to the scarcity of water in Cyprus, its high cost and the increasing demands for fresh water for other purposes, alternative water resources must be found and innovative approaches will help solve the problem. Thus, there is a need to develop more efficient irrigation methods to save water as well as to use some unconventional water resources for agricultural purposes. Irrigating with secondary-treated sewage effluent which will be available soon in considerable quantities (15–20 million m^3/year) as a product of biological treatment plants, is such an alternative innovative approach of particular interest to Cyprus.

The use of effluent may increase agricultural productivity not only because of its quantity but also because of its high nutrient content (Hershkovitz and Feinmesser, 1967; Noy and Feinmesser, 1977; Sollenberger, 1981), and it may further improve with the use of advanced, efficient irrigation methods such as trickle irrigation systems (Overman and Evans, 1978; Solomon and Keller, 1978; Oron, Ben-Asher and DeMalach, 1982). It has been demonstrated (McCaslin and Lee-Rodriquez, 1977) that sewage effluent is even superior to recommended commercial chemical fertilizer treatments. Surprising results have been obtained, particularly from calcareous soil known to be deficient in micronutrients, especially iron (Fe). Adding to this, in arid and semiarid areas it is the cheapest and, in some cases, the only water available for irrigation, so it appears that treated sewage effluent is an important potential source of irrigation water that cannot be ignored.

Apart from the expansion of irrigated agriculture, eventual environmental pollution will also be minimized because the treated effluent will have to be disposed of, and its use for irrigation is considered by far the best option under our conditions. In addition to these, the establishment of sewage treatment plants at a number of housing estates and the planned major sewage treatment projects at Larnaca and Limassol pose a pressing need for research on the use of such treated effluent for irrigation. For these reasons, treated municipal effluent has recently

* Agricultural Research Institute, Nicosia, Cyprus

received increased research attention by the Agricultural Research Institute, with the main objective of finding ways for reusing this 'valuable' effluent for irrigation rather than disposing of it in other ways. However, the widespread application of sewage effluent is not without serious problems and is sometimes limited by certain economic, technical and public health uncertainties about the impacts of such use. There is concern that the quality of foodstuffs, land resources and groundwater resources may be partially impaired due to residual contaminants present in the treated effluent.

Trace heavy metals, soluble salts and other toxic constitutents in the effluent may be dangerous because of phytotoxicity and biomagnification – the tendency for many plants to absorb and accumulate some toxic substances which, in food and fodder crops, could present a hazard to human and animal health. Salinity, however, is presumably the most important single water quality factor which affects both soil and crop production. Therefore, an important factor to be considered before use is that the application of sewage effluent should be such as to prevent a build-up of soluble salts and toxic materials in the soil, affecting thereafter crops and polluting groundwater (Jacobs, 1977; Westcot and Ayers, 1984).

A key problem that has been of even more concern is the survival of pathogens and viruses in sewage effluent and their effects on the crops being irrigated. This is of particular concern when the crop is to be consumed raw by humans. Pathogenic protozoa, bacteria, viruses and other organisms can enter plant tissues in a variety of ways and may even be found in the edible portion of the plant. However, irrigating such crops with high quality effluent, in order to avoid contamination, might not be the best solution because of the high cost of treating such effluent. A balanced approach, combining treatment methods capable of providing acceptable purification of wastewater, the use of advanced methods of irrigation (trickles and minisprinklers), with restriction of crops to those presenting a low level of public health risk (industrial and fodder crops), as suggested by Shuval (1974), appears to be the most prudent policy and therefore of priority in our research in order to achieve the maximum social and economic benefits from wastewater reuse. Trickle irrigation has the advantage of lower possibility of contamination (Romanenko, 1970) and, although Goldberg (1976) demonstrated that with such a system even untreated sewage is absolutely safe, research under our local conditions is needed since a serious problem associated with trickle irrigation is emitter clogging, caused by chemical and biological build-up in the minute water passageways (Oron, Shelet and Turzynski, 1979).

The experimental project undertaken by the Agricultural Research Institute is the first step for evaluating, under Cypriot conditions, the agricultural use of treated sewage for irrigation and fertilization, and furthermore to provide information concerning soil, underground water and plant contamination, with contaminants present in the effluent, such as pathogens, viruses, NO_3-N and heavy metals. The latter, although not necessarily a serious problem everywhere, will be thoroughly examined, because soil once contaminated with heavy metals, according to our present-day technology and knowledge, cannot be cleaned. Such information will be helpful for extending the use of such effluents for irrigating crops intended particularly for animal feeding. Yet, for the maximum possible benefit to be obtained, it is still absolutely necessary that the study of the impacts of sewage effluent on soils, plants, animals, humans and the environment be approached in an integrated way for conditions in Cyprus, which lacks experience in these matters. Only then can the results obtained be evaluated and act as guidelines for accepting and expanding the reuse of sewage effluents. In line with these ideas, intensive interdisciplinary experimental work has been designed in Cyprus to be undertaken at two locations where secondary-treated sewage effluent is already available.

This chapter reports some of the preliminary results obtained during the first year (1984) of an ongoing study using secondary-treated effluent to irrigate cotton. The objective of the phase of the study reported here was to study the effect of effluent, supplemented with nitrogen, on soil and crop yield. Results from the 1985 experiments concerning the influence of municipal effluent on the soil, N-nutrition, growth, yield and lint quality of drip irrigated cotton will be presented. The influence of effluent on growth, N-nutrition and yield of sunflower and sudax (*Sorghum vulgare x S. sudanensis*) will also be presented.

Materials and methods

A long-term irrigation experiment was initiated in 1984 on a calcaric lithosol soil (35% clay, 28% silt, 37% sand, 26% $CaCO_3$) to study the influence of treated municipal effluent on the soil and on N-nutrition, growth and yield of drip-irrigated crops. At the beginning of the experiment the surface 60 cm of soil contained 0.48 g/kg total N, 7.3 mg/kg of $NaHCO_3$-extractable P, and 270 mg/kg of exchangeable K. The mean values of NO_3-N to a depth of 120 cm are given in *Table 19.1*. In 1984 the test crop was cotton (*Gossypium hirutum* cv. Sindos 1980), whereas during 1985, besides cotton, sunflower (*Helianthus annunus* cv. HS-52) and sudax (*Sorghum vulgare x Sorghum sudanensis*) were included in the test crops.

The experimental site is near Nicosia, with an arid climate typical of the eastern Mediterranean basin in general. In dry years, the area receives less than 200 mm of total annual rainfall between November and April and during the dry months of the year regular irrigation of crops is needed. The mean monthly maximum temperatures reach as high as 35°C in July–August with evaporation during these months reaching 10 mm/day (Metochis, 1977).

Table 19.1 Residual soil NO_3-N as influenced by sewage effluent (S) and fresh well water (F) at four nitrogen levels*

Depth (cm)	*Initial*	*N rates* (mg/l)							
		0	*30*	*60*	*90*	*0*	*30*	*60*	*90*
		Effluent				*Fresh water*			
0–15	13	4	6	15	25	3	3	3	10
15–30	12	2	7	18	30	2	4	4	15
30–45	16	3	6	20	35	4	5	5	23
45–60	19	2	8	18	38	6	6	6	25
60–75	23	17	14	25	40	10	17	22	23
75–90	24	35	32	28	42	29	24	25	20
90–105	23	24	21	20	20	37	30	21	19
105–120	23	22	18	17	22	15	16	12	18

* Each value on the table is the arithmetic average of four soil samples

Table 19.2 Chemical composition of fresh water and secondary-treated effluent

Water	*EC_w* (dS/m)	*SAR*	*pH*	*Chemical composition* (mol ±/m³)							
				Ca	*Mg*	*Na*	*K*	*HCO_3*	*CO_3*	*SO_4*	*Cl*
Fresh	3.0	11.3	8.8	1.8	5.4	21.7	0.2	4.4	1.6	7.5	15.5
Sewage	2.6	12.0	9.2	2.2	3.1	19.6	1.2	3.7	3.6	5.3	13.4

Two sources of irrigation water are used, secondary-treated sewage effluent and fresh well water (*Table 19.2*), and four concentrations of N($N_0 = 0$, $N_1 = 30$, $N_2 = 60$ and $N_3 = 90$ mg N/l) are applied continually as ammonium nitrate in the irrigation water. An activated sludge treatment plant adjacent to the experimental site supplies effluent averaging 20–30 mg N/l, 15–20 mg P/l and 23–27 mg K/l. The biochemical oxygen demand (BOD_5), chemical oxygen demand (COD) and suspended solids (SS) were on the average during the 1984 irrigation season 40, 75 and 55 mg/l, respectively. The most probable number of coliform counts was variable. There are no heavy metals or organic toxicants at concentration levels sufficient to present a problem, because wastewater is derived from a residential community with no industrial inputs. A split plot design with four replications was used for all crops, with either wastewater or well water assigned to the main plots and N concentrations to the subplots, consisting of three rows 30 m long. Spacing between rows was 1.0, 0.8 or 0.7 m for cotton, sunflower or sudax, respectively. Planting of cotton in 1984 was rather late (early June) and many bolls, especially of treatment N_3 irrigated with effluent, were not mature by mid-November when harvesting was concluded. Fertilizer treatment began 3 weeks after emergence and plants were irrigated at a pressure of 1.2 atm three times a week, based on evaporation from a class A pan (Epan). For cotton, in 1984 a total of 595 mm of effluent or well water were applied in 32 irrigations during the season.

Soil samples were taken from each main plot to a depth of 120 cm in 15 cm increments prior to initiation of the experiment, to measure soil background conditions. Thereafter samples were taken the same way below the tricklers semiannually (before planting and after harvesting). Leaf samples were also taken during the growing period.

Results and discussion

Soil NO_3-N

The total amount of N applied to drip-irrigated cotton with a cumulative amount of 595 mm of water was 0, 179, 357, or 536 kg N/ha for the N_0, N_1, N_2 or N_3 treatments, respectively.

Analysis of soil samples collected to a depth of 120 cm before planting and below the tricklers after the first irrigation season showed significant main treatment effects on soil NO_3-N concentration (*see Table 19.1*). The roots were apparently very active in removing N, as evidenced by the soil depletion of NO_3 that occurred in the treatments supplied with low N concentrations. With the well water, in particular, there was even depletion from the soil profile of NO_3-N with the fertilizing concentration of 60 mg N/l. Alternatively, the restricted downward movement of NO_3 together with the depletion of N obtained in the N_0, N_1 and N_2 treatments irrigated with the well water suggest that concentration of N up to 60 mg/l was inadequate for cotton irrigated with well water. Contrary to this, with sewage effluent the 60 mg N/l were sufficient and the soil profile was not depleted of NO_3-N. Significant NO_3 build-up occurred below 60 cm only in plots irrigated with effluent supplied with the highest N level. With the well water there was a slight NO_3 accumulation in the zone of 0–60 cm with the highest N level, which in general was very similar to that which occurred with the sewage effluent supplied with 60 mg N/l. Therefore, fertilizer N was used very efficiently with sewage effluent when continually applied up to 60 mg N/l, whereas with the well water, there was efficient use of N up to 90 mg/l. Evidently, the same results indicate that N naturally occurring in the effluent is a valuable nutrient source, which with fresh water should be given as fertilizer. However, the NO_3 accumulation at greater depths, when effluent is supplied with excess of N, poses a potential groundwater pollution hazard.

Soil salinity

The initial and final salinity profiles corresponding to each N level (*Table 19.3*) indicate a significant effect of both water and N levels on soil salinity. Apparently, the 595 mm of water applied to all treatments provided greater leaching in the plots fertilized with the lower N levels, because the plants in the N_0 and N_1 treatments were not sufficiently supplied with N and grew less. Because of this, more water was extracted by the plant roots at the higher N levels, due to their extensive vegetable growth, resulting thereafter in more salt accumulation in the surface soil profile, probably because less water penetrated deeper into the soil profile, out of the rooting zone. As a result, there was a general increase in soil salt build-up with N which, however, was significantly higher with the fresh well water. With the latter water, irrespective of the N treatments, the salt build-up to a depth of 45 cm, as indicated by the mean values in *Table 19.3*, was approximately twice as much as occurred with the treated effluent, although the difference in EC_w between the two waters was only 0.4 dS/m (*see Table 19.2*).

The significantly higher increase in soil salt build-up when well water was used is due mainly to the higher concentration of Na and Cl in soils irrigated with this water (*Table 19.4*). Furthermore, with the well water, the concentration and

Table 19.3 EC_e in dS/m as influenced by treated effluent and fresh well water at four nitrogen levels

Depth (cm)	*Initial*	*N rates* (mg N/l)									
		0	*30*	*60*	*90*	*mean*	*0*	*30*	*60*	*90*	*mean*
		Effluent					*Fresh water*				
0–15	5.6	2.6	2.2	3.7	2.8	2.2	4.5	7.7	7.9	6.6	6.7
15–30	4.4	2.1	2.4	3.5	2.8	2.7	3.8	6.4	7.3	7.0	6.1
30–45	4.9	2.1	2.8	5.1	4.1	3.5	2.7	4.6	5.9	6.9	5.0
45–60	3.8	2.2	2.8	5.2	4.8	3.7	2.8	3.0	6.4	4.4	4.1
60–75	2.6	2.9	2.8	5.2	4.2	3.7	2.5	3.7	6.0	4.3	4.1
75–90	2.3	2.8	3.3	4.0	4.3	3.7	3.4	4.7	4.7	4.2	4.3

Table 19.4 Chemical composition of soil saturation extract as influenced by fresh water and treated effluent

Depth (cm)	*SAR*	*Chemical composition* (mol (±)/m³)							
		Ca	*Mg*	*Na*	*K*	*HCO₃*	*CO₃*	*SO₄*	*Cl*
Initial									
0–15	12.4	14.6	6.2	40.0	0.6	1.2	–	16.4	35.0
15–30	11.0	12.8	4.8	32.6	0.4	1.6	–	17.4	25.0
30–45	10.3	14.8	6.2	33.5	0.2	1.2	–	14.0	36.5
45–60	9.8	10.8	4.0	27.0	0.1	1.0	–	11.2	27.0
60–75	9.4	6.4	2.6	20.0	0.1	1.0	–	10.8	13.0
75–90	9.6	5.0	2.2	18.3	0.1	1.0	–	7.2	14.5
Fresh									
0–15	15.6	12.4	6.0	47.2	0.6	4.0	–	13.1	46.4
15–30	14.1	11.7	7.0	43.0	0.5	2.6	–	15.9	45.1
30–45	14.6	8.4	5.8	38.9	0.2	2.0	–	14.3	36.1
45–60	10.4	9.8	4.9	28.2	0.1	2.8	–	13.9	27.2
60–75	11.1	8.2	4.0	27.3	0.1	1.6	–	18.4	28.5
75–90	11.0	11.0	4.2	26.4	0.1	1.2	–	9.6	32.5
Sewage									
0–15	8.3	4.0	3.0	15.6	0.7	4.0	–	8.0	10.0
15–30	12.3	4.8	2.4	23.4	0.4	3.2	–	8.8	13.5
30–45	13.8	7.6	3.8	33.0	0.1	2.4	–	10.2	27.0
45–60	11.9	8.0	3.6	28.7	0.1	2.4	–	10.8	24.0
60–75	11.5	8.4	3.2	27.8	0.1	2.4	–	8.4	26.0
75–90	11.0	8.4	4.0	27.3	0.1	2.2	–	15.0	20.5

fraction of Na became progressively higher relative to that of Ca and Mg, with a consequent increase in soil solution SAR (*see Table 19.4*). In the 1984 irrigation season, hydraulic conductivity was not evaluated, but it has been observed that, although irrigation was with tricklers, there was surface runoff indicating a probable decrease in water permeability. Contrary to this, with the sewage effluent there was a certain leaching of Na deeper into the soil leading to lower, generally acceptable, SAR at the surface 15 cm. Because of this, better infiltration of water was observed with the effluent.

In general, there was an improvement in soil salinity with the effluent (*see Table 19.4*). Highly soluble salts were leached from the soil surface (0–30 cm) to deeper layers and an EC_e was developed close to that of the effluent used. Therefore, although the initial EC_w of the effluent was relatively high (2.6 DS/m), soil salt build-up below the tricklers was prevented and was kept below the salinity level that cotton can tolerate. This suggests that the fraction of 0.55 of Epan used for irrigating cotton was sufficient when effluent was used.

Leaf analysis

Mean NO_3-N, P and K concentrations of laminae and petioles are given in *Table 19.5*. Evidently N and P in the treated effluent provided fertilizer benefits to the crop. P was, in general, of higher concentration in both laminae and petioles in the

Table 19.5 Chemical composition of cotton laminae and petioles as influenced by fresh well water (F) and sewage effluent (S) at four nitrogen levels

Water	*Constituent*					
	NO_3–N		*P*		*K*	
	F (mg/kg)	*S* (mg/kg)	*F* (g/kg)	*S* (g/kg)	*F* (g/kg)	*S* (g/kg)
Laminae						
N_0	333c	349d	3.12a	3.50a*	17.6c	18.0b
N_1	454b	541c*	3.28a	3.66a*	20.8b	23.0a
N_2	551a	594b*	3.27a	3.61a*	23.7a	23.3a
N_3	620a	701a*	2.94b	3.22b*	24.9a	25.0a
Petioles						
N_0	1561c	1624c	1.77a	1.82bc	47.6b	43.4b
N_1	3522b	3593b	1.84a	2.02a*	52.4a	50.5a
N_2	5536a	5712a*	1.78a	1.90b*	53.6a	53.8a
N_3	5624a	5957a*	1.55b	1.70c*	51.5a	53.0a

Means within columns followed by the same letter are not significantly different at 0.05 level of probability based on Duncan's Multiple Range Test.

* Indicates differences at 5% level between F and S for each element at the same nitrogen level

treatments irrigated with effluent. K was not affected, most probably because of the high exchangeable K of the soil (270 mg/kg). The apparently insignificant difference in the NO_3-N concentration in petioles of the N_0 and N_1 treatments is due to the dilution factor, because plants at these N levels irrigated with effluent grew taller with more vegetable growth and gave higher yields than did cotton irrigated with well water. It may be concluded that the nutrients in treated effluent are valuable, and some fertilizer amounts may be saved by the use of effluent for irrigation.

Crop yields

The seed cotton yield obtained in the first season averaged 3.19 and 3.66 tonnes/ha for the fresh and sewage effluent waters, respectively (*Table 19.6*). Yield was higher with the effluent, particularly at the lower rates of N, which shows the value

Table 19.6 Yield in kg/ha of lint cotton irrigated with fresh water or sewage effluent at four nitrogen levels (4000 plants/ha)

Irrigation	*N rates* (mg N/l)			
	0	*30*	*60*	*90*
Fresh	2775b*	3075b	3503b	3398a
Sewage	3585a	2788a	3900a	3360a

* Means within rows followed by the same letter are not significantly different at 0.05 level of probability based on Duncan's Multiple Range Test.

of treated effluent as a source of N. The relatively low yield obtained in 1984 with all treatments may be attributed to spacing (2.5 plants/m^2) but also to the delay of about 45 days in planting (early June). During 1985, yield is expected to increase in relation to that of the 1984 season, because plant density is about five plants/m^2 and planting was in time (29 April). A decrease in yield at the highest N level with both waters might have resulted also from late planting. Results from this year's experiments will be valuable for explaining both the low yield obtained in general in 1984 and also the decrease in yield at the highest N level.

General aspects

By using trickle irrigation, to avoid exposure to wind-blown effluent of the people working at the experimental site, and with certain although minimum precautionary measures (using gloves, and wearing rubber boots) no adverse health effects resulted. Also, there was no unacceptable odour from the storage pond where the effluent was collected before irrigation, or even during irrigation.

A serious problem, however, associated with trickle irrigation in arid zones, particularly when applying effluent, is emitter clogging caused by chemical and biological build-up in the minute water passageways (Oron, Shelet and Turzynski, 1979). Under our conditions and with the quality of effluent used, by installing a gravel filter no severe filtering problem occurred and there was no unusual emitter clogging.

Acknowledgements

The authors thank Dr C. S. Serghiou, the Director of ARI, for his constant support during this work, and the assistance of the professional and technical staff of the Soils and Water Use Section.

References

GOLDBERG, S. D. (1976) *New Techniques in the Reuse of Effluent.* Report. Washington DC. World Bank

HERSHKOVITZ, S. Z. and FEINMESSER, A. (1967) Utilization of sewage for agricultural purposes. *Water and Sewage Works,* **114,** 181–184

JACOBS, L. W. (1977) *Utilizing Municipal Sewage Wastewaters and Sludges on Land for Agricultural Production.* East Lansing, Michigan: Michigan State University Publication, no. 52

McCASLIN, B. D. and LEE-RODRIQUEZ, V. (1977) *Effect of Using Sewage effluent on Calcareous Soils.* Scientific Paper no. 20. Alberquerque, New Mexico: New Mexico State University, pp. 598–614

METOCHIS, C. (1977) *Potential Evaporation of Lucerne. Technical Bulletin 21,* Cyprus: Agricultural Research Institute

NOY, J. and FEINMESSER, A. (1977) *The Use of Wastewater for Agricultural Irrigation. Water Renovation and Reuse.* New York: Academic Press, pp. 73–92

ORON, G., SHELET, G. and TURZYNSKI, B. (1979) Trickle irrigation using treated wastewaters. *Journal of Irrigation and Drainage Division, ASCE,* **105,** 175–186

ORON, G., BEN-ASHER, J. and DeMALACH, Y. (1982) Effluent in trickle irrigation of cotton in arid zones. *Journal of Irrigation and Drainage Divsion, ASCE,* **108,** 115–126

OVERMAN, A. R. and EVANS, L. E. (1978) Effluent irrigation of sorghum X.3, sudangrass and Kenaf. *Journal of the Environmental Engineering Division, ASCE,* **104,** 1061–1066

ROMANENKO, N. A. (1970) On helminth eggs in crops grown on sewage farms. *Hygiene and Sanitation, USA,* **35(11),** 257–259

SHUVAL, H. I. (1974) DIsinfection of wastewater for agricultural utilization. *7th International Conference on Water Pollution Research.* Paper No. 7c(ii) pp. 9–13

SOLLENBERGER, G. (1981) Farming with higher-priced fertilizer: it's back to basics. *The Furrow,* **6,** 12–15

SOLOMON, K. and KELLER, J. (1978) Trickle irrigation uniformity and efficiency. *Journal of the Irrigation and Drainage Division, ASCE,* **104,** 293–306

WESTCOT, W. D. and AYERS, S. R. (1984) Irrigation water quality criteria. In: *Irrigation with Reclaimed Municipal Wastewater Guidance Manual,* edited by G. S. Pettygrove and T. Asano. Report No. 84-1 wr. Sacramento, California: California State Water Resources Control Board

Chapter 20

Mexican experience in using sewage effluent for large scale irrigation

Nicolas Sanchez Duron*

Introduction

The Number 3 Mezquital Irrigation District of Mexico was established in 1945 and is located in the southwestern part of the State of Hidalgo, between 19°53′ and 20°30′ latitude north and between 98°59′ and 99°38′ longitude west. It is roughly north of Mexico City.

The Irrigation District has a mean annual rainfall of 494 mm, a large part of which falls during the months of July, August and September. Average temperature is 17°C, with a minimum of −3°C, and therefore 18 days with frost per annum, during the wintertime. May is the warmest month of the year.

Under these climatic conditions irrigation is indispensable in order to achieve productive agriculture. This has been possible in the Mezquital area largely through the use of Mexico City's sewage effluent for irrigation; otherwise it would not be possible to carry out the type of agriculture that now exists in this area, since the annual rainfall is not only limited but also poorly distributed over the growing season.

The availability of irrigation water and the existence of favourable climatic conditions for agriculture and animal husbandry have definitely contributed towards solving the social and economic problems in this area.

Although the use of sewage water for irrigation in the Mezquital Valley began in 1886, it was not until 1945 that the Ministry of Hydraulic Resources (Secretaria de Recursos Hidraulicos) established the Number 3 Irrigation District. Historically, in 1671, the external drainage channel of Nochistongo was excavated in order to control the flow of water stored in Zumpango Lake. The water is conducted by the Cuautitlan River in the State of Mexico to the Tula River in the State of Hidalgo. In 1890, the old tunnel of Tequixquiac was built to connect the Grand Canal and the conducting river and to be the most important outlet for the drainage and sewage water flowing out of the Valley of Mexico. In 1943, the new Tequixquiac tunnel was constructed in order to complement the capacity of the old Tequixquiac tunnel.

More recently, in 1962, the western drainage outlet was put into service to conduct the sewage water and rainfall runoff from the Valley of Mexico into the Nochistongo drainage channel. In 1967, the construction of a new deep drainage tunnel, 49.742 km long and 6.5 m in diameter with a capacity of 200 m^3/s, was

* Ministry of Agriculture and Water Resources, Mexico, DF

initiated and completed in 1975. This new deep tunnel partially replaces the Grand Canal which is now of insufficient capacity due to the population growth in the urban area of Mexico City.

Water quality

In general, the sewage effluent (mixed with rainfall runoff) and clean water from local dams are used for irrigation in the Mezquital Valley. As can be seen in *Table 20.1*, not only the values for turbidity, pH and electrical conductivity vary for the different waters, but also their colour, odour and sediments are variable. However, they are relatively uniform throughout the year. The values for turbidity, colour, odour and sediments correspond to individual samples. The pH and conductivity values are based on an average of 15 determinations. The pH values are slightly above neutrality and the values of electrical conductivity occasionally reach a maximum of 2.250 mmhos/cm, which indicates a salinity hazard. However, the major part of the crops grown in the district are tolerant to the existing range of electrical conductivities.

Table 20.1 Characteristics of irrigation waters

Source	*Turbidity*	*Colour*	*Odour*	*Nature of sediments*	*pH*	*Electrical conductivity (EC)* (mmhos/cm)
Central outlet (sewage + rainfall runoff)	26.25	clear	odour-free	–	7.50	0.590
Tequixquiac tunnel (sewage + rainfall runoff)	224.25	yellow	septic	organic	7.55	1.500
Requena Dam (clean water)	3.00	clear	odour	none	7.20	0.360
Endho Dam (clean water)	10.00	clear	septic	–	7.05	0.700
Irrigation canals (with mixture of all available water)	104.30	clear; whitish-yellowish; greenish-yellow	odour-free	none	7.05	1.470

Values shown in *Table 20.2* indicate variations between 259 and 1063 ppm for dissolved solids; the values for grease content vary between 2.26 and 23.00 ppm. Only a single value is reported for organic carbon content. However, evidence does exist that the organic matter counteracts the detrimental effects of exchangeable sodium content in the soils. The organic matter improves the physical conditions of the soil by acting as a source of energy that benefits the metabolism of soil microorganisms, which induce aggregate stability of the soil particles, increase soil porosity and diminish the values of apparent density.

Table 20.2 Solids, grease, carbon and sulphonates content of irrigation water

Source	*Dissolved solids* (ppm)	*Grease* (ppm)	*Organic carbon* (ppm)	*Alkyl benzene sulphonates* (ppm)
Central outlet (sewage + rainfall runoff)	410	2.26	–	7.30
Tequixquiac tunnel (sewage + rainfall runoff)	–	–	–	–
Requena Dam (clean water)	259	–	–	2.79
Endho Dam (clean water)	442	4.66	–	15.95
Irrigation canals (mixture of all available water)	1063	23.00	6.85	15.52

Table 20.3 Other characteristics of the water used in this irrigation district

Source	*Effective salinity (ES)*	*Possible salinity (PS)*	*Sodium adsorption relation (SAR)*	*Residual carbonate (RSC)*	*Possible percentage of sodium (PPS)*	*Biochemical oxygen demand (BOD)*	*USDA water classification*
Central outlet (sewage + rainfall runoff)	1.87	1.60	1.13	0.01	86.09	0.34	C_2S_1
Tequixquiac tunnel (sewage + rainfall runoff)	9.57	6.60	5.07	2.74	88.81	2.01	C_3S_2
Requena Dam (clean water)	1.75	1.39	1.14	0.00	69.71	0.39	C_2S_1
Endho Dam (clean water)	2.78	2.50	0.72	0.00	42.44	0.28	C_2S_1
Irrigation canals (mixture of all available water)	8.89	6.48	3.62	0.52	75.19	0.66	C_3S_2

Most crops cultivated in the area are tolerant to the range of values of alkyl-benzene sulphonates shown in the table. However, other organic chemical products may cause new problems of contamination in the future.

With regard to exchangeable cation content, it is pertinent to point out that the Ca and Mg contents of the water, 5.93 and 15.19 me respectively, are above the contents of Na and K, 3.63 and 7.00 me respectively, while the exchangeable anions HCO_3 (3.47 me/l) and Cl (5.50 me/l) predominate over the CO_3 (1.88 me/l), SO_4 (2.41 me/l) and NO_3 (4.16 me/l) contents. These data indicate that the sewage effluent will have little effect on soil salinity, if it is properly managed.

Factors with a favourable effect are the length of the conveyance of the sewage effluent before it is used for irrigation, the existing good internal soil structure that favours drainage conditions, as well as the high calcium content of the regional soils and subsoils. All these contribute to preventing the accumulation of high concentrations of soluble salts and of exchangeable sodium in the soil, which could otherwise be harmful to crops.

All the values reported in *Table 20.3* show a variation that can be attributed to the diverse proportions of sewage effluent mixed with rain runoff and clean water available in the dam storage systems, or to the conditions prevailing in the irrigation canals. However, the residual sodium carbonate content is low. The USDA classification of the water falls between medium and high salinity hazard, and the values for sodium adsorption ratio (SAR) also indicate the danger of sodification. Therefore, theoretically, this water should not be used for irrigation purposes without special management practices. It is a fact, however, that this water has been used for irrigation successfully for many years without serious problems. Possibly the explanation for this lies in the long distance that the sewage effluent runs before being used, and consequently its prolonged exposure to oxidation.

Detailed economic studies will be required, and also evaluation of the diverse social factors, to establish the feasibility and necessity for adequate sewage treatment, taking into consideration the nature of the natural resources involved.

Effect of heavy metals and trace elements

The eight elements studied were found to exist in concnetrations above those tolerated in irrigation water; they are shown in *Table 20.4*. The high levels of elements found, combined with the presence of pathogenic microorganisms, could represent serious problems in the Irrigation District. However, it has been found that these heavy metals precipitate as insoluble compounds and the concentrations of soluble forms of Mn, Fe, Cu and Zn in the soil are within permissible limits.

Table 20.4 Elements contained in irrigation waters

Element	*Limit of tolerated concentration* (me/l)	*Concentration found in water* (me/l)
As	0.10	0.49
B	0.75	2.83
Cd	0.01	0.03
Cu	0.20	0.35
Cr	0.10	0.21
Fe	5.00	12.90
Mn	0.20	0.32
Mo	0.01	1.44

Even though it is recognized that the accumulation of toxic elements may continue, it is believed that a dynamic equilibrium has been reached in the soils. For example, the majority of the regional crops and animals are tolerant or semitolerant to the present boron content in the irrigation water.

Effects of human health

Contamination with faecal materials is common in Mexico, and constitutes a serious aspect of air pollution, not only in the Valley of Mexico but also in the Mezquital Valley. Consequently, the situation prevailing in the Mezquital Valley is not an exception. The existing natural conditions plus the proximity in which the people live, together with all kinds of domestic animals, could favour the propagation of different kinds of diseases among the people. However, restrictions have been established by the Ministry of Health (Secretaria de 'Salubridad', now of 'La salud') regarding the consumption of fresh vegetables produced in this region, and these are applicable to: celery (*Apium graveolens L.*), watercress (*Nasturtium officinale L.*), coriander (*Coryandrum sativum L.*), cabbage (*Brassica oleracea L.*), lettuce (*Lactuca sativa L.*), carrot (*Daucus carota L.*), and other vegetables. Such restrictions are also applied to diverse fruits in the effort to avoid many diseases. However, it has not been possible to eliminate the occurrence of intestinal infections in spite of the application of existing preventive measures.

The vegetable and fruit crops consumed raw present the greatest health hazards to man and facilitate the spread of parasitic infections by viruses, fungi, bacteria, protozoa, etc., found in the soils and the water used for irrigation.

In this region, 90% of the houses have safe drinking water (the other 10% obtain their safe drinking water from public taps). As well as eletricity, however, only 10% have sanitary drainage, 5% are equipped with domestic latrines and the remainder (85%) use the open fields as toilets. Under these conditions, it is common to find widely spread infections of *Enterobacter escherichia coli* and *Enterobacter aerogenes*. Since almost all houses have running potable water, it is recommended as a common practice that all vegetables are washed thoroughly with running water before they are eaten or that they are treated with solutions of commercial disinfectants that have a base of chlorine salts.

The parasitic agents that most frequently contaminate water and soil, and cause disease, are:

- parasitic pathogenic microorganisms, excreted by humans and transmitted by direct contact with contaminated water or soil;
- parasitic pathogenic microorganisms of animals, transmitted to humans by direct contact with water and soil contaminated by detritus from infected animals;
- native parasitic pathogenic microorganisms existing normally in some soils which are transmitted to humans by contact with those soils.

The inhabitants of the Number 3 Irrigation District are living in a habitat that is unfavourable to human health due to the presence of pathogenic microorganisms existing in the local agricultural and animal products. Not only rapid demographic growth but also the bacterial and chemical pollution of the water and soils, and the overgrazing of pasturelands, are responsible for the existing environmental contamination. On the other hand, the regional use of sewage effluent has been an important factor in the production of food crops that are used not only locally but also contribute to the satisfaction of national needs.

In other words, the use of sewage has permitted raising the standard of living where previously only poverty prevailed, providing occupation to many people, markedly increasing the land values due to irrigation of the regional dry-land areas and reducing farmer immigration, by producing higher crop yields and improving farm family incomes. Under these circumstances, the regional endemic diseases are considered a matter of secondary importance by the farmer population. Consequently, it is not enough to determine which programmes can have a marked influence in improving the sanitary conditions in a given region, it is also necessary that the whole population acquire adequate consciousness that such programmes are in their interest and need the permanent support of the people themselves to be effective.

Sewage treatment

No special treatment is given to the sewage that flows toward the Mezquital Valley. It is assumed that during the course of its lengthy transit to this region, as well as in its passage through several waterfalls, thorough oxidation occurs and there is relatively good homogenization of the water.

Effect on soil fertility and productivity

The yields of the majority of crops harvested in the agricultural year are now greater than the yields obtained 10 years ago, except for pasture. On this basis, it is believed that fertility conditions, if measured on the results of productivity, are better than in past years.

During the present year the population of the 16 municipalities that constitute the Number 3 Irrigation District is 300 000 inhabitants. During the agricultural year 1983–84, 52 175 ha were harvested, with a production of 2 226 599 tonnes of food crops, with a value of 8 269 454.074 pesos (or approximately US$ 33 077 816).

The Tepatepec and Progreso soil series are *in situ* soils and represent approximately 89% of the total area of the Mezquital region; the Actopan and Lagunillas soils series form the remaining 11%. The different geological conditions that prevail in the region known as the Valley of Mezquital can be grouped as follows:

- the sediments of clastic materials that have fluvial and alluvial origins, local lens of volcanic ash and limestone. Considering their physical and lithological characteristics they are classified as having high permeability;
- the rocks constituted principally by clastic materials, with lens of volcanic ash and lava spreads, that have medium permeability;
- the tertiary formations that show up almost everywhere in the area, integrated by lava spreads with ash associations of basaltic and andesitic composition, volcanic rocks with a composition that varies between riolite and basalt, with local deposits of clay, silt and limestone; the formations are considered to be impermeable;
- towards the north and southwest of the area, limestone and marble limestone predominate. These formations have low permeability.

It is believed that the high content of organic matter and plant nutrients in the sewage have improved the physical and chemical properties of the shallow soils in the Number 3 Irrigation District.

The generally good conditions for soil drainage, the calcareous nature of the regional soils, their abundance of calcium carbonate and the light and medium textures prevalent, have given rise to a prosperous and stable agriculture in the region. It is also necessary to consider another important factor that may have contributed to the success obtained. The excess irrigation water applied to different crops, mainly to alfalfa, which is rotated with other crops, has increased the soil organic matter and systematically leached the soils, preventing the accumulation of soluble salts.

The volumes of water used for irrigation in the last five agricultural years have been the following:

1979–80 1 200 915 500 m^3
1980–81 1 018 657 500 m^3
1981–82 1 418 098 100 m^3
1982–83 1 127 740 000 m^3
1983–84 1 125 593 000 m^3

The principal sources of the water have been the dams, the Salado river, the central outlet or deep drainage of the metropolitan area of Mexico City, the springs of the Cerro Colorado and the Lagunilla drain. Sewage water constitutes approximately 80% of these volumes.

The continuous utilization of these lands for crop production has prevented farmers from abandoning them. The agricultural development and the crops grown have provided food for both the people and animals living in the local area as well as in Mexico City. On these soils, there is animal husbandry without a high degree of organization, the main species being sheep, cows, goats, swine, horses and

Table 20.5 Crops, areas harvested and yields in irrigation district number 3

Crops		*Area harvested* (ha) *and yield* (kg/ha)			
		1970–71	*1975–76*	*1980–81*	*1983–84*
Corn (maize)	Harvested (ha)	17 914	21 023	17 907	18 371
	Yield (kg/ha)	3 938	3 896	4 566	4 581
Beans	Harvested (ha)	1 266	1 222	1 646	1 028
	Yield (kg/ha)	1 259	1 768	1 521	1 430
Wheat	Harvested (ha)	7 293	2 634	2 005	399
	Yield (kg/ha)	1 919	3 119	3 225	3 134
Alfalfa	Harvested (ha)	12 708	15 206	20 339	19 515
	Yield (kg/ha)	95 300	89 154	91 175	96 481
Oats	Harvested (ha)	2 998	691	1 002	2 489
	Yield (kg/ha)	18 150	19 898	32 470	25 348
Barley	Harvested (ha)	–	832	1 812	1 268
	Yield (kg/ha)	–	19 620	19 939	16 823
Pastures	Harvested (ha)	13	11	65	109
	Yield (kg/ha)	142 500	107 000	44 276	93 832

donkeys. The Health Department has recommended that none of the vegetables produced in this region be consumed raw unless previously washed with adequate germicides and that the milk always be pasteurized.

The main crops harvested in the Number 3 Irrigation District are shown in *Table 20.5*. For the years shown in the table, foodcrops represent 54, 50, 42 and 37.8% of the harvested area. Although the yields of maize and beans can possibly be increased, they are already higher than the national average for these crops. In the case of wheat, it is possible to improve the yield markedly, since the average national yield is now close to 4.0 tonne/ha. The forage crops represent 32, 34, 45 and 44.7% of the total harvested area in the Mezquital Valley for the same groups of years. Safflower is at present being investigated experimentally to determine the best varieties, dates of planting, rates of planting, number and frequency of irrigations, etc.; other oil crops, such as rape and sunflower are also under research. The vegetable crops, such as Italian squash, chili peppers, tomatoes, husk tomatoes etc., cover approximately 10% of the harvested area. Food, oil, forage and vegetable crops comprise 93% of the regionally harvested crops, and flower and fruit crops the remaining 7%.

Conclusions

The development of agriculture and animal husbandry in the region has been a very important factor in keeping the farmers on their lands, by giving economic occupation to the regional rural population. Favourable social and economic conditions, as well as the characteristics of the regional soils, including good drainage, and the fact that the hydrological basins of the Valley of Mexico and the Mezquital Valley complement each other, combined with climatic conditions encouraging agricultural activities, have permitted the rapid development of the Number 3 Irrigation District. In previous years, local farmer associations were responsible for the distribution of the sewage and rainfall that drain out of the Valley of Mexico, and which have been utilized in the Mezquital Valley since the end of the last century. The most serious problems in this Irrigation District are related to pathogenic microorganisms, heavy metal contamination and the salinity hazard.

Recommendations

To utilize the available sewage effluent for agricultural and animal production, constant supervision and interdisciplinary studies are required. Such activities can be developed under the coordination of a responsible agency, such as the Minsitry of Agriculture and Hydraulic Resources. At the same time, technicians must be trained in the use of the proper analytical equipment which must be available at a strategic site in the Irrigation District. Such equipment should be used for physical, chemical and biological analyses of the water, soils and plants, as well as of animals and animal products produced in this region.

Every effort must be made to make people continuously aware of the existing problems and to participate in the measures required by health programmes that will be of benefit to them. To achieve the desired results, it is necessary to continue research related to human health in the region, recognizing that the Ministry of

Health has strictly prohibited the consumption of raw vegetables which come in contact with the soils and irrigation water in this district. Finally, it is considered necessary to evaluate socially, technically and economically the need for adequate treatment of the sewage effluent used for irrigation.

Bibliography

ARTEAGA, I. de H., ZAVALA, J. T., SCHETTINO, P. M. S. and JIMENEZ, C. P. (n.d.) *Determinacion de la contaminacion fecal en frutas y verduras de mercados de la Ciudad de Mexico.* Mexico, DF

CAMARGO, I. O. (1985) Diversas informaciones verbales e informes y materiales proporcionados en las oficinas del distrito de riego No. 03 en Mexquiahuala, HGO

DURON, N. S. (1980) *Perspectivas y evolucion agicola de Mexico.* Mexico DF

DURON, N. S., TORRES, E. O, ZAPATA, R. V. y GONZALES, R. C. (1962) Secretaria de Agricultura y Ganaderia. Traduccion: 'Department of Agriculture – USDA'. Instituto Nacional de Investigaciones Agricolas. *Diagnostico y Rehabilitacion de 'Suelos salinos y sodicos'.* Mexico, DF

HERNANDEZ, J. T. (1981–82) Instituto Mexicano del seguro Social. Sistema IMSS – Coplamar. Delegacion regional Hidalgo. Zona II Pachuca 1. *Estudio de Comunidad. Unidad medico rual 101.* Colonia Morelos, Municipio de Mexquiahuala, HGO

MARQUEZ, H. M. (n.d.) *Tratamiento por aplicacion al suelo. Una solucion viable al manejo de las aguas residuales del area metropolitana del valle de Mexico.* Director ejecutivo de Eco-Ingenieria SA Mexico, DF

MATA, G. F., (1981) Secretaria de Agricultura y Recursos Hidraulicos. Direction General de Estudios – Subdireccion de Agrologia. *Clasificacion de la capacidad del uso de la tierra segun el SCS – USDA.* Mexico, DF

MORENO, J. C. (1980) Secretaria de Agricultura y Recursos Hidraulicos. Comision daguas del valle de Mexico. *Irrigacion de tierras agricolas con aguas residuales en el valle de el Mezquital, HGO.* Trabajo presentado en el Simposio Internacional sobre Renovacion de Aguas Residuales Municipales para Reuso en Sistemas Agricolas e Industriales. Hacienda Cocoyoc, Mor

Chapter 21

Planning and strategy of wastewater and wastewater reuse in Jordan

Saqer El-Salem* and Mahmoud Talhouni†

Introduction

During the last few years sewerage system networks in Jordan have been rapidly developing and expanding. In the year 2000, Jordan will have 36 treatment plants serving about 65% of the total population of the Kingdom. The wastewater effluent from all these plants will be about 60 million m^3/year of treated wastewater.

As water in Jordan is a scarce commodity, the reuse of municipal and industrial wastewaters will be one of the major sources of water for irrigation purposes in the near future. For this reason, it has been stipulated that the quality of the treated water must achieve public health and agricultural requirements. By using this effluent for restricted irrigation, an area of about 30 000 ha can be irrigated, allowing the production of about 6000 tonnes of barley and about 32 000 tonnes of alfalfa.

General information

Population

The population of the East Bank of Jordan (*Figure 21.1*) according to the 1979 Census was 2.15 million, with a net growth rate of about 3.8%/year. The largest percentage of the population is urban, with about 50–75% living within a 50 km radius of the centre of Amman, the capital of Jordan.

Area

The total area of the Kingdom is 94 800 km^2. Only 9% of this area receives more than 200 mm of rainfall per year, while 1% receives about 500 mm where there are rocky and steep mountains (*Table 21.1*).

About 93% of the total cultivable area depends on natural rainfall and the remaining percentage produces about 40% of the total agricultural production, which is approximately 70% of the gross value of agricultural production (Department of Statistics, 1979, 1985).

* Director, Treatment Plants Department, Water Authority of Jordan (and Project Manager of Waste Stabilization Pond Project)
† Secretary General, Water Authority of Jordan

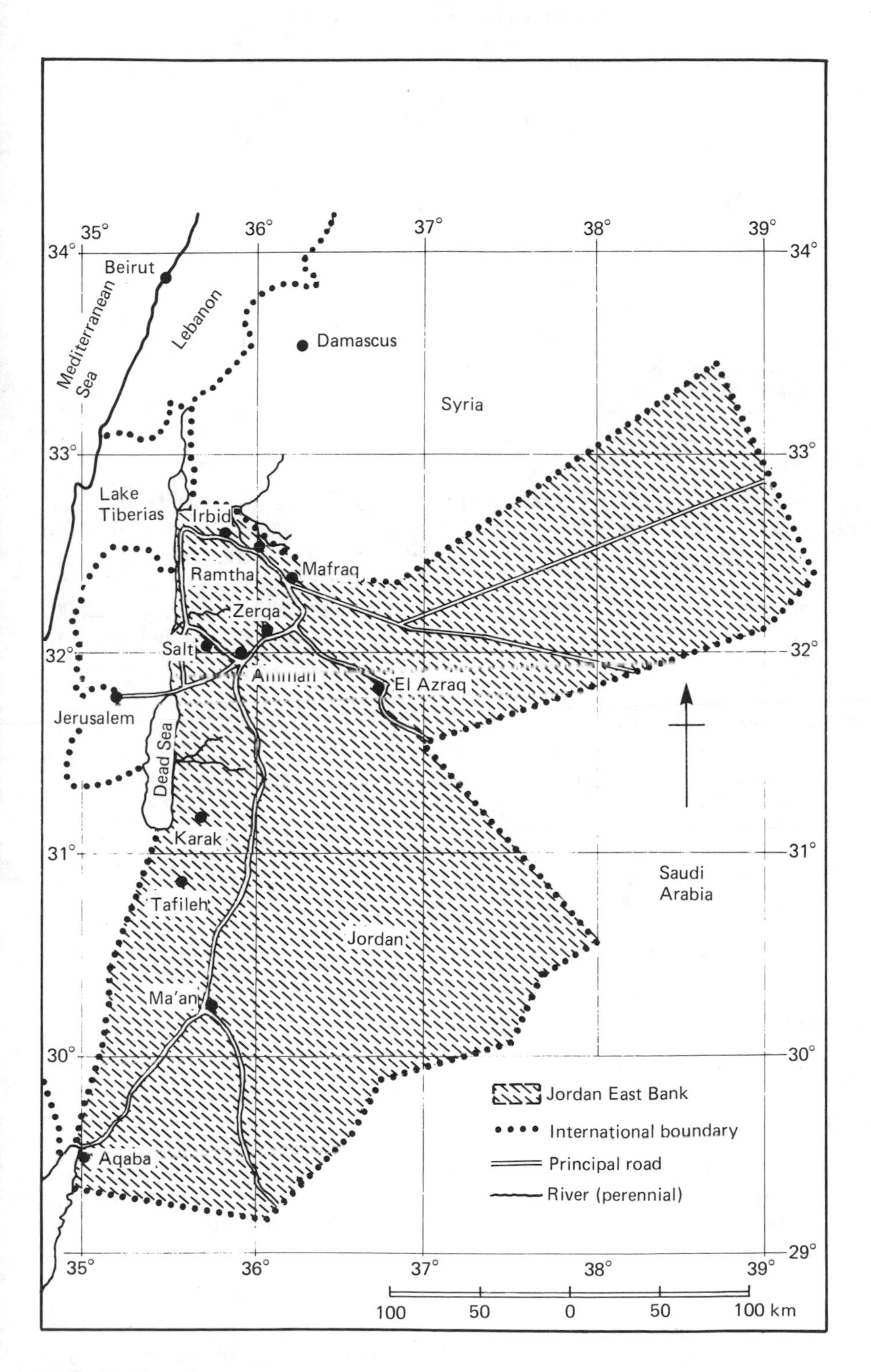

Figure 21.1 Location map

Table 21.1 Land areas classified by rainfall, East Bank

Area	*Average rainfall* (mm)	*Area* (Md*)	*Per cent*
Arid desert	100	75.0	81.1
Desert	100–200	9.6	10.3
Marginal	200–300	5.3	5.7
Semiarid	300–800	1.7	1.8
Semihumid	500	1.0	1.1
Total		92.6	100.0

Source: National Planning Council, Jordan (1976)
* Md = million dunums; 1 dunum = 1000 m^2

Water resources

The total annual surface water resources of the Kingdom are about 880 million m^3 in addition to 220 million m^3 of groundwater extracted, thus making a total in the order of 1100 million m^3. Projected demands for these water resources are summarized in *Table 21.2* and it is evident that irrigation consumes 70% of the total demand and that the surface water resources will be exhausted by the year 2000. Consequently, reuse of treated municipal and industrial wastewater will have to become one of the major sources of water for irrigation purposes in the near future.

Table 21.2 Water demand (millions of cubic metres)

Year	*Industrial and domestic*	*Irrigation*	*Total*
1985	120	401	521
1990	187	537	724
2000	268	606	874

Source: Water Authority of Jordan Studies (1985)

Water supply

About 80% of both the urban and rural populations are supplied with water through house connections from wells supplemented by springs and, in the near future, treated surface water. The remainder are supplied either by private wells or by tankers operated privately or by the Water Authority of Jordan (WAJ).

The average consumption is given in *Table 21.3* and it can be seen that the water consumption rate has been kept constant after 1990, on the assumption that services will be improved and losses in the distribution network will be reduced to 20% in the future (Water Authority of Jordan Studies, issued in April 1985).

Sanitation services (sewerage)

At present, only four cities – Amman, Salt, Aqaba and Jerash – have sewage collection and treatment systems. They serve about 30% of Jordan's population.

Table 21.3 Average water consumption rate (per caput litres/day)

Year	*Amman*	*Urban*	*Rural*
1985	88	73	48
1990	94	77	56
2000	94	77	67

Sewerage systems are under construction in Irbid, Zerqa and Ruseifa and are ready to start in Ramtha, Maddaba, Mafraq, Ma'an, Tafileh, Anjara, Ajloun, Ein Janneh, Karak, Soof, Mokhayyam Soof, Wadi Esse'neh, Swaileh and Jbaiha. These systems will provide services for another 15% of the country's population.

The design details for sewerage systems and treatment plants for all cities, towns and villages with a population of more than 8000 were prepared before the end of 1986 and the construction of the systems will be executed during the following 5 years. In other areas not connected to a sewerage system, septic tanks and cesspits are common.

Institutional responsibilities

The establishment of a National Water Authority (WAJ) was projected in the 5 year plan for economic and social development, 1981–85. The Authority was to assume the responsiblity for defining water consumption in the long run for domestic, industrial and agricultural uses, based on projected demands.

The Water Authority of Jordan was established by Temporary Law number 34 published in the *Official Gazette* on 15 December 1983 and commenced functioning in mid-January 1984. It comes under the Office of the Prime Minister and carries full responsibility for all water and sewerage systems and related projects all over Jordan. The Water Authority of Jordan, while exercising these responsibilities, is charged with the following tasks:

- set up a water policy that reserves the rights of the Jordanian Kingdom in all its water resources including development, maintenance and utilization of the resource, in order to raise the social, economic and sanitary standards within the Kingdom;
- survey the different water resources, conserve them, determine ways, means and priorities for their use and their implementation;
- develop the potential water resources in the Kingdom, increase their capacity and improve their quality, and put forth programmes and plans to meet future water needs by providing additional water resources from inside or outside the Kingdom, including the use of water treatment and desalination;
- regulate and advise on the construction of public and private wells, investigate water resources, drill exploratory, reconnaissance and productive wells, license well drilling, drilling rigs and drillers;
- study, design, construct, operate, maintain and administer public sewerage projects, including collecting, rectifying, treating, disposing and other processes related to water uses;
- draw up terms, specifications and special requirements in relation to the preservation of water and water basins, protect them from pollution and

ascertain the safety of water and sewerage structures, public and private distribution and disposal networks;
- carry out theoretical and applied research and studies regarding water and public sewerage to achieve the Authority's objectives, including the preparation of approved water quality standards for different uses, and of technical specifications concerning materials and construction in order to apply the findings to the Authority's projects in coordination with other concerned departments; and to publish the final findings and standards so as to generalize their application by all means available to the Authority;
- carry out or participate in planning, studying, designing, implementing, operating, maintaining and administering all water and sewerage projects or their supplemental projects in any area.

Prior to the establishment of the Water Authority of Jordan, different administrative bodies were responsible for water supply, water resources investigation and sanitation, including the Amman Water and Sewerage Authority, the Jordan Valley Authority, the Natural Resources Authority and the Water Supply Corporation. The latter and the Ministry of Municipal and Rural Affairs were both responsible for the urban and rural areas outside Amman and the Jordan Valley. The Natural Resources Authority was responsible for developing, monitoring, gauging and master planning the water potential of the whole Kingdom. All these main agencies now come under the umbrella of the Water Authority of Jordan.

As for agriculture, two bodies are responsible.

The Ministry of Agriculture, Agricultural Directorate

The Directorate is staffed by agricultural engineers whose function is to provide assistance and extension help to farmers. They also have links with Faculty of Agriculture researchers at the University of Jordan, who keep them informed of research activities in the district and occasionally help in coordinating field days. There is an Agricultural Research Station and a Seed Increase Farm, one called the Ramtha Agriculture Station and the other the Marrow Agriculture Station. Their main purpose is to propagate certified seeds, but they also carry out research into cereals and legumes for the different research organizations and the programmes of Jordan. Field days are held in these stations and farmers are invited to attend.

Jordan Cooperative Organization (JCO)

The JCO has a higher degree of contact with farmers than any other organization in Jordan. In 1982, there were 39 311 members of different societies, of which 16 062 belonged to agricultural societies, and there are 169 of these. Although the major function of the JCO is to extend seasonal or medium-term loans to members, it also provides agricultural supplies for farmers regardless of membership. Cooperative societies also assist farmers by hiring out cooperatively owned machinery and by the handling or processing of agricultural produce.

Impact of the lack of wastewater systems on surface and groundwater quality

Groundwater forms a significant portion of the total water resources (about one-third). Its protection for continuous usefulness is thus most important.

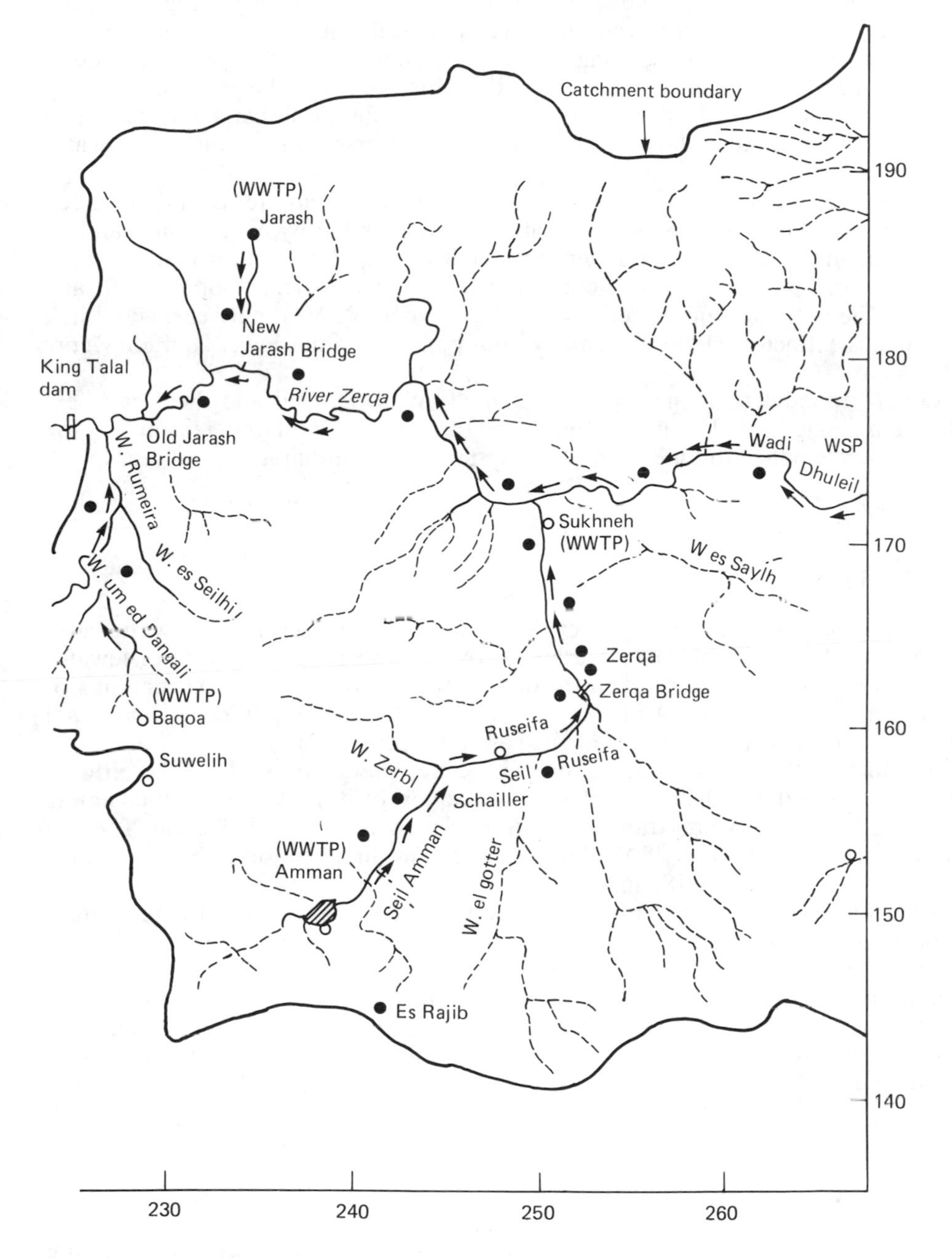

Figure 21.2 Wadi Zerqa basin at King Talal Dam site (scale 1:250 000). WWTP = Wastewater treatment plant; WSP = Waste stabilization ponds. →→ Direction of flow from WWTPs and WSP. ● Ground water wells

Wastewater management must take into consideration the recharge of the aquifers and the movement of groundwater. There is now great concern regarding the discharge of relatively large quantities of cesspool effluent into the ground, most of which has an impact on underlying aquifers. Cesspool effluent must be reduced in the future and completely eliminated finally. This will affect the potential capacity of the upper aquifers. Wastewater discharges into water bodies and upper aquifers must be monitored and controlled to avoid adverse impact on these water resources.

It has been reported in different studies that the upper aquifer quality has been strongly affected by human activities and is characterized by high nitrates and an increasing mineral and bacterial density. The lower aquifers in the Amman–Zerqa area underlie the most heavily populated areas, in the vicinity of Amman and Zerqa. The potential aquifer recharge is of the order of 5 Mm^3 per year, about half of which is extracted within Amman by wells penetrating this lower aquifer (*Figure 21.2*).

Other groundwater basins, such as Baq'ah, Wadi Dhuleil and the Dead Sea, although less affected, must be protected and are part of the reason for the government's policy to build new sewer systems in all urban and rural areas.

Existing reuse

From the aforementioned, it is evident that indirect recycling of wastewater effluent is already taking place, although there is no planned direct wastewater reclamation and reuse currently in Jordan. However, there is indirect recycling of wastewater effluent from Amman, Salt and Jerash via the Seil Zerqa and Wadi Shu'eb, respectively, principally for irrigation (*Figure 21.2*).

In Amman, the sewered area drains to the treatment works where the effluent discharged into the wadi mixes with the surface flow of storm water and then enters the King Talal Reservoir. After the detention time it is used for irrigation. The live storage of the reservoir is 48 Mm^3 and the current direct run-in from the sewage treatment plant is about 18 Mm^3.

Water discharged to septic tanks in unsewered areas soaks away into the ground. A portion may emerge as springs along the wadis in the cities (such as Amman, Zaraq, Irbid, Sweileh, Wadi Essier and Baqa'ah), particularly during wet conditions, but most of it returns to the ground, mixing with the natural recharge to the groundwater and travelling with the general groundwater movement along the wadis. It is likely that a large proportion of this returns to groundwater and is abstracted by the numerous borehole pumps along the wadi beds.

Potential volume of reclaimed water for reuse

As stated earlier (p. 261), during the next 5 years all cities will be generally sewered, using water as a means of transporting sanitary wastes to the point of treatment and/or reuse. At this time, large flows of treated wastewater effluent will be available either to be discharged to wadis or transported to agricultural land for reuse. To protect this effluent from any mixing with other water in the wadis or

Table 21.4 Wastewater treatment plants, areas served and flow quantities of effluent from these plants for the year 2000

Number	*Location of plant*	*Year of operation*	*Cities, towns and villages served*	*Flow* (m^3/day)
1	Ain Ghazal	1969	Amman	110 212
2	Khirbet Essamra*	1985	Greater Amman	
3	El-Baq'ah	1986	El-Baq'ah and Sweileh	6 276
4	Wadi Esseir	1986	Wadi Esseir	2 212
5	Jawa	1986	Jawa, Tayybeh, Qwesmeh and Umelhiran	1 750
6	Irbid 1	1986	Irbid	10 220
7	Irbid 1-A	1988	Irbid area	2 717
8	Irbid 2	1988	Irbid area	6 252
9	Ramtha	1986	Ramtha	2 214
10	Mafraq*	1986	Mafraq	1 737
11	Kufranjeh	1986	Kufranjeh, Ajloun, Anjarah	1 705
12	El-Karak	1986	El-Karak	978
13	Madaba*	1986	Madaba	2 304
14	Ma'an*	1986	Ma'an	832
15	Tafileh	1986	Tafileh	1 023
16	Aqaba	1985	Aqaba	2 025
17	Salt	1981	Salt	2 994
18	Jerash	1984	Jerash	816
19	Al-Azraq	1987	Al-Azraq	111
20	Na'ur	1987	Na'ur	446
21	Mahis	1987	Mahis and Fuheis	763
22	Es-Sukhneh*	1987	Es-Sukhneh	324
23	Sahab	1987	Sahab	980
24	Khirbet Essouq	1988	Kirbet Essouq, Ejwaideh and Abu Alamda	747
25	Um Qasir	1988	Um Qasir and Moqableen	217
26	Zahar*	1988	Zahar, Kofor Yuba and Jamhah	392
27	Tayybeh	1988	Tayybeh and Deir Ess'eneh	590
28	Dair Abu Sa'id	1988	Dair Abu Sa'id, Kofor Elma and Ashrafiyyeh	1 006
29	Kofor Khall	1987	Kofor Khall, Sakhrah, Ebbien and Ebbellien	834
30	Mokhayyam Ghazzeh	1987	M. Ghazzeh, Sakeb, Reimoo, Kitteh and Nahleh	1 319
31	Shooneh Shamaliyyeh*	1987	Shooneh Shamaliyyeh	578
32	Talbiyyeh	1987	Talbiyyeh and Jizeh	339
33	Mo'tah	1987	Mo'tah, Mazar Janoobiyyeh and Adnaniyyeh	490
34	Husseiniyyeh	1987	Husseiniyyeh	74
35	Mazra'ah*	1987	Mazra'ah	95
36	Wadi Moosa	1987	Wadi Moosa	445
			Total	166 017

* Effluent reduced 10% due to evaporation reckoned to take place in Waste Stabilization Ponds System

Notes

1. Flow quantities were calculated from the *per caput* water consumption rates (given in *Table 21.3*) for projected populations in the year 2000.
2. Flow quantities were calculated assuming:
   only 80% of population to be served in the areas with sewerage system network;
   only 75% of the water consumed in Amman, Zerqa, Irbid and urban areas will be discharged as raw sewage
   only 50% of the water consumed in rural areas will be discharged as raw sewage
3. Populations for the year 2000 are projected based on growth rates shown in *Table 21.5* (information from Department of Statistics and World Bank study, 1982).

illegal discharges to them, it was decided that the best way to achieve this was to use the effluent where it is produced.

The potential volume of reclaimed water from each city is computed on the basis of projected population by the year 2000 as shown in *Table 21.4*. Clearly, not all of this water can be used for irrigation because wastewater is generated with more or less uniformity from month to month, whereas irrigation water needs vary widely from season to season. Assuming that use of winter flows would be infeasible for Amman and the northern cities area, during this period the water could be stored in storm-water reservoirs or discharged to wadis. If all flows from May to the end of September for Amman and the northern cities area, and for the whole year for the southern cities and Azraq, were usable, a calculation can be made of how much land could be irrigated by the given flow. A generalized computation of this nature is presented in *Table 21.6* for each drainage area (treatment plant).

Table 21.5 Population projections

Population centre/band	*Assumed annual growth rate* (%)	
	1979–90	*1990–2005*
Amman municipality	5.0	4.0
Zerqa municipality	4.0	3.5
Irbid municipality	4.5	3.5
Other cities with 1979 population over 15 000	3.5	3.0
Towns with 1979 population 3000 to 15 000	3.0	2.5
Villages with 1979 population below 3000	2.0	1.72

Table 21.6 Area of agricultural land that could be irrigated with wastewater effluent

Number	*Treatment plant*	*Area* (ha)	*Number*	*Treatment plant*	*Area* (ha)
1	Khirbet Essamara	20 113.7	20	Mahis	139.2
2	El-Baq'ah	1 145.4	21	Es-Sukhneh	59.0
3	Wadi Esseir	403.7	22	Sahab	178.8
4	Jawa	319.4	23	Khirbet Essouq	136.3
5	Irbid 1	1 865.2	24	Um Qasir	39.6
6	Irbid 1-A	495.8	25	Zahar	71.5
7	Irbid 2	1 141.0	26	Tayybeh	107.7
8	Ramtha	404.0	27	Dair Abu-Sa'id	138.6
9	Mafraq	317.0	28	Kofor Khall	152.2
10	Kufranjeh	311.2	29	Mokhayyam Ghazzeh	240.7
11	El-Karak	178.5	30	Shooneh Shamaliyyeh	105.5
12	Madaba	420.5	31	Talbiyyeh	61.9
13	Ma'an	151.8	32	Mo'tah	89.4
14	Tafileh	186.7	33	Mazralah	17.3
15	Aqaba	396.6	34	Husseiniyyeh	13.5
16	Salt	456.4	35	Wadi Moosa	81.2
17	Jerash	149.0			
18	Al-Azraq	20.3			
19	Na'ur	81.4		Total	30 325.0

Table 21.6 was computed using the following simplifying assumptions:

- effluent will not be stored from the beginning of September to the end of April;
- evaporation at treatment plants with the waste stabilization pond system is taken as 10% of the total flow, while for other systems of treatment no losses are considered;
- areas of land to be irrigated are calculated for barley, which needs 200 mm/dunum;
- rainfall quantities are not taken into consideration;
- calculations were made for wastewater effluent expected in the year 2000.

Effluent quality for reuse in agriculture

Agricultural water quality requirements fall into two main categories: (a) pertains to the crops, soils and agricultural works, and (b) to safety and the public health of the consumers of the products. The first group is concerned mainly with the chemical characteristics of the water, while the second is concerned essentially with biological parameters, in addition to heavy metals and certain stable organic compounds.

Crop and soil concerns

The common potential problems relate to salinity, permeability, toxicity and miscellaneous factors such as excess nutrients. Potable water quality in Amman and the southern and northern areas, given in *Table 21.7* show that these waters have fairly low salinity. Normal increases in salinity through the one-use cycle are not expected to degrade water quality to the level that would further reduce crop yields. Wastewater resulting from the one-use cycle of these water supplies will have a somewhat higher sodium absorption ratio (SAR) than potable water but will probably remain in the safe range, as shown in *Table 21.8*. Sodium and chloride levels are generally low in water supplies (*see Table 21.7*) and are not expected to affect the quality of wastewater for reuse (*see Table 21.8*).

Public health concerns

Wastewater treatment processes can reduce pathogen levels substantially, asssuming that the design and the process is selected taking this fact into consideration.

Wastewater treatment

Treatment plants will have the greatest environmental significance so they must conform with public health and agricultural water quality requirements. Not only the human habitat must be protected but also the groundwater resources.

The lack of sewerage systems creates a dispersed public health problem because ground and surface waters become polluted. Surface water pollution can also concentrate that problem at one point. To conserve water in Jordan, in each final feasibility study for treatment plants, it was considered necessary to prepare a

Table 21.7 Test results of potable water samples taken in 1984 (yearly average)

Source	NO_3 (mg/l)	*pH*	*SAR*	HCO_3 (me/l)	SO_4 (me/l)	*Cl* (me/l)	*K* (me/l)	*Na* (me/l)	*Mg* (me/l)	*Ca* (me/l)	*TDS* (mg/l)	*EC* (mmhos/cm*)
AGTP/Amman	9.20	7.73	2.52	2.82	0.67	2.00	0.08	3.00	1.05	1.57	379.8	0.586
Irbid	31.15	7.58	2.73	2.36	0.98	8.46	0.10	5.45	3.78	2.88	822.0	1.285
Salt	50.56	7.32	0.59	3.95	0.25	1.27	0.13	1.06	1.68	3.57	423.0	0.640
El-Karak	71.45	7.65	1.47	3.20	0.94	1.75	0.45	1.85	1.36	3.33	473.0	1.120
Aqaba	12.20	8.00	0.85	1.51	0.14	0.95	0.00	0.90	0.41	1.78	217.5	0.340

* at 25°C

Notes

1. Results averaged for the following months:
   Amman: January, July, August, September and October
   Irbid: August and October
   Salt: May, June, July and September
   El Karak: April and August
   Aqaba: April and August
2. Figures are calculated from the results obtained from the Water Authority of Jordan laboratory records.

Table 21.8 Test results of wastewater effluent samples from treatment plants taken 1984 (yearly average)

Source	*NO_3* (mg/l)	*pH*	*SAR*	*HCO_3* (me/l)	*SO_4* (me/l)	*Cl* (me/l)	*K* (me/l)	*Na* (me/l)	*Mg* (me/l)	*Ca* (me/l)	*TDS* (mg/l)	*BOD* (mg/l)	*SS* (mg/l)	*COD* (mg/l)	*NH_4* (mg/l)	*PO_4* (mg/l)	*EC* (mmhos/cm*)
AGTP/Amman	3.11	7.80	4.92	9.37	1.67	5.46	0.79	8.03	2.55	2.72	1164	85	109	321	78.0	36	1.82
Salt	7.28	7.70	3.45	7.96	1.53	3.94	1.08	5.50	2.12	2.90	964	49	91	244	–	36	1.51
Jerash	18.20	8.00	2.34	4.48	1.21	3.36	0.64	4.43	2.74	4.00	756	32	32	132	8.15	–	1.21
Aqaba	–	7.00	4.21	5.75	0.56	3.21	0.20	5.40	1.86	1.42	748	200	–	867	–	–	1.17

* at 25°C

Notes

1. Results averaged as follows depending on results available:
   - AGTP/Amman: average of 12 months
   - Salt: BOD_5 and TDS as average for 12 months, other results as average for 6 months
   - Jerash: average for 5 months
   - Aqaba: one grab sample test result
2. Figures are calculated from the results obtained from the Water Authority of Jordan laboratory records.

separate financial feasibility study to justify the reuse for effluent as the primary, or as a supplemental source, of irrigation water. The sewage treatment plants have been designed as wastewater reclamation plants to process raw sewage, through full secondary treatment, to produce an effluent suitable for restricted agricultural irrigation of crops that are eaten cooked, or for irrigation of continuous grazing areas. The treatment process should be biological or biological and physical, it should not be energy intensive, it should require a minimum chemical input, and, therefore, it should optimize the local topography and climate to help in the process. It is essential that the specifications are established for a safe reclamation plant. Effluent that will be adequate for agricultural irrigation will also support fish life and will meet quality standards for indirect recharge of groundwater.

Through an order issued on 25 January 1983, it was established by the Prime Minister's Office that all treatment plants must have quality standards to secondary treatment level with a mean of 30 mg/l BOD_5 mg/l–30 mg/l SS. The standards are:

30 days average BOD_5: 30 mg/l
7 days average BOD_5: 45 mg/l
30 days average SS: 30 mg/l
7 days average SS: 45 mg/l .

The Water Authority of Jordan has also decided to add polishing ponds at post-secondary treatment to achieve substantial reductions in the pathogen level, followed by post-treatment disinfection if necessary. Treatment in the secondary stage would include screening, grit removal, primary settlement, biological oxidation and secondary settlement.

Additionally, management measures will be implemented to minimize the negative impact on the various water quality problems, for example:

- selecting food or fodder crops that are not oversensitive and can effectively utilize the nitrogen in the effluent (high yield crop production);
- reducing commercial, chemical and organic fertilizer application rates to compensate for the nitrogen fertilizer value of the sewage-originated irrigation water;
- the type of irrigation system will be controlled and designed to minimize the pathogenic contamination of the produce;
- selecting the agricultural area as near as possible to the treatment plant;
- selecting the type of crops that do not have direct contact in the human consumption chain, such as alfalfa, wheat, fodder, fibre crops, trees, barley, clover and maize.

The final disposal of the effluent is considered as an integral part of the design. A significantly large area of suitable land must be available in the vicinity for irrigation, and the area will form a unit to which only trained workers have access.

Theoretical yield and benefits

The area of agricultural land that could be irrigated with wastewater effluent during an irrigation season (shown in *Table 21.6*) will give a production of barley and alfalfa as shown in *Table 21.9*.

Table 21.9 Production of barley and alfalfa obtained from land irrigated with wastewater effluent*

Number	*Treatment plant*	*Production* (tonnes)		*Number*	*Treatment plant*	*Production* (tonnes)	
		Barley	*Alfalfa*			*Barley*	*Alfalfa*
1	Khirbet Essamra	3862	21 455	20	Mahis	27	148
2	El-Baq'ah	220	1 221	21	Es-Sukhneh	11	63
3	Wadi Esseir	77	431	22	Sahab	34	191
4	Jawa	61	341	23	Khirbet Essouq	26	145
5	Irbid 1	358	1 990	24	Um Qasir	8	42
6	Irbid 1A	88	529	25	Zahar	14	76
7	Irbid 2	219	1 217	26	Tayybeh	21	115
8	Ramtha	78	431	27	Dair Abu Sa'id	35	196
9	Mafraq	61	338	28	Kofor Khall	29	162
10	Kufranjeh	60	332	29	Mokhayyam Ghazzeh	46	257
11	El-Karak	34	190	30	Shooneh Shamaliyyeh	20	112
12	Madaba	81	448	31	Talbiyyeh	12	66
13	Ma'an	29	162	32	Mo'tah	17	95
14	Tafileh	36	199	33	Mazra'ah	3	18
15	Aqaba	76	423	34	Husseiniyyeh	3	14
16	Salt	105	583	35	Wadi Moosa	16	87
17	Jerash	29	159				
18	Al-Azraq	4	22				
19	Na'ur	16	87		Total	5816	32 345

* Production rate for barley in the year 2000 = 192 kg/dunum
Production rate for alfalfa in the year 2000 = 4000 kg/dunum

These rates are taken from Abdel Hamid Shouman Establishment(1982).

Conclusions

It is evident from the previously mentioned information and tables that the reuse of treated wastewater in Jordan will be an important element of the government economic, agricultural and water policy. By reusing this water from the various cities and areas, the agricultural land area will be increased and have a better geographical distribution. As a result of using treated wastewater for irrigation, a large quantity of potable water can be saved and, additionally, ground and surface water can be protected against pollution.

This chapter shows that, by using wastewater, 5816 tonnes of barley can be produced, which was about 38% of the total amount imported into the country during the year 1983, and about 32 345 tonnes of alfalfa can also be produced. These quantities can be obtained if the land is used only once for each crop. The value of these products, according to 1985 prices, is about 4.5 million Jordanian dinars, which is nearly equal to the value of 1983 imports of concentrated materials used for preparation of forage.

Bibliography

ABDEL HAMID SHOUMAN ESTABLISHMENT. (1982) *Analytical Review on Food Problems in the Arab World*
DEPARTMENT OF STATISTICS (1979) *Housing and Population Census, Jordan.* Amman: DOS
DEPARTMENT OF STATISTICS (1985) *External Trade Statistics, Jordan.* Amman: DOS

HUMPHREYS, H. AND SONS (1978) *Water Use Strategy for North Jordan.* Prepared for Government by H. K. of Jordan

JAMES M. MONTGOMERY, CONSULTING ENGINEERS, INC., BROWN AND CALDWELL CONSULTING ENGINEERS, in association with CONSULTING ENGINEERING CENTRES (1981–83) *Feasibility study of the municipal water distribution improvement and sewerage and storm water drainage systems in Madaba, December 1981; and in Tafileh, March 1983.* Prepared for the National Planning Council, Jordan

JAMES M. MONGOMERY CONSULTING ENGINEERS, INC. and DMJM INTERNATIONAL (1982) *Final master plan report and engineering design and economic analysis report for wastewater disposal for Greater Amman Area, February.* Prepared for the National Planning Council, Jordan

JONZEY AND PARTNERS CEB, in association with ENGINEERING SCIENCE INC. (1983) *Kufranja wastewater reclamation facility design calculations, August.* Prepared for the Water Authority, Jordan

MRM CONSULTING ENGINEERS CO. LTD., in association with TRAVERS MORGAN RKL (1985) *Final pre-feasibility report, March.* Prepared for the Water Authority, Jordan

SAQER EL-SALEM (1983) *Reuse of wastewater.* Prepared for the National Conference on Potable Water and Environmental Rehabilitation, June, Ministry of Health, Jordan

SAQER EL-SALEM (1984) *Wastewater treatment and possible alternatives for use in Jordan.* Short Course on Environmental Engineering, November, Jordan University

WATER AUTHORITY, JORDAN (1985) Reports and Records

WATER AUTHORITY LAW (1983) H. K. of Jordan

Chapter 22

Plans for reuse of wastewater effluent in agriculture and industry in the Kingdom of Saudi Arabia

Misfer S. Kalthem* and Ahmed M. Jamaan†

Introduction

The Kingdom of Saudi Arabia, being an arid country, lacks perennial rivers, and groundwater constitutes the main source of natural water supply, out of which agricultural irrigation consumes a major share. Rainfall in the Kingdom is low and, therefore, recharge to the deep sedimentary aquifers is almost insignificant (Ministry of Agriculture and Water, 1977, 1980). Because of extraction of groundwater to meet the demand in various sectors, the non-renewable aquifers are showing signs of a significant water level decline. Wastewater effluent can supplement these demands to a certain degree in non-domestic sectors.

The Kingdom's policy is to utilize all available treated municipal wastewater in the most beneficial manner for several purposes, among which the agriculture sector is afforded top priority (Kalthem and Tabaishi, 1980). The importance of reclaimed wastewater is due to the great need for new water resources to meet the increasing water demands for agriculture and landscape irrigation, and for industrial uses. Moreover, it is needed to reduce the excessive groundwater abstraction and for possible recharge of aquifers.

The purpose of this paper is to outline the present activities and future plans for reclaimed wastewater reuse in agricultural irrigation in the Kingdom of Saudi Arabia, giving specific examples from the Riyadh, Madinah and Qassim regions. This does not mean, however, that these are the only reuse projects in Saudi Arabia. Wherever there is a wastewater treatment plant, reuse planning of some scale is taking place, such as for the Jeddah, Makkah, Hail and Dammam areas (Ministry of Agriculture and Water, 1984a).

Wastewater availability

It is expected that about 674 Mm^3/year of reclaimed wastewater will be available by the year 2000 in the Kingdom. This is, however, a relatively small quantity compared with the water demand for agriculture and industry (*Table 22.1*). The

* Director, Water Research and Studies Division, Water Resources Development Department, Ministry of Agriculture and Water, Riyadh, Saudi Arabia
† Hydrogeologist, Water Resources Development Department, Ministry of Agriculture and Water, Riyadh, Saudi Arabia

Table 22.1 Summary of water demands, 1990 and 2000

Item	*Demand* (Mm^3/year)		
	1398$_H$ (1978)*	*1410$_H$* (1990)*	*1420$_H$* (2000)*
Urban networks:			
Water supplied	196	979	1643
Treatment losses	2	49	164
Subtotal	198	1028	1807
Agriculture	3171	3684	5119
Independent industries	18	74	182
Other domestic use	113	54	27
Out of town parks	–	35	106
Subtotal	3500	4875	7241
Wastewater available for reuse	35	368	674
Total	3465	4507	6567

Source: Ministry of Agriculture and Water (1979)
*H = Hegira

main urban areas alone are expected to provide more than 70% of the total available reclaimed wastewater in the year 2000 (*Table 22.2*).

Reuse in the Riyadh area is at present confined to Dirab (57 000 m^3/day), Diraiyah (53 000 m^3/day) and, very soon, an additional 80 000 m^3/day will be used in the Ammariyah area. The three areas together constitute about 4000 ha of land for irrigation (Ministry of Agriculture and Water, 1981). With the proposed expansion of the Manfuha sewage treatment plant to 600 000 m^3/day and construction of another treatment plant to the east of Riyadh, the total wastewater expected to be available in the year 2000 will be about 800 000 m^3/day. Additionally, 20 000 m^3/day of reclaimed wastewater is used by Riyadh Refinery (Kalinske, 1979; Al-Marshoud and Javed, 1982).

The present wastewater flow from the Madinah treatment plant is about 27 000 m^3/day, which is expected to be increased to 54 000 m^3/day in 1990 and 141 000 m^3/day in the year 2000, with an estimated land area under irrigation of 3100 ha (Ministry of Agriculture and Water, 1984b).

Table 22.2 Estimated reclaimed wastwater

Rank of the population centre	*1990* (m^3/day)	*2000* (m^3/day)
National	833 000	1 074 000
Regional	175 778	285 200
District	–	152 200
Rural cluster	–	334 100
Total	1 008 778 m^3/day	1 845 500 m^3/day
Total	368 Mm^3/a	674 Mm^3/a

In the Qassim area, the total quantity of wastewater from the five proposed municipal treatment plants will be 98 700 m^3/day which, it is estimated, will irrigate 823 ha of land (Ministry of Agriculture and Water, 1984c).

Wastewater effluent quality

Although irrigation with municipal wastewater is in itself an effective form of wastewater treatment (land treatment), some degree of treatment must be provided for raw municipal wastewater before it can be used for agricultural or landscape irrigation. The purpose of treatment is to provide: (a) protection for the environment and public health; (b) prevention of nuisance conditions during storage and application on land; (c) prevention of possible damage to crops and soils; and (d) protection of water resources, especially shallow groundwater, from pollution (Pettygrove and Asano, 1984).

The Ministry of Agriculture and Water, realizing the importance of providing an adequate degree of wastewater treatment to assure reliable public health protection against any adverse effects from contaminants and disease transmission during reuse, in 1982 proposed draft Wastewater Regulations (Ministry of Agriculture and Water, 1982). At present, these are under revision to be formally submitted for issuance as a Royal Decree. These Wastewater Regulations, in their revised form, envisage a minimum of tertiary treatment in order to attain the quality level for unrestricted irrigation (*Table 22.3*).

Table 22.3 Maximum contaminant level in treated wastewater for unrestricted irrigation

Parameter	*Maximum contaminant levels (MCL)**	*Parameter*	*Maximum contaminant levels (MCL)**
BOD_5	10.0	Lead	0.1
TSS	10.0	Lithium	0.07
Aluminium	5.0	Manganese	0.2
Arsenic	0.1	Mercury	0.001
Beryllium	0.1	Molybdenum	0.01
Boron	0.5	Nickel	0.02
Cadmium	0.01	Nitrate	10.0
Chlorides	280	Selenium	0.02
Chromium	0.1	Zinc	4.0
Cobalt	0.05	Oil and grease	Absent
Copper	0.4	Phenol	0.002
Cyanide	0.05	pH	6.0–8.4
Fluoride	2.0	Coliforms (MPN/100 ml)†	2.2
Iron	5.0	Turbidity (NTU)	1.0

* Expressed in mg/l, except as noted

† The reclaimed wastewater shall be considered adequately disinfected for unrestricted irrigation if the geometric mean of the coliform organisms in the water does not exceed MPN 2.2/100 ml as determined from the bacteriological test results of the last 7 days, and the number of coliform organisms does not exeed MPN 23/100 ml in any sample.

Improvement in wastewater quality

The quality of reclaimed wastewater depends upon the nature of the raw wastewater and the treatment offered. The major characteristics of raw wastewater originate in the municipal water supplies. Previously, the major part of the water

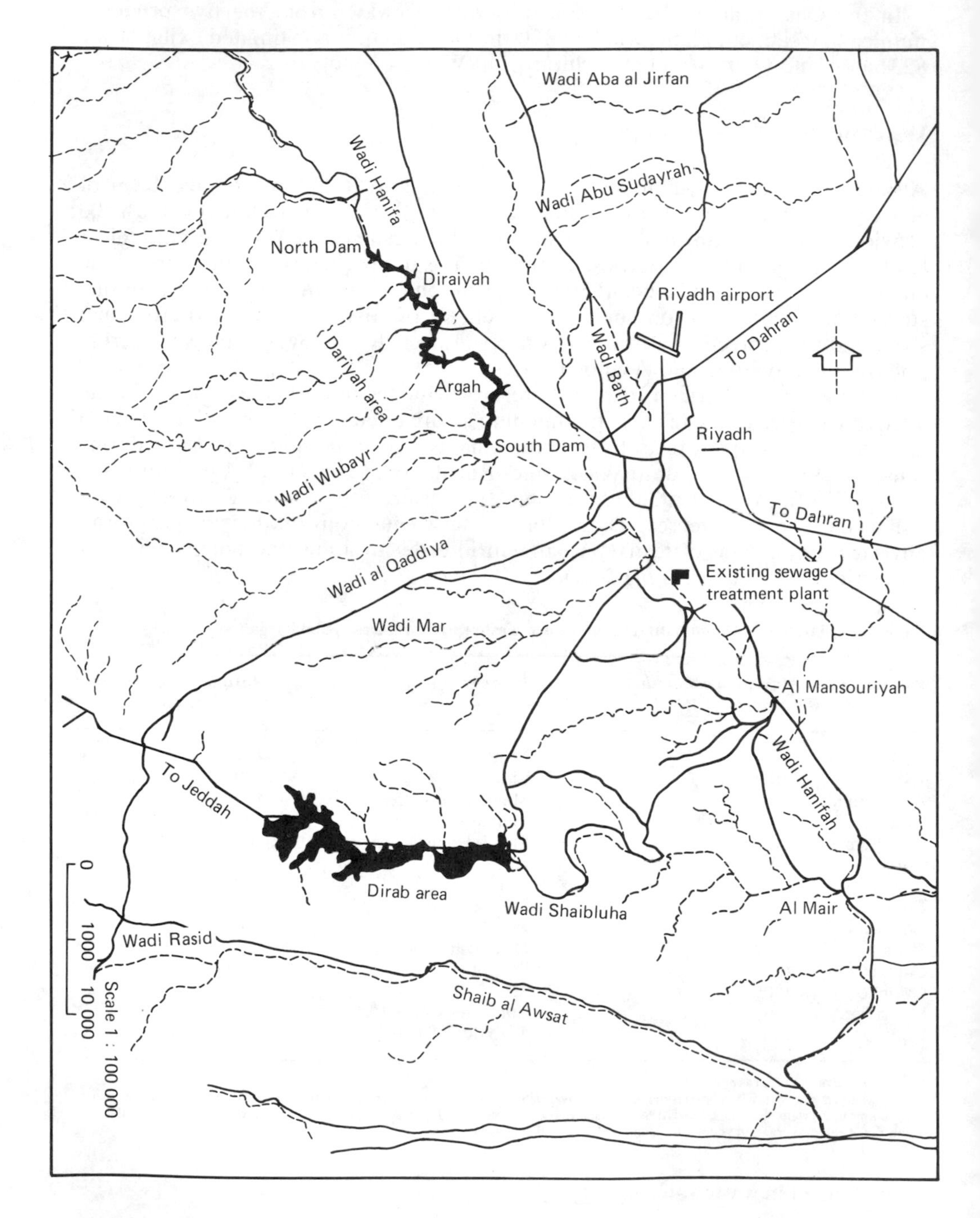

Figure 22.1 Map of Dirab and Diraiyah area

supplies for municipal use in the urban centres was derived from wells, where total dissolved solids (TDS) were rather high being usually more than 1000 mg/l. However, this situation has gradually changed, since various portions of water supplies are now desalinated water with low dissolved solids, in the range from 350 to 500 mg/l. This means that wastewater is now expected to improve from the salinity viewpoint and may prove to be more useful for agricultural applications; it may even improve the salt content of the groundwater in areas where such applications are taking place, as in the Riyadh, Madinah and Qassim areas. The quality of wastewater effluent is also expected to improve further as a result of new wastewater treatment plants with facilities for tertiary treatment.

Monitoring of wastewater quality

Monitoring of wastewater effluent quality is an important function for maintaining the level of treatment provided in the treatment plant. In addition to sampling and analysis for some of the parameters listed in *Table 22.3*, continuous monitoring systems are presently being considered for Madinah and Qassim which will measure parameters such as TOC, turbidity, DO, pH, NH_4 and residual chlorine (Ministry of Agriculture and Water, 1984b and c).

Utilization of wastewater effluent in agriculture

Riyadh

The agricultural sector is earmarked as the first priority to receive treated wastewater and the government initiated the first reclaimed wastewater reuse pilot projects in Dirab and Diraiyah near Riyadh in 1981–82 (*see Figure 22.1*). The Riyadh wastewater treatment plant, for the moment, has a capacity of 200 000 m^3/day and the biological treatment is plastic media trickling filters. This treatment plant produces an effluent with BOD_5 and SS of 10–20 mg/l (Shammas, 1983).

The cultivated areas under reclaimed wastewater irrigation in Diraiyah and Dirab are estimated at 850 and 2000 ha, respectively. In Diraiyah the areas of farm holdings are relatively small and irregular, with the average area of each farm around 12 ha, but in Dirab the farms are between 60 and 70 ha.

In Diraiyah, the cultivable alluvial soils consist of deep, rather homogeneous, fine, sandy loams, practically all of which are already utilized for cultivation. In Dirab, the best available soils are the deep alluvial soils in the northern part of the wadi. These alluvial soils are composed of typical flood plain alluvium with alternating layers of coarse and fine textured soil.

In Diraiyah, the irrigation method mainly utilized is surface irrigation practised as basin, furrow or border irrigation, whereas in Dirab, mechanical cultivation is practised with sprinkler or drip irrigation. The present cropping pattern in Diraiyah area is estimated to be as shown in *Table 22.4*.

Diraiyah receives 53 000 m^3/day of treated wastewater pumped to a stand pipe at the top of a hill upstream of Hanifa Dam and then flowing to the reuse area by gravity (*Table 22.5*). During the high demand summer season, the farmers supplement the irrigation water by pumping groundwater from shallow alluvium and fractured limestone as well as deep Minjur sandstone aquifers. The total dissolved solids in the well waters is around 1500 mg/l. The relatively hot

Table 22.4 Cropping pattern in Diraiyah

Crop	*Area* (ha)	*Percentage of total area*
Date palms with undercrops and scattered fruit trees	680	80.0
Vegetables and fodder crops	50	5.9
Fallow	20	2.4
Houses, gardens, roads, non-cultivated areas, etc.	100	11.7
Total	850	100.0

Table 22.5 Present reuse of wastewater in Riyadh

Area	*Quantity* (m^3/day)	*Nature of use*
Dirab	57 000	Irrigation of 2000 ha
Diraiyah	53 000	Irrigation of about 800 ha
Ammariyah	80 000 (being designed)	Irrigation of 1200 ha
Sub-total	190 000	4000 ha
Riyadh Refinery	20 000	Industrial use (cooling, washing, etc.)
Total	210 000	

groundwater cools down before it reaches the farm area (Ministry of Agriculture and Water, 1981).

In Dirab, the present cropping pattern in the area is estimated to be as shown in *Table 22.6*; 57 000 m^3/day of treated wastewater is pumped to a hill in Dirab, some 35 km away from the treatment plant, from where it flows by gravity to the reuse area through a distribution network. In summer, irrigation water is supplemented by pumping groundwater from shallow wells of 150–200 m deep. Water quality in the shallow wells is three times inferior to that of the effluent. In the low demand season, the excess reclaimed wastewater is allowed to flow down to open land in the wadi where it helps recharge the shallow limestone aquifer.

Table 22.6 Cropping patterns in Dirab

Crop	*Area* (ha)	*Percentage of total area*
Date palm	10	0.5
Fruit (olives, citrus)	5	0.25
Vegetables	50	2.5
Fodder crops (alfalfa, millet, sorghum, barley)	300	15.0
Wheat	1400	70.0
Experimental area (Ministry of Agriculture and King Saud University Research Station)	100	5.0
Fallow, houses, roads, etc.	135	6.75
Total	2000	100.0

Another wastewater reuse project is to start in Ammariyah. About 80 000 m^3/day of treated wastewater will be pumped to the area for irrigation. The total agricultural land available is of the order of 1200 ha. Most of the new farms are sizeable. The soil characteristics and all other agricultural requirements are similar to Diraiyah.

Madinah

The city of Madinah, located about 150 km inland from the Red Sea coast, receives desalinated sea water. The local basaltic and sub-basaltic aquifer supplements the domestic water requirements and is simultaneously blended with the distilled sea water to bring TDS down to 500 mg/l. At present, treated wastewater is 27 000 m^3/day and is expected to be around 100 000 m^3/day by the year 1995, to irrigate about 1750 ha of land during the phase I period. By the year 2000, in phase II, 141 000 m^3/day is expected to irrigate 3100 ha of land (*Table 22.7*). The wastewater treatment plant is equipped with trickling filter units now and will be using the conventional activated sludge process followed by filtration. *Figure 22.2* shows the area.

Table 22.7 Present and future wastewater availability in Madinah and Qassim areas

Years	*1984* (m^3/day)	*1990* (m^3/day)	*2000* (m^3/day)	*Required farm area in 2000* (ha)
Madinah	23 000	54 000	141 000	3100
Qassim				
Buraydah	5 000	35 700	57 344	495
Unayzah	6 000	11 529	22 400	171
Ar Rass	–	6 615	11 424	96
Bukariyah	–	2 249	3 584	29
Riyadh Al Khabra	–	2 495	3 965	32
Total for Qassim area	11 000	58 588	98 717	823

The soil and land classification of the entire reuse area of 3100 ha in the Madinah area can be categorized as:

Class A soil: 20%
Class B soil: 30%
Class C soil: 50%

Soil texture may be described as mainly sand, loamy sand and sandy loam.

Qassim region

Wastewater reuse possibilities in selected townships, namely Buraydah, Unayzah, Ar-Rass, Al-Bukariyah and Riyadh Al-Khabra, have been investigated and found to be technically feasible to develop agriculture with the reclaimed wastewater.

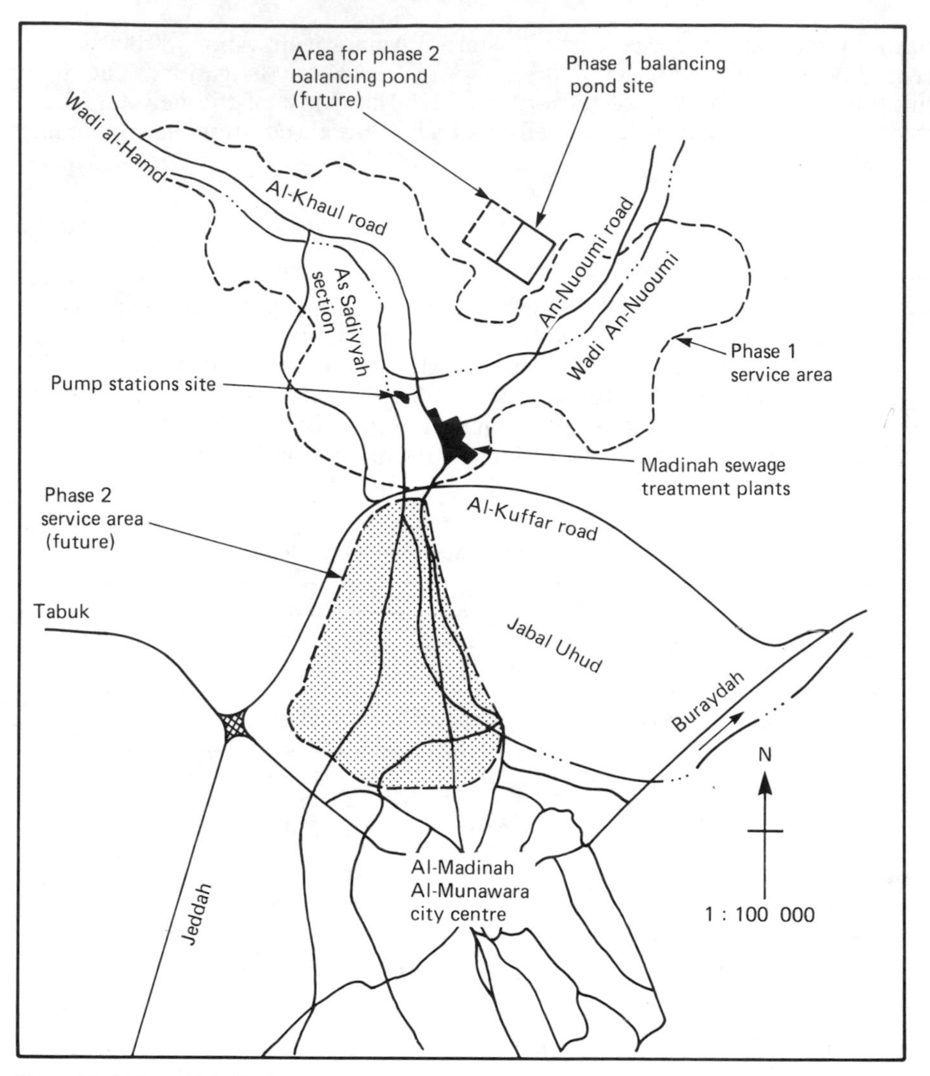

Figure 22.2 Map of Madinah area

Wastewater availability and land areas to be irrigated are given in *Table 22.7*, and *Figure 22.3* shows the area.

Sludge utilization

In most of the municipal wastewater treatment plants in the Kingdom, sludge treatment is provided. During this treatment, sludge is normally digested anaerobically and then dried on sand beds. The quality of the dried sludge from Madinah is presented as an example in *Table 22.8*. The average annual sludge production rate is about 3500 dry tonnes/year in Madinah, and the density of air-dried sludge is 650–700 kg/m^3 (Ministry of Agriculture and Water, 1984d).

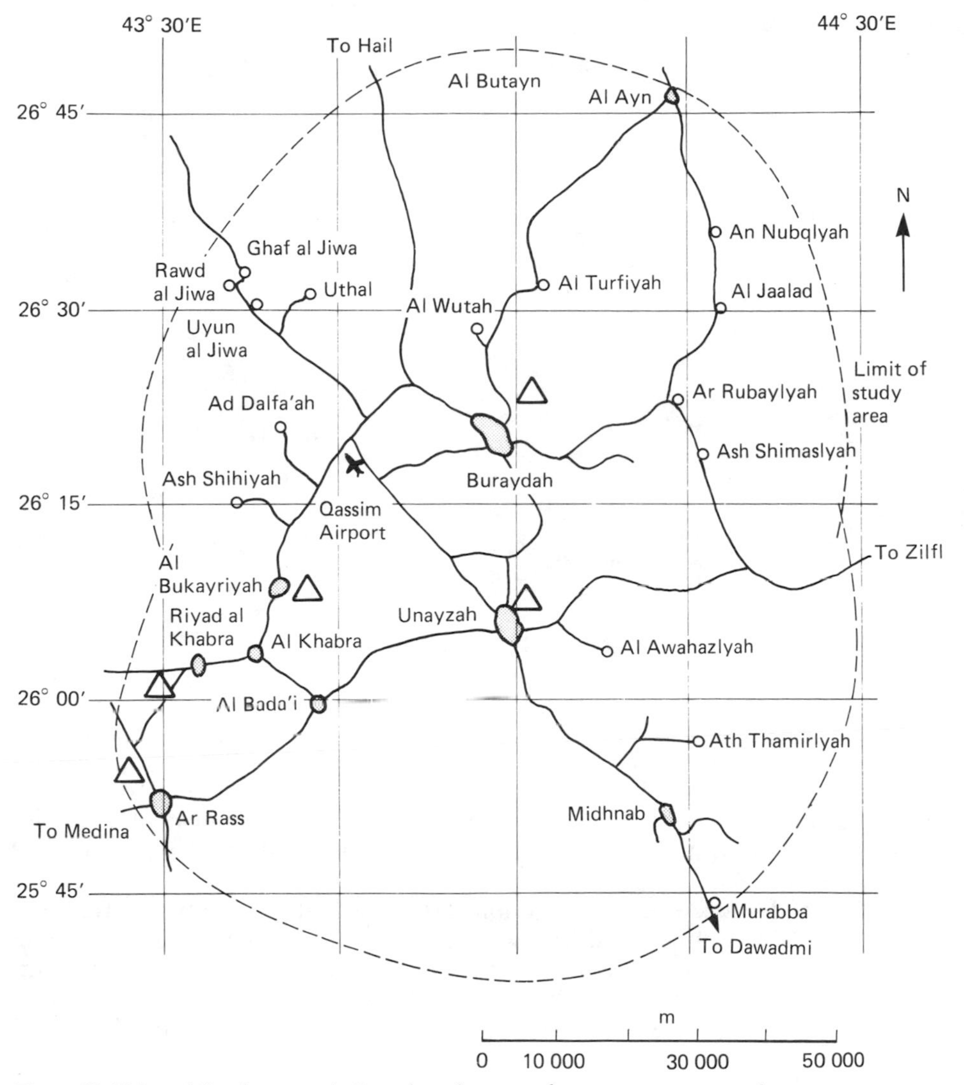

Figure 22.3 Map of Qassim area. △, Location of proposed sewage treatment works

Dried sludge is now commonly used all over the Kingdom by farmers as fertilizer and soil conditioner. After drying, it is usually piled up in treatment plants, and farmers pick it up in their own trucks.

The Ministry of Agriculture and Water is also carrying out research to develop guidelines for the proper application of sludge on land (Shammas and Shankiti, 1985). According to the preliminary results, mineralization of sludge nitrogen proceeds at a rather slow rate. Consequently, there is a depletion of soil nitrogen for at least a month, and, therefore, it is necessary to apply supplementary fertilizer with the sludge for at least a period of 1 month. It has been recommended that sludge should be applied at least 1 month before planting, along with supplemental nitrogen fertilizer.

Table 22.8 Sludge quality for Madinah

Constituents	*Quantity**
Total solids content (%)	94.4
Volatile solids content (%)	61.7
Total Kjeldahl nitrogen	5223
Ammonia nitrogen	2060
Total phosphorus	2236
Potassium	1850
Arsenic	0.7
Cadmium	0.3
Chromium	1.0
Copper	88
Lead	9.5
Mercury	0.5
Molybdenum	1.4
Nickel	0.5
Selenium	0.6
Zinc	515

* All units are in mg/kg of sludge except as indicated

Utilization of wastewater effluents in industry

From a total of $674 \times 10^6\,m^3$/year of reclaimed wastewater which will be available in the year 2000, only 10% will be utilized by industries. According to the Ministry's allocation guidelines, reclaimed wastewater rather than groundwater resources are to be allocated to independent industries located in the inland regions. Additional advanced treatment is usually provided by individual industries according to their own process requirements; Riyadh Oil Refinery is a good example of this.

The advanced wastewater reclamation plant belonging to Riyadh Refinery receives, on average, 20 000 m^3/day of secondary-treated wastewater effluent from Riyadh Municipal Treatment Plant (Kalthem and Tabaishi, 1980). In the plant, inflow is first treated physically and chemically through aeration and chemical clarification, followed by two stages of recarbonation, filtration and disinfection to produce general utility water which is used mainly in cooling towers, for firefighting and floor washing and as a desalter make-up water. Further treatment of the general utility water through activated carbon adsorption columns and reverse osmosis units produces high quality feed water for the high-pressure boilers of the Refinery.

Desalination of sea water is preferred, in every respect, to the advanced treatment of reclaimed wastewater in obtaining high quality water for industrial use. Therefore, industries located in the coastal regions meet their water demand through sea water desalination, as is the case in Jubail and Yanbu industrial cities.

Acknowledgement

The authors wish to express their gratitude to Mr Mustafa Noory, Director General of the Water Resources Development Department, Ministry of Agriculture and Water, Riyadh, for his encouragement and guidance during the preparation of this chapter. They wish also to thank Dr A. Gur, Mr A. R. K. Javed and Dr R. A. Khan of the Water Resources Development Department, Ministry of Agriculture and Water, for their assistance in the preparation of this chapter. They further would like to thank Mr Muwafaq Sughair, Director of Riyadh Wastewater Treatment Works, for his kind cooperation in providing necessary information.

References

AL-MARSHOUD, F. and JAVED, A. R. K. (1982) Wastewater reuse in the Kingdom of Saudi Arabia. *Proceedings of the Symposium of Water Resources and Utilization in the Kingdom of Saudi Arabia, Riyadh*

KALINSKE, A. A. (1979) Reclamation of wastewater treatment plant effluent for high quality industrial reuse in Saudi Arabia. *Water Reuse Symposium I.* Denver: Colorado: AWWA Research Foundation.

KALTHEM, M. and TABAISHI, S. (1980) Augementation of groundwater with wastewater effluent in Wadi Hanifa Basin. *Symposium on Water Resources Development,* Madrid, Spain

MINISTRY OF AGRICULTURE AND WATER (1977) *Al-Hassa Development Project: Groundwater Resources Study and Management.* BRGM

MINISTRY OF AGRICULTURE AND WATER (1979) *National Water Plan* BAAC and Water Resources Development Department (unpublished)

MINISTRY OF AGRICULTURE AND WATER (1980) *UMM Er Radhuma Study – Draft Final Report.* GDC International Ltd.

MINISTRY OF AGRICULTURE AND WATER (1981) *Feasibility Study, Reuse of Wastewater in the Diraiyah and Dirab Areas.* VBB

MINISTRY OF AGRICULTURE AND WATER (1982) *Draft National Wastewater Regulation.* Riyadh: MAW

MINISTRY OF AGRICULTURE AND WATER (1984a) *Draft Fourth Five-year Plan, Section-II.* Riyadh: Water Sector

MINISTRY OF AGRICULTURE AND WATER (1984b) *Draft Preliminary Design Report: Investigations and Engineering Design for Wastewater Reuse in Madinah Area.* CH2M-Hill International

MINISTRY OF AGRICULTURE AND WATER (1984c) *Draft Preliminary Design Report: Investigations and Engineering Design for Wastewater Reuse in Qassim Area.* John Taylor and Sons

MINISTRY OF AGRICULTURE AND WATER (1984d) *Master Plan – Phase I Investigation and Engineering Design for Wastewater Reuse Planning in Madinah Area.* CH2M-Hill

PETTYGROVE, G. S. and ASANO, T. (1984) *Irrigation with Reclaimed Municipal Wastewater. A Guidance Manual.* Report No. 84–1. California State Water Resources Control Board

SHAMMAS, A. T. and SHANKITI, A. (1985) Towards the establishment of guidelines for the land application of treated sewage wastes. *First Symposium on Environmental Protection and its Administration in the Kingdom of Saudi Arabia*

SHAMMAS, N. K. (1983) Wastewater treatment in the capital city of Riyadh. *Symposium on Water Resources in Saudi Arabia, Vol. 2.* Saudi Arabia: King Saud University

Chapter 23

Treated sewage effluent for irrigation in Kuwait

Agriculture Affairs and Fish Resources Authority, Kuwait

Introduction

The State of Kuwait is situated at the north-western corner of the Arabian Gulf. Its area is 17 818 km^2 of which 12% is arable land; however, the actual area under cultivation is only about 1% due to the scarcity of water and other limiting factors. Dry farming is impossible due to high evaporation rates and low annual rainfall. The climate of Kuwait is desert tropical, with average maximum temperature of 44°C and the hottest months are July and August. Average minimum temperature is 8°C, with an absolute minimum temperature of −4°C. The coldest months are December and January. Rain falls between November and March, with a yearly average of 100 mm. Relative humidity normally fluctuates between 27 and 63%, reaching 100% as maximum and 5% as minimum. The evaporation rate is very high, averaging 14.1 mm/day and may attain 21.1 mm during summer months (class 'A' pan). North-west winds prevail most of the year, cold in winter and hot in summer. In July and August, south-east winds laden with humidity sweep over Kuwait from the Arabian Gulf. Sand storms blow between March and August but may occur at any time during the year. The hinterland of Kuwait is undulating desert rising gradually towards the east with some low ridges and hills. Land elevation above sea level varies between 0 and 300 m.

The population of Kuwait in mid-1985 was estimated to be 1 900 000. Although the agricultural sector in Kuwait does not contribute more than 0.2% to the total national income, it plays an important role in the economy of the country by attempting self-sufficiency in vegetable production and animal products, in addition to promoting landscaping and increasing recreational parks.

The availability of irrigation water at a reasonable cost is one of the most important factors which affects the development and expansion of agriculture in any country, especially in arid zones. As this subject is very important to Kuwait, many investigations and studies have been carried out in this connection for about a quarter of a century. Desalinated seawater used in the houses is rather expensive and it was decided to reuse the wastewater for agricultural purposes instead of returning it to the sea, resulting in pollution and harmful effects on marine life.

For a long time, the untreated sewage effluent has been used to irrigate forestry projects which are far from inhabited areas. The wastewater given secondary treatment in the Giwan Treatment Plant has been used since 1956 to irrigate some plantations in the experimental farm and, after its completion, effluent from Al Ardiya Plant was also used.

As the reuse of treated sewage effluent is of public concern regarding the health of human beings and animals, Kuwaiti authorities attached great importance to conducting studies and making contacts as widely as possible, to benefit from the experiments and experience of other countries in this field , especially in respect of the precautions to be taken. Construction of treatment plants and preparation of the infrastructure in agricultural fields where the treated water was to be used were implemented simultaneously with various studies on the subject. The methods of utilizing the treated sewage effluent were studied by several international consultants who submitted their reports and recommendations*. These studies were reviewed by a committee from FAO and the appropriate technical staff of the Agriculture Authority. WHO was also contacted for advice and they recommended getting in touch with specialists in this field.

Studies were continued on several aspects by health and scientific committees within the country and by international organizations. Many projects throughout the world were visited, in South Australia, United States and the Federal Republic of Germany, to see what others had achieved and to discuss all the anticipated problems. A report was prepared on these visits and contacts.

At the invitation of the Ministry of Public Health, a group of consultants from WHO visited Kuwait from 13 to 18 March 1982 and submitted a report on the utilization of treated sewage effluent. The conclusions of their report can be found in the appendix to this chapter.

Sewage treatment facilities in Kuwait

As a result of these studies and contacts, the government of Kuwait decided to take advantage of modern technology for treating sewage so as to utilize the treated effluent for agricultural purposes, which undoubtedly costs only a fraction of the cost of desalination of seawater.

At present, three sewage treatment plants have been constructed, one at Ardiya, which is the largest of the three, one at the coastal villages and one at Jahra. The design average flow of the Ardiya Treatment Plant is 150 000 m^3/day, that of the coastal villages is 96 000 m^3/day and Jahra 65 000 m^3/day. Failaka Island is also provided with a small treatment plant having a capacity of 10 000 m^3/day. It consists of waste stabilization ponds and is expected to produce wastewater of similar quality to the three mainland treatment plants.

Several sewage pumping stations have been constructed and serve to convey raw sewage to the treatment plants. Raw sewage from domestic premises is conveyed by pressure and gravity mains to these pumping stations from whence it is pumped to the respective plant for treatment; and the effluent produced is pumped for agricultural purposes.

Construction of a data monitoring centre (the name given to the treated effluent pumping, storage and administrative centre) has been completed and the facilities provided include:

- two 170 000 m^3 storage tanks which receive treated effluent pumped from Ardiya and Jahra sewage treatment plants;

---

* AHT-Agrar Hydrotechnic, Federal Republic of Germany; FMC Corporation, United States and Grontmij, The Netherlands.

- pumping station/valve house;
- administrative building which is the headquarters of the operation and maintenance staff who manage the eflfuent project. It includes chemical and bacteriological laboratories for monitoring treated effluent quality at all stages of use and application, for monitoring the effect of treated effluent irrigation on the agricultural soil, etc. Telemetric signals are received here giving basic information regarding gravity and pumped effluent flows throughout the system;
- workshops for maintenance and stores.

The effluent, which is already treated, is further upgraded by means of tertiary treatment consisting of chlorination, rapid sand filtration and final chlorination. Initially, raw sewage is received at the Ardiya Sewage Treatment Plant where it is treated by a modified conventional method – the two-stage activated sludge process. The quality of the effluent can be summarized as BOD_5 concentration of 40 mg/l and total suspended solids of 40 mg/l, which are the standards required up to the tertiary processes. Sewage from the other areas is received by their respective treatment plants, where the treatment is based on the extended aeration system and the quality is within the standard range of 40 mg/l BOD_5 and 40 mg/l suspended solids concentration. Treated effluent from the three treatment plants, after secondary treatment, is subjected to a third stage of treatment (that is, tertiary treatment) consisting of chlorination, rapid sand filtration and final chlorination.

Each treatment plant is equipped with an up-to-date laboratory where chemical and biological tests are carried out, such as BOD_5, COD, sludge measurements, TDS, microbiological examination, and the results obtained reflect the quality of the final effluent.

Waste sludge passes to pre-thickeners and then to drying beds or to the digesters, where it remains for a period of about 25–30 days. It is then transferred to the thickeners and finally to the drying beds. This sludge, which is used as a soil conditioner and fertilizer, is supplied to government and private agricultural farms.

At present, no industrial waste is permitted to be connected to the sewerage system except after pretreatment.

The total cost of producing 4545 litres of tertiary-treated effluent is 454 fils*.

Agricultural and forestry irrigation

The government strategy for implementation of the Effluent Utilization Project is to give highest priority to the development of irrigated agriculture by intensive cultivation in enclosed farm complexes, together with environmental forestry in large areas of low-density, low water-demand tree plantations. The ultimate project design provides for the development of 2700 ha of intensive agriculture and 9000 ha of environmental forestry.

The following facilities are provided for agricultural irrigation systems:

- two 170 000 m^3 effluent storage tanks;
- irrigation pumping station: maximum output 2400 l/s, with the installation of 12 pumps each at 90 m;

June 1985: US$1 = 0.3030 Kuwaiti dinar (1 KD = 1000 fils)

For forestry irrigation the systems include:

- *storage tanks:* treated effluent is gravitated or pumped to the forestry storage tanks;
- *irrigation pumping stations:* each forestry storage tank has an associated pumping station. Within the pumping station are pressure filters with 75 μm stainless steel screens. In addition, chlorine dosing facilities to dose up to 2 mg/l of chlorine are incorporated. Each pumping station has an associated fertilizer injection system.

Treated effluent is supplied via control points to blocks of forestry. Pressure-regulating valves, to reduce pressure, and 'Y' screen filters of 125 μm are included at the control point. The header mains downstream of the control point feed 12.5 mm polythene drip lines fitted with pressure-compensating drip emitters (two per tree) discharging 4 l/h operating over a 0.7–3.5 bar inlet pressure range.

Crops and vegetables

Treated sewage effleunt is at present used at the experimental farm and the irrigation project to irrigate the following:

- *Fodder plants:* alfalfa, elephant grass, Sudan grass, field corn (maize), vetch, barley, etc.
- *Field crops:* field corn (maize), barley, wheat and oats
- *Fruit trees:* date palms, olive, zyziphus and early salt-tolerant vines (sprinklers must not be used for fruit trees)
- *Vegetables:* potatoes, dry onions, garlic, beet and turnip can be irrigated by any method;
  vegetables which are to be cooked before consumption, such as egg plant, squash, pumpkin, cabbage, cauliflower, sweet corn, broad beans, Jews mallow, Swiss chard, etc., can be irrigated in any way but without sprinkling;
  vegetables which are eaten raw, such as tomatoes, water melons and other melons, can be irrigated with tertiary-treated sewage effluent by drip irrigation with soil mulching

The yield of green alfalfa is 100 tonnes/ha per year. Total production from the agricultural irrigation project, where the treated sewage effluent is mainly used, has reached 34 000 tonnes of vegetables and green fodder plants, including dehydrated alfalfa and barley straw. With this production, a reasonable quantity of some varieties of vegetables is made available to the local market, the total demand for green alfalfa for animals is met and some of the needs for dehydrated fodder are satisfied.

Health aspects of the reuse of sewage effluent for irrigation

An efficient monitoring system must be provided to conduct daily analyses for heavy metals and bacteria.

The proposed guidelines for tertiary treated effluent to protect health are:

- suspended solids 10 mg/l
- BOD_5 10 mg/l

• COD 40 mg/l	
• disinfection	chlorine residual of about 1 mg/l after 12 h at 20°C
• coliform bacteria	
for forestry, fodder and crops not eaten raw	10 000/100 ml
for crops eaten raw	100/100 ml
Not to be used on salad greens or strawberries	

As regards heavy metals, cadmium is the only one that might cause a health problem in food grown on treated sewage effluent, therefore it will be monitored and crops will be analysed for cadmium. It is also planned to measure cadmium in the kidneys of animals fed on forage irrigated with treated sewage effluent.

Agricultural labourers who deal with the sewage effluent are medically controlled as a pre-employment measure and given periodic 6 monthly examinations and periodic vaccinations. It has been observed that no outbreaks of infectious diseases have occurred since this procedure began in 1976.

Appendix: Conclusions of the Report submitted by WHO consultants who visited Kuwait from 13 to 18 March 1982

(1) The plan for reuse of effluent from the tertiary treatment of sewage in agriculture and forestry is devoid of significant public health risk.
(2) The proposed agricultural and forestry applications will not produce public health risk even if the standards are occasionally not met.
(3) If the tertiary treatment plant reliably meets the standards for more than 2 years, then the water could be considered for more demanding applications, including horticulture but not for use as potable water.
(4) Reliable operation of the secondary treatment plant with properly trained operators is necessary for the success of the proposal.
(5) The proposed chlorination step before the filters will not improve the hygienic quality of the effluent but some chlorine at this point may protect the filters.
(6) Nitrification is important in order to obtain high levels of disinfection at reasonable levels of chlorine.
(7) Odour control is important for public acceptance of the project. It is also one indication of good operation.
(8) Sulphide production in the sewers not only produces objectionable odours but leads to crown corrosion and interferes with plant operation.
(9) The provision of flow measurement devices at the Ardiyah plant is inadequate and existing devices are of doubtful reliability.
(10) A system of analytical quality control with intercalibration among different laboratories is an essential part of the collection of reliable data.

Chapter 24

The use of treated sewage effluent for irrigation: case study from Kuwait

R. O Cobham* and P. R. Johnson†

Introduction

Background

Until about 10 years ago, sewage effluent in the Middle East was largely an unsavoury and hazardous product of the human energy chain. For the most part it was discharged either to the sea or to wadis or onto unwanted and remote areas of desert. In a few cases it was used for irrigating agricultural crops, as well as for both productive and amenity trees. These functional uses invariably entailed health hazards for consumers, local residents and, not least, the work force involved in disposal and use.

The oil wealth of the Middle East has increasingly generated hitherto unimaginable opportunities for the development of cities, public amenities and domestic conveniences. Development has brought with it a growing awareness of the need to utilize scarce resources to the full. This applies particularly in the case of water resources. The introduction of reticulated water and sewage systems has yielded a new resource of considerable potential and value: treated sewage effluent in place of the embarrassing waste product of yesteryear. Not only have advances in the technologies of sewage treatment, distribution and application improved the qualities of urban life, but they have opened up opportunities for agriculture and afforestation in formerly austere areas of desert. In place of the occasional green oasis, it is now not uncommon in certain areas for lush plains of fodder and other crops to stretch to the horizon.

Aims

The purpose of this paper is to describe the main planning and management activities involved in seeking to make the best use of this new resource. It is based on general experience of effluent treatment and usage projects throughout the Middle East and upon one in particular, namely the Kuwait effluent utilization project.

Public health considerations and treatment standards associated with the use of wastewater are already well documented (Banks, 1981; Aikman, 1983; Beynon,

* Representing the Overseas Development Administration, London
† John Taylor and Sons, Consulting Engineers

Cobham and Matthews, 1984; Cowan and Johnson, 1984; Anon., 1985). Thus, this chapter purposely concentrates on the planning, land use, application and management aspects.

The choice of after uses

The value placed upon treated effluent is understandably high, since the ability of engineers to reduce health hazards has opened up a whole range of usage options; at least for discussion, even if not for selection and implementation. *Table 24.1* is interesting in that there was a large number of reuse options which the Kuwait project was required to investigate.

Table 24.1 The range of after uses investigated and initially selected

Potential after uses		*Kuwait project*
Agriculture:	cash and fodder crops	0
Afforestation:	commercial	0
Afforestation:	environmental protection/ sand stabilization	0
Beautification/landscaping in protected/ semiprotected areas		0
Ground water recharge		+
Industrial use		X
Recreation:	ponds and lakes	X
Livestock watering		X
Aquaculture		X

Key: 0 signifies initial selection as a chief component of the draft reuse plan
+ signifies initial selection as meriting further investigation under a pilot scheme
X signifies rejection for this particular project

The planning and implementation process

Planning to achieve the best use of treated sewage effluent is a complex and at times controversial activity. It entails many different skills, notably all forms of engineering, including public health and irrigation, surveying, hydrogeology, pedology, agriculture, aquaculture, forestry, horticulture, urban and rural planning, landscape architecture, economics, management and, not least, common sense.

The procedures involved in preparing plans which enable optimum use to be made of the treated effluent are similar to those involved in most forms of resource planning (Cowan and Johnson, 1984; Cobham, 1983; Cobham *et al.*, 1984). They are summarized in *Figure 24.1*. Within this broad outline of the overall planning and implementation process, there is a multiplicity of factors requiring investigation and appraisal. An attempt has been made to summarize the main physical, social and economic components in *Figure 24.2*.

Experience has shown that, arising either directly or indirectly from these components, there are a number of key issues or tasks upon which the ultimate success of an effluent reuse scheme is likely to depend. These are as follows:

- the organizational and managerial provisions made to administer the resource, to select the reuse plan and to implement it;

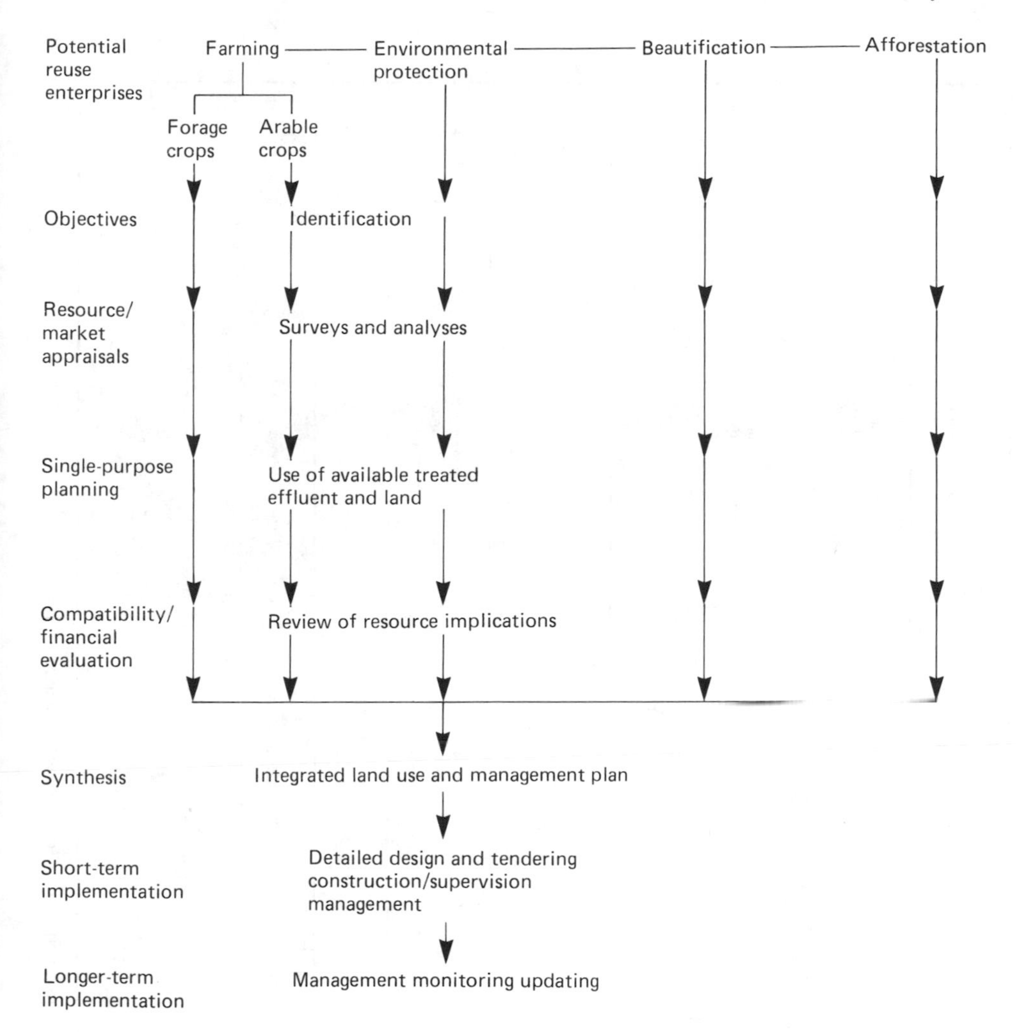

Figure 24.1 An outline of the resource planning and implementation processes

- the importance attached to public health considerations and the levels of risk taken;
- the choice of single-use or multiple-use strategies;
- the criteria adopted in evaluating alternative reuse proposals;
- the level of appreciation of the scope for establishing a forest resource.

A discussion of each of these follows later in this chapter. First, however, in order to provide a firm basis for such discussion, a brief description of the Kuwait reuse plan follows.

Kuwait case study

Background

Some 14 years ago, the first major sewage (secondary stage) treatment works were commissioned in Kuwait at Ardiyah. The initial supplies of treated effluent were

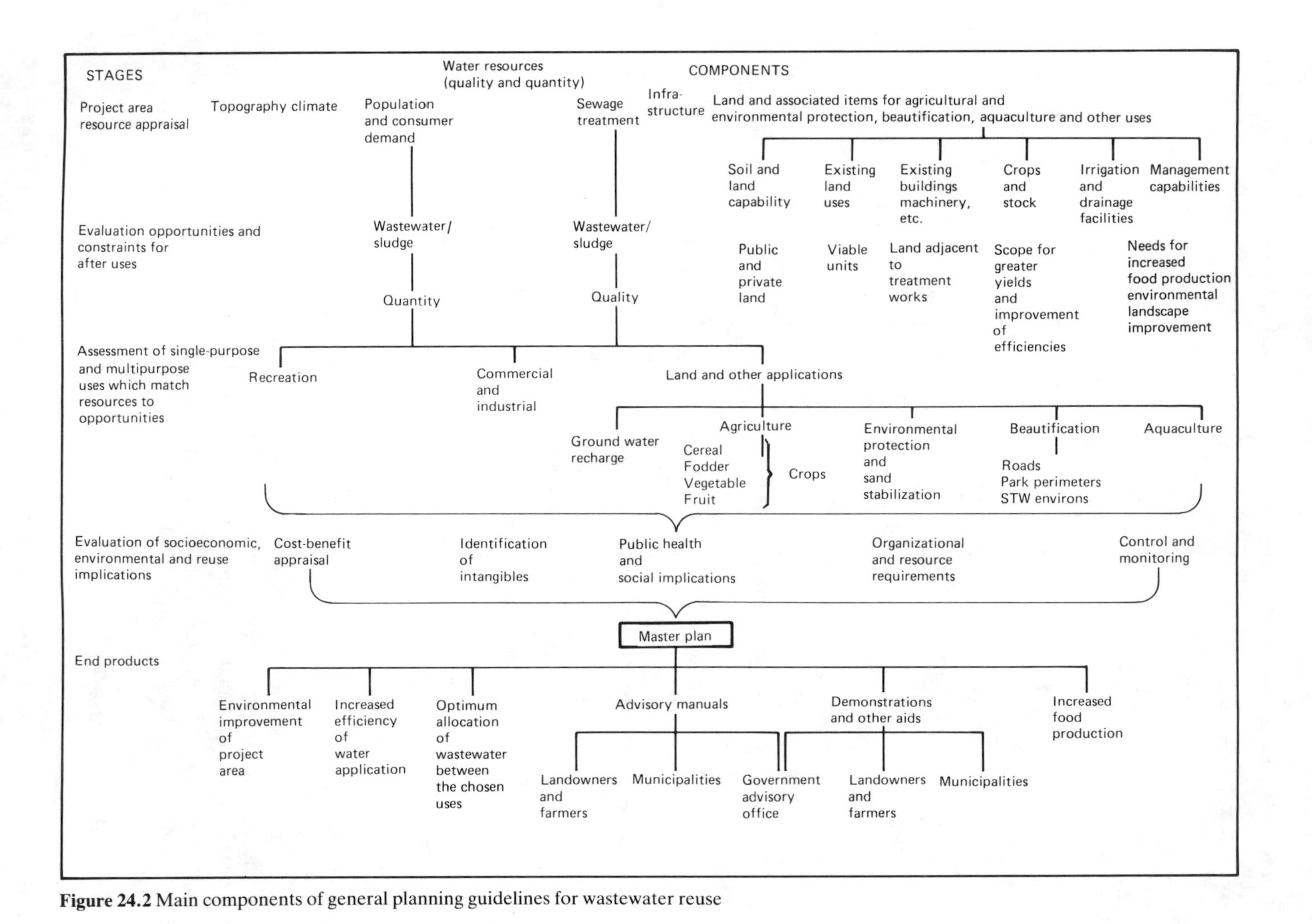

Figure 24.2 Main components of general planning guidelines for wastewater reuse

distributed to the experimental farm of the Department of Agriculture at Omariyah. These trials were undertaken using potable water, brackish water and treated effluent in order to compare crop yields. In 1975 an 850 ha farm was established by the United Agricultural Production Company (UAPC), especially for the purpose of utilizing the effluent. The directors of this close shareholding company represented the main private organizations involved in Kuwait agriculture, in particular the local dairy, poultry and livestock farming organization. In 1975 only part was under cultivation: forage (alfalfa) for the dairy industry was the main crop, using side-roll sprinkler irrigation, but aubergines, peppers, onions and other crops were also grown on an experimental basis, for which semiportable sprinklers and flood-and-furrow techniques were used. Today the whole of the farm is under cultivation.

Simultaneously came plans to expand the sanitary sewer network and associated treatment facilities, in order to service the burgeoning population of Kuwait city and its hinterland. Three new sewage treatment works, one serving the coastal villages, another at Jahra in the west and one for Failaka Island were designed.

Until the late 1970s agriculture in Kuwait was severely limited; the total cropped area even in 1976 only amounted to 732 ha. The country relied heavily upon food imports and the imports of both fresh and dried alfalfa were considered to be unnecessarily high.

In late 1977 the Ministry of Public Works initiated the preparation of a Master Plan for effective use of all treated effluent in Kuwait, covering the period up to the year 2010. The objective of the project is to make Kuwait as self-sufficient in certain food products as possible by then.

Approximately 27 million m^3/year of effluent is currently being used, which will rise to 125 million m^3/year by the year 2010, constituting one of the most ambitious schemes for agricultural reuse in the Middle East.

Land available and limiting factors

The project area was approximately 16 000 ha of land for possible reuse, of which some 1800 ha were thought to be suitable for intensive agriculture. The remainder was considered to have forestry or amenity potential. In all, some 24 sites were proposed. One of the first tasks was to assess the designated sites in terms of their relative development merits for single or multiple use. This led both to modification and confirmation of designated sites and to the identification of additional areas with development potential. The two major constraints for all irrigated development in Kuwait are the presence of 'gatch' (a hard impermeable calcareous layer at varying depths) and the high-percolation-rate/low-water-holding capacity of the soils. Thus, each of the sites was examined in detail and ranked according to its use potential. This was assisted by a low-density soil survey, which confirmed the generally superior nature of the soils closest to the western sewage works, compared with those in the vicinity of the works for the coastal villages. The areas designated for agricultural development generally contained soils which were better than those present on land designated for amenity/'afforestation' purposes. Deep to moderately deep sands through to loamy sands of a well-drained and non-saline nature were found on the proposed agricultural sites. The incidence of both shallow soils and of 'gatch' was low in the designated agricultural areas.

Table 24.2 Strategy options for effluent reuse

A Master plan based on single-purpose use	*B Master plan based on multi purpose use*		
		Use priorities	
1 *Allocation exclusively to agriculture*	*First*	*Second*	*Third for 'seasonal surplus'*
1.1 Field grazing by sheep 1.2 Zero grazing by livestock, based on high forage–low concentrate diet 1.3 Zero grazing by sheep, based on high concentrate diet 1.4 Zero grazing by dairy cattle, based on high forage diet 1.5 Zero grazing by dairy cattle, based on high concentrate diet	1 Forage production for dairy enterprises based on high concentrate diet; vegetables grown in rotation	Forestry	Extra forage or vegetables or forestry irrigated at subsistence rates in some months
2 *Allocation exclusively to vegetables* 2.1 Production on extensive scale 2.2 Production on intensive scale	2 Forage production for dairy enterprise based on low concentrate diet; vegetables grown in rotation	Forestry	Extra forage or vegetables or forestry irrigated at subsistence rates in some months
3 *Allocation exclusively to forestry*	3 Forestry, including a modest 'maximum production' area	Forage production based on high concentrate diet; vegetables grown in rotation	Extra forestry irrigated at subsistence rates in some months
3.1 'Maximum production' forestry (625 trees/ha – 25 m^3 wood/ha year) 3.2 'Environmental protection' forestry (125 trees/ha – 5 m^3 wood/ha year)	4 Forestry, including a substantial 'maximum production' area	Forage production based on low concentrate diet; vegetables grown in rotation	Extra forestry irrigated at subsistence rates in some months

Reuse options

Public health considerations substantially reduced the number of options which realistically could be evaluated in preparing the reuse Master Plan (John Taylor and Sons in association with Cobham Resource Consultants, 1978). Thus the sites originally designated as green grass amenity areas for informal public recreation were almost completely deleted.

Four types of use were identified as being 'safe' candidates meriting further investigation, namely:

- agriculture and horticulture on an extensive field scale, involving the production of 'safe' crops. The latter were defined as those either requiring cooking prior to human consumption or those with edible parts not requiring and receiving irrigation for a significant period prior to harvesting. These constraints limited the crops to alfalfa, potatoes, onions and garlic;
- intensive production of the three 'safe' vegetable crops under protected or fully controlled environmental conditions;
- commercial afforestation for fuel and charcoal;
- green belts of trees and shrubs adjacent to highways, designed so as to minimize sand blow and to enhance the appearance of the visual corridor. The need, for health reasons, to protect such areas against access by both grazing animals and members of the public was recognized.

Identification of Master Plan options embraced the four single-purpose uses on the one hand and a host of multiple-use combinations on the other, as indicated in *Table 24.2*. Evaluation of the various options entailed a considerable amount of survey and resource budgeting work, the main items in addition to the soil and water quality assessments already outlined, being:

- the estimation of the annual, monthly and daily water demands of the 'safe' crops, using the most appropriate irrigation techniques;
- an appraisal of the agricultural, horticultural and charcoal sectors of the Kuwait economy to determine the relative contributions which could be made by each of the 'safe' crops to domestic consumption, import savings, etc.;
- the matching of the predicted annual, monthly and daily irrigation demands with the available supplies, so as to achieve optimum physical and financial utilization throughout the year.

A further example of the multiple reuse strategies identified and evaluated in the course of typical Middle East reuse projects is provided in *Table 24.3*.

Technical and economic appraisals

From the various supply and demand studies undertaken, it was clear that development of the agricultural economy should be based primarily upon growth of the existing agribusiness sector. The need to develop an efficient, intensive forage production system as a basis for expanding the national dairy industry was recognized to be a top priority task. The potential for developing both vegetable and domestic charcoal production was also revealed.

Attention focused on the choice of irrigation methods based on their relative efficiencies, as well as upon the capital expenditure and staffing implications of each of the options. The systems and methods chosen for this project are

Table 24.3 Strategy options reviewed

Strategy A Limited multipurpose strategy geared primarily to the establishment of environmental protection areas on a large scale	Tertiary treated wastewater allocated between priority uses, as follows: Priority 1 Landscaping the sewage treatment works and its approach road Priority 2 Creating, where appropriate, a nursery to grow the large numbers of plants required Priority 3 Development of green belts planted with trees and shrubs for environmental protection and improvement, principally in sand dune areas; these are areas of type 1 planting* Priority 4 Establishment of environmental protection planting type 2* Priority 5 Utilization of seasonal surplus quantities of wastewater to establish and sustain environmental protection planting type 3*
Strategy B Multipurpose strategy based upon landscaping and environmental protection planting	Tertiary treated wastewater allocated between priority uses, as follows: Priority 1 Landscaping the sewage treatment works and its approach road Priority 2 Creating, where appropriate, a nursery to grow the large numbers of plants required Priority 3 Beautifying important highways and roads, particularly ring roads Priority 4 Establishing the structure planting for important public recreation facilities such as rural and urban parks
Strategy C Multipurpose strategy involving allocation primarily to agriculture	Tertiary treated wastewater allocated between priority uses, as follows: Priority 1 Landscaping the sewage treatment works and its approach road Priority 2 Growing commercial fodder crops on government owned and managed farms Priority 3 Beautifying important highways and roads, particularly ring roads Priority 4 Protecting highways, treatment works, dwellings and farm land against wind blown sand by the establishment of green belts consisting of environmental protection planting type 1 Priority 5 Establishment of environmental protection planting type 2 areas Priority 6 Utilization of seasonal surplus quantities of wastewater to establish and sustain areas of environmental protection planting type 3

* In the interests of ensuring that under each of the strategies maximum benefit is gained from the total wastewater supplies available, three different types of environmental protection planting have been proposed, namely:

Environmental protection planting type 1, irrigated throughout the year at the 'thrive' rate; the total application of wastewater amounts to 4000 ppm/ha year.

Environmental protection planting type 2, irrigated at the 'thrive' rates in the five winter months and either at, or as near as possible to, the 'survival' rates for the summer months.

Environmental protection planting type 3, irrigated at the survival rates for as many months as possible, which usually means solely for the period November to May inclusive.

Table 24.4 The choice of irrigation systems and methods

Enterprise	*System*	*Method: options*
Forage	Sprinkler	Hand moved; self-propelled rain guns; centre pivots; solid set; *side roll*
Extensive vegetables	Sprinkler	As above
Maximum production forestry (4 m × 4 m planting grid)	Drip	Pressure compensating emitters
Environmental protection forestry (4 m × 4 m planting grid)	Drip	Pressure compensating emitters
Agricultural wind breaks	Drip	Pressure compensating emitters
Intensive vegetables (1 ha unit, 100 m × 100 m)	Drip strip	Twin walled strip of either 0.08 mm (one crop life) or 0.3 mm thickness (two to three crop life); plus hand operated sprinkler system for leaching

Note

Provision was made for the following in the design of the irrigated network: onsite storage, pumping plant, filtration and distribution systems, fertilizer injection. The master plan contained recommendations concerning the provision of appropriate management, supervisory and training inputs.

summarized in *Table 4.4*. Sensitivity analyses were also undertaken, covering changes in yields, costs and prices.

Strategy assessments: land and effluent use

It was recognized that in theory it might at some future stage be necessary for national economic or strategic reasons to attach overriding importance to one particular land use, such as milk production. However, it was deemed inappropriate to adopt any of the single-purpose options, on account of the poor use which would be made either of effluent supplies or of the designated land areas. In the case of the option involving exclusive use for 'environmental protection' forestry, the area which could be irrigated with the available effluent far exceeded that designated for the project.

Experience gained from other resource planning projects indicated the advantages of adopting a mix of reuses. These included greater flexibility, increased financial security, more efficient use of the wastewater throughout the seasons, etc. These factors pointed to the need to search for a multiple-use strategy. Identification of the best combination of uses was strongly influenced by the hierarchy of predicted financial performances for the individual uses, as shown in *Table 24.5*.

Table 24.5 Indices of relative financial performance (during the first 10 years)

Enterprise (100 ha units)	*Indices*
'Maximum production' forestry	1
Extensive vegetables in rotation with barley hay	6
Extensive vegetables in rotation with alfalfa	11
Dairying based on alfalfa and a high concentrate diet	97

However, the results of the soil and other surveys, coupled with efficiency considerations, suggested that the adoption of a multipurpose strategy would only be appropriate in the case of the effluent treated by the northern/western located works. For the designated sites surrounding both the coastal village and Failaka Island works, there was a strong case for concentrating reuse on one or more types of forestry, irrigated at either optimum or subsistence rates. It was recognized in the case of the northern/western works, that it would be best to confine agricultural development initially to one rather than all of the major designated sites.

Distribution strategies

Various engineering strategies were considered for the distribution of effluent, both to the designated sites and between the works. Evaluation of these led to the conclusion that it would be desirable to transfer the seasonal surpluses available from the coastal village works to the western sites designated for agricultural and forestry uses. Careful evaluation of the four multipurpose strategies pointed to the superiority of strategy B1, as judged by the efficiency of use and the scope for developing agricultural potential. Since only reuse for forestry was feasible in the case of the coastal village sites, it seemed sensible to select a use strategy for the northern/western treatment works which gave priority to agriculture. Both the significant underuse of the UAPC farm infrastructure and the good financial prospects for both dairying and vegetable production pointed to strategy B1 being the one which would best serve the national interest.

The Master Plan

Table 24.6 displays the overall plan proposed in terms of both distribution system and land use. The recommendations submitted for the western and northern sites provide an insight into a main part of the Master Plan:

First priority should be devoted to developing an integrated system of forage (used in a high concentrate ration dairy enterprise) and extensive vegetable production on the UAPC farm, so that full utilization is made of its existing and potential facilities as soon as possible. The utilization should be based on:

- modern irrigation techniques;
- strengthening of shelter belts;
- provision of adequate effluent storage facilities to cope with temporary problems or breakdowns in the system;
- trial of different irrigation equipment under local commercial conditions;
- investigation of the relative merits of vegetable production on intensive and extensive scales;
- improvement of both management and technical husbandry skills.

Second priority should be given to developing fresh forage/hay production in rotation with vegetables on the other agricultural sites to the point where at least two, if not three, vegetable storage units are justified.

Once the first two priorities have been achieved, then as much prime and subsistence Environmental Protection Forestry as possible should be planted. Provided that the trials concerning commercial timber production yield positive results, it is recommended that an area of at least 213 ha of Maximum Production Forestry should also be included.

Table 24.6 The Master Plan – land use implications

Western and northern sites	*1980* (ha)	*2010* (ha)
Agriculture		
Forage: dairy enterprise	70	670
Forage: open market sale	149	589
Horticulture		
Extensive vegetables	50	200
Forestry		
'Maximum production' forestry	20	213
'Environmental protection' forestry at recommended irrigation rate	3808	7826
'Environmental protection' forestry at subsistence irrigation rate	401	–
Others		
Existing trial sites, vegetable areas	46	46
Subtotal	4544	9544
Coastal village sites		
Forestry		
'Maximum production' forestry	52	787
'Environmental protection' forestry	1673	1673
Subtotal	1725	2460
Failaka Island		
'Environmental protection' forestry	176	284
Total	6445	12288

The resource implications of the Master Plan were assessed. These included the tree nursery production required for the new forestry areas, the infrastructural requirements including boundary walls and roads, the irrigation equipment and machinery needs for the forage, vegetables, etc.

Implementation: construction, operation and maintenance

The construction of the coastal and western treatment plants began in 1979 and was thus underway while the effluent utilization report and designs were being produced. The construction works for effluent utilization began in mid-1981. Civil engineering works for effluent utilization began in mid-1981. Civil engineering construction work to the value of KD 33.4 million was completed by the autumn of 1983, while mechanical and electrical works to the value of KD 11.0 million took until the end of 1983. Delays in provision of the permanent power supplies to all 12 sites deferred the start of commissioning of the project until 1985. Full-scale operation of the project began in late 1984.

Present and future management

While physical progress to date has largely matched the expectations, many of the organizational and management arrangements still remain to be finalized.

Review of key considerations

Introduction

The case study provides a basis for discussion of some of the key factors. Reference was briefly made to the five aspects at the end of the third section and a review of each in turn now follows.

Organization and management commitment

Initially, the scope and success of any resource planning and implementation project depends in large measure upon the administrative skills which are brought to bear. The potential uses of the treated effluent span a wide range of both urban-based and rural-based interests, at least at regional and local levels. Consequently, it has been considered desirable that policy decisions concerning:

- the allocation of the effluent between competing uses;
- the investment in supporting resources, especially the recruitment and training of the managerial and technical staff, required to administer each main component of a reuse plan; and
- the maintenance of quality standards

should be taken by a Consultative Council comprising a group of senior officials representing each of the main interests. The executive functions should be delegated to a specially constituted committee under the leadership of a project director.

If the aim is, and it should be, to ensure that optimum use is made of the treated effluent from a national standpoint, then 'desiderata' become necessities. In short, more than just nodding acceptance of the type of organizational structure recommended and displayed in *Figure 24.3* is required, not only to ensure that the optimum plan is chosen and implemented, but that it is achieved.

Unfortunately the same level of priority displayed in establishing feasibility studies does not appear to have been given, so far, to setting up the required organizational framework and procedures.

Finally, and ultimately, the most important ingredients in any effluent reuse scheme is the level of supervision at all stages of the process. This starts at the treatment works and proceeds to the performance of the irrigation equipment right through to quality control of the resulting products, be they primarily of commercial or environmental value.

As with most major development projects the creation of appropriate management, monitoring and public relations procedures is as important as the construction works. This is a subject area where much pioneering work remains to be done if the endeavours of consulting engineers and land-use experts are to achieve their full potential.

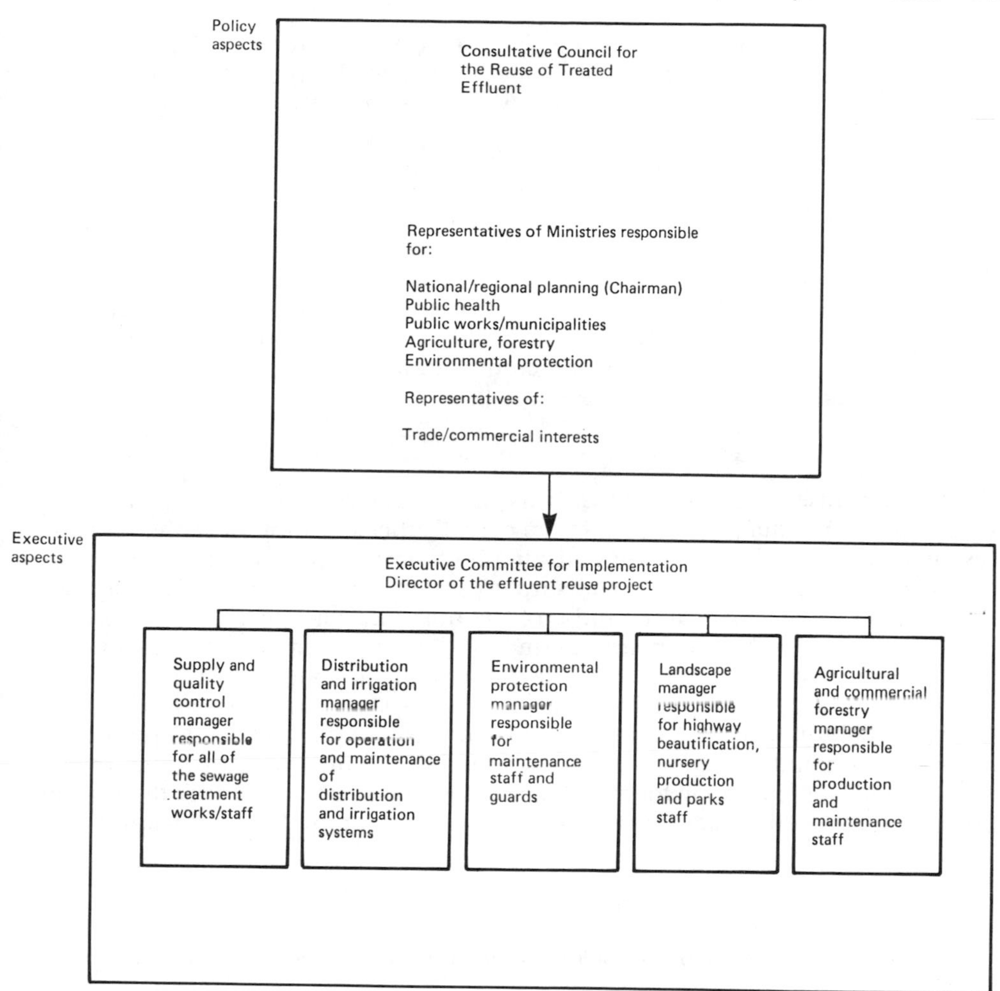

Figure 24.3 Typical organizational and management structures required

Public health risks

The importance accorded to safeguarding public health varies from country to country and from time to time. This is a factor which has to be recognized when preparing reuse plans. Likewise the fact that practice does not always match national regulations or guidelines. Despite any discrepancies between *de facto* and *de jure* considerations, it is always the latter which provide the basis for consultants' work.

As a general rule, consultants take great care to provide minimum risk proposals, recognizing that the ultimate decision may be influenced by political and financial considerations. In Kuwait, the consultants' recommendations to exclude all amenity uses and to restrict agricultural use in effect to four 'safe' crops was upheld.

In the end, the responsibility for not abusing the resource rests with each individual member of the public. If all the recommended precautions involving direct site supervision and the use of both fencing (to protect the public from gaining entry to areas of tree and shrub planting irrigated with treated effluent) and notices of the type used in Abu Dhabi (to warn the public of effluent irrigated green areas) are ignored, it can be concluded that responsibility should rest with the individual. However, in spite of the underlying logic, such simplistic views are not always tenable, since one individual can be the source of an epidemic. In the end, a decision has to be taken as to what constitutes reasonable levels of risk and public protection. The decision is not made any simpler by the fact that it is usually extremely difficult to trace the precise route of viral epidemics.

In view of there being none or only poor records, the adoption of conservative design principles by consultants would appear to be the only safe and responsible course of action. Until such time as records indicate otherwise, standards should not be relaxed.

The choice of single-use or multiple-use strategies

While single-use strategies undoubtedly have financial merit, they do have significant shortcomings in relation to the overall efficiency with which the valuable resource is itself used. The lower utilization of the single-purpose strategies stems directly from the fact that seasonal surpluses are generated, since the irrigation requirements in summer exceed those of the other scasons quite substantially. In the absence of complementary uses during these periods, unproductive disposal of the seasonal surpluses is inevitable. In the case of the multiple-use master plan adopted for Kuwait, the effective utilization of any such surpluses was achievable through establishment of seasonally irrigated areas of environmental protection forestry, using indigenous species. A comparison of the levels of both treated effluent and land utilization for multiple-use and single-use plans under the Kuwait project indicates the superiority of the multiple-use approach, as shown in *Table 24.7.*

Table 24.7 Comparison of utilization efficiencies achievable under single-purpose and multipurpose strategies

Kuwait case study	*Proportion of total Effluent used*		*Proportion of total designated land used*	
	1980 (%)	2010 (%)	*1980* (%)	*2010* (%)
1 Single-purpose strategies				
Dairying based on alfalfa	63	63	3	12
Vegetables: extensive scale	44	44	8	30
Vegetables: intensive scale	49	49	12	46
'Maximum production' forestry	64	64	13	52
'Environmental protection' forestry	63	36	61	100
2 The preferred multipurpose strategies				
Overall multiple use	100	100	34	68
Total agricultural sites	N/A	N/A	9	34
Total forestry sites	N/A	N/A	46	83

N/A = not applicable

Reuse strategy evaluation criteria

The criteria used to test the feasibility and suitability of potential reuse strategies are extremely important. Invariably, project appraisals concerning economic or financial components are not easy and straightforward. Inclusion of environmental considerations, as is the case with reuse effluent projects, does not ease the task.

Table 24.8 displays the criteria used to evaluate effluent reuse strategies. They represent a mixture of quantitative and qualitative yardsticks, thus reflecting the many dimensions of reuse projects. While a strong case can be made for using a diverse array of criteria, in the final analysis a value judgement is required by the decision-takers. One of the lessons to be learnt from the case study and indeed other effluent reuse projects is the need to establish the criteria and their relative 'weights' at an early stage.

Table 24.8 Strategy selection criteria

(1) Utilization of wastewater for multipurpose use
(2) Provision of flexibility
(3) Recognition and satisfaction of the individual and genuine needs of each individual townships
(4) Preservation and development of distinct and separate identities for each of the townships
(5) Generation of significant and prestigious community benefits
(6) Implementation and subsequent management of the project to be simple and sustainable
(7) Achievement of cost-effectiveness criteria
(8) Provision of valuable socioeconomic contributions in terms of increases in local skills, raw materials for local industries and recreation facilities
(9) Integration with activities and initiatives of existing institutions and with urban landscape plans
(10) Inclusion of both high priority elements common to all townships (for example, beautification of sewage treatment works and environmental protection planting areas near to the sewage treatment works)

The scope for afforestation

As yet, the scope for utilizing effluent as a means of establishing potentially commercial, as distinct from environmental protection, forestry has barely been realized. In Kuwait, certainly, the potential for reducing charcoal inputs is appreciated. However, the fact that in many arid countries there is virtually even less of a tradition to practise commercial forestry than 'high tech' farming is largely responsible for the relatively reticent approach towards forestry. To some extent this is surprising. Annual productivity levels achieved in irrigation sand areas are estimated to range between 5 and 25 m^3/ha for *Prosopis* and *Tamarix*. In practice, the considerable yield variations depend upon irrigation levels in particular, but also upon stocking and survival rates as well as the incidence of illegal cutting.

The results of the appraisal for afforestation in one project are interesting:

> The establishment of environmental protection planting could possibly produce some tangible economic benefits to be offset against the costs. Preliminary investigations indicate that the demand for the sorts of timber products, with the possible exception of fence posts, which could be produced, namely fuel wood, poles and charcoal, is limited. Historically, timber was an important resource in the local economy, as evidenced by the concern expressed as recently as 20 years ago about the depletion of supplies and over-exploitation. However, changes in availability and in the relative costs of alternative fuels for cooking and heating, together with the introduction of new building materials and techniques, appear

to have led at best to a static demand and at worst to a declining market for most types of locally produced timber.

Even charcoal, although still apparently sold in reasonable quantities despite being more expensive than alternative petroleum based fuels, is mainly imported. Nonetheless, despite this declining market considerable quantities of similar wood products are imported each year. This suggests that there may be opportunities for using new and improved techniques of timber production and processing. If, in future, market trends changed significantly, or if for any other reason it was thought desirable, there is no doubt that timber could be produced. Given adequate water, very high yields could be achieved. Yields of up to $10\,m^3$/ha are possible, on the basis of both the proposed planting densities of 400 trees/ha and irrigation at a rate of $8000\,m^3$/ha year. Lower yields, but still giving some timber return could be obtained as a useful by-product. This could be achieved simply by utilizing the surplus water available for much of the year to irrigate the proposed environmental protection areas at higher rates.

Alternatively, if desired, greater emphasis could be placed on timber production as an objective from the outset. This would involve establishing much smaller areas, which would be adequately watered throughout the year and could possibly be planted with better timber species.

Charcoal production would enable maximum advantage to be taken of the potentially high growth rates which can be achieved in the harsh conditions which prevail. There are indications that there is a growing demand for charcoal for mainly recreational uses. It is important to note that the quality of charcoal available in local markets is high. However, *Eucalyptus* and *Prosopis* are both species from which good quality charcoal can be obtained.

In order to compete with imports, careful processing of domestic produce would be necessary. In many countries the use of steel kilns, producing high yields of clean charcoal, is well established. Under such a project no wood for processing could be expected for at least 10 years after planting.

Pursuit of production as well as environmental improvement objectives would involve extra costs. The labour involved in harvesting and processing operations would be the main extra expense involved. Harvesting would bear an approximately 10 year cycle. Subsequently vigorous coppice regrowth should occur. However, it might be necessary to interplant in some cases after harvesting. Since interplanting could involve up to 50% of the area, this could also add, but not significantly, to the costs.

The current and likely future demand for building materials offers a potential market for timber. Although recent trends in the national construction industry are moving away from traditional techniques and materials, there may be possible uses for locally produced hardwood timber, if only for scaffolding and fence posts.

The use of treated effluent for afforestation, offering environmental and possibly recreational benefits as well, is considered by the consultants to be a subject which merits an R and D programme in its own right.

Pointers for the future

Experience to date has shown that the costs of both effluent reuse and desalination are high. Consequently, it is very important that optimum use should be made of all treated water.

Unfortunately, in the agricultural sector it has been observed (Bowen-Jones and Dutton, 1983) that 'the price gap between food demand and supply is only met by the large direct and indirect subsidization of almost all (domestic) agricultural activities. This in turn does little to encourage the efficient use of any resource other than that of Government financial support'.

Against this background there are five main pointers, which emerge from the experience to date of using treated effluent for agricultural and environmental purposes:

- provided sewage effluent is treated to remove elements harmful to plant growth, it is a valuable resource capable of generating acceptable crop yields;
- there are public health and operator risks due to the possible presence of viruses, which suggest that continual vigilance is required;
- the type of treatment is important in the interests of ensuring that the effluent can be applied efficiently, prevention of the build-up of algae being particularly important in the case of drip irrigation equipment;
- ultimately, the level of success achieved depends upon attention to detail at every stage between the treatment works and the 'downstream' products; thus
- management and supervision skills along with the provision of training facilities need to be of a consistently high standard.

Experience suggests that the potential for utilizing this new resource in a very wide range of outlets needs to be investigated under R and D projects. The terms of reference for such projects should, it is suggested, include provision for long-term experimental, development, training and demonstration work.

References

AIKMAN, D. I. (1983) Wastewater reuse from the standpoint of irrigated agriculture. *The Public Health Engineer,* **11**(1), 35–41

ANON (1985) Wastewater – for the right crops. *Arab World Agrobusiness,* **1**(1)

BANKS, P. (1981) Effluent reuse: how the Gulf States balance benefits and risks. *World Water*

BEYNON, R. B., COBHAM, R. O. and MATTHEWS, J. R. (1984) Kuwait effluent utilization project: concept and implementation. Paper 12, *Third Arab Water Technology Conference, Dubai*

BOWEN-JONES, H. and DUTTON, R. (1983) Agriculture in the Arabian Pensinsula, The Economics Intelligence Unit, *Special Report No. 145*

COBHAM, R. O. (1983) Professional integration in place of ineptitude. *Proceedings of the British Ecological Society and Landscape Institute Conference, Manchester*

COBHAM, R. O., MATTHEWS, J. R., McNAB, A. STEPHENSON, E. and SLATTER, M. J. S. (1984) Demonstration farms: agricultural landscapes. *CCP170,* Countryside Commission, Manchester

COWAN, J. P. and JOHNSON, P. R. (1984) Reuse of effluent for agriculture in the Middle East. *Reuse of Sewage Effluent,* Chapter 7. London: Thomas Telford Ltd

TAYLOR, J. and SONS, in association with Cobham Resource Consultants (1978) *Draft Master Plan Report: Utilization of Treated Sewage Effluent in Agriculture,* Volumes 1–4, Government of Kuwait, Ministry of Public Works

Chapter 25

Irrigation of Damascus plain (the Ghouta) with polluted water from the river Barada

M. N. Al-Rifai*

Introduction

Ernest Renan (1823–92) the French philosopher, historian and scholar of religion, first visited Damascus in the 1860s and later immortalized the Ghouta (the plain of Damascus) in the passage:

> The impression conveyed by this richly cultivated countryside, these delightful orchards, crisscrossed by streams and bearing the choicest fruits, is one of calm and happiness . . . from earliest antiquity to our own times, the whole of this zone which surrounds Damascus with coolness and well-being has had only one name, has inspired only one dream: God's Paradise.

In 1985, the area is still looked upon as the premier picnic site for the Damascene to escape from the heat and noise of the city in the spring and summer. However, it would seem that in the future, the Barada and its derivatives, which are increasingly transformed into open sewers, will change to a large extent this charming picture of the Ghouta.

The siting of the City of Damascus, which is the capital and most important city in Syria, was determined by the existence of the river Barada which flows eastwards out of the hills of the Anti-Lebanon and into the plain of Damascus (*Figure 25.1*). The manmade derivative canals of the Barada (*Figure 25.2*), which have been developed over the centuries, fan out to form the oasis east of Damascus, the Ghouta, and provide irrigation water for the numerous fruit and vegetable farms. As the City of Damascus, now considered as probably the oldest existing city in the world, expanded, so small satellite villages developed in the Ghouta area and the irrigation network was expanded.

The river Barada

The principal sources of the river Barada are springs in the Anti-Lebanon mountains, which emerge to form a small lake in the Zebdany valley, 45 km north-west of Damascus. The water flows through the steep-sided Barada valley until its confluence with the Ein El Figeh spring, which contributes about two-thirds of the total Barada flow.

---

* Professor of Irrigation and Sanitary Engineering, University of Damascus, Syria

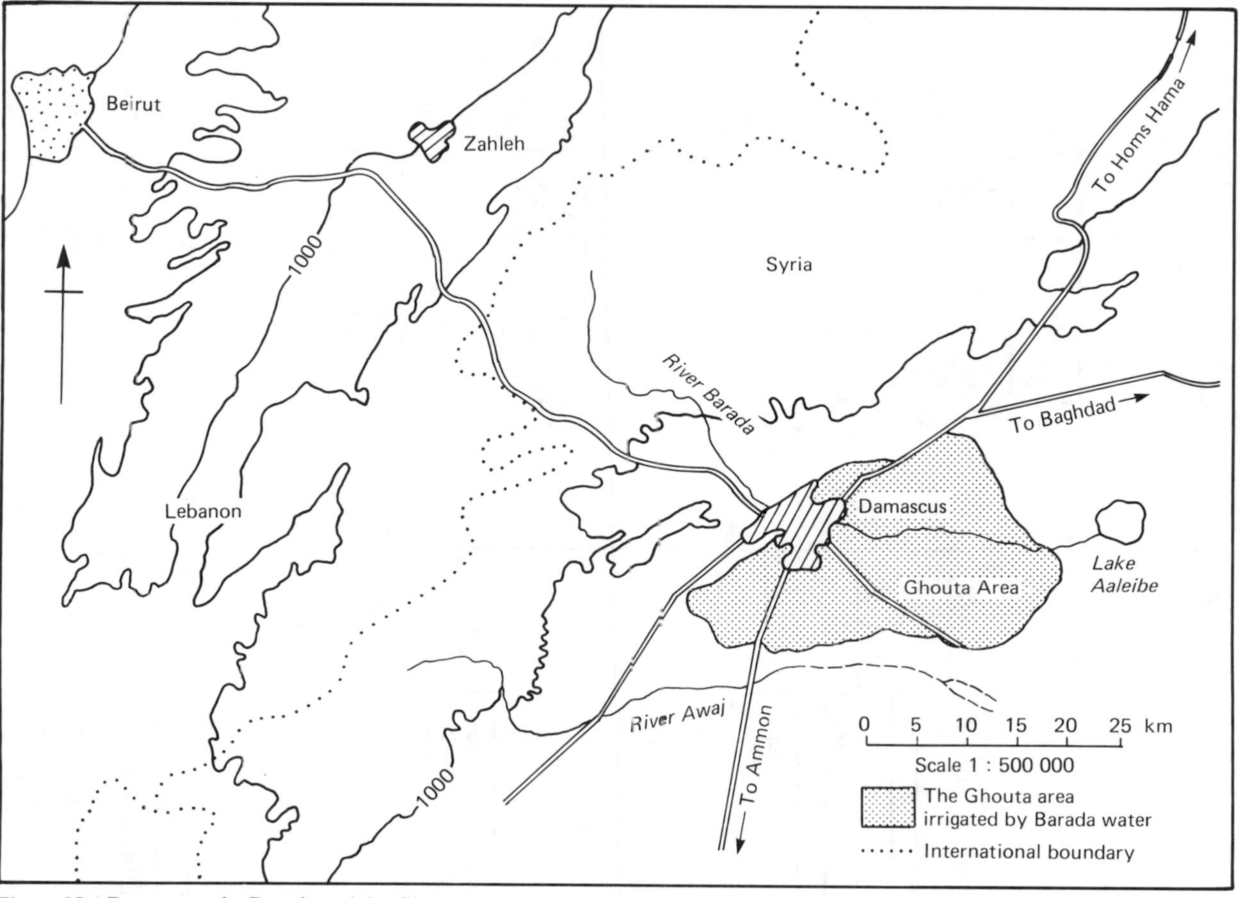

Figure 25.1 Damascus, the Barada and the Ghouta

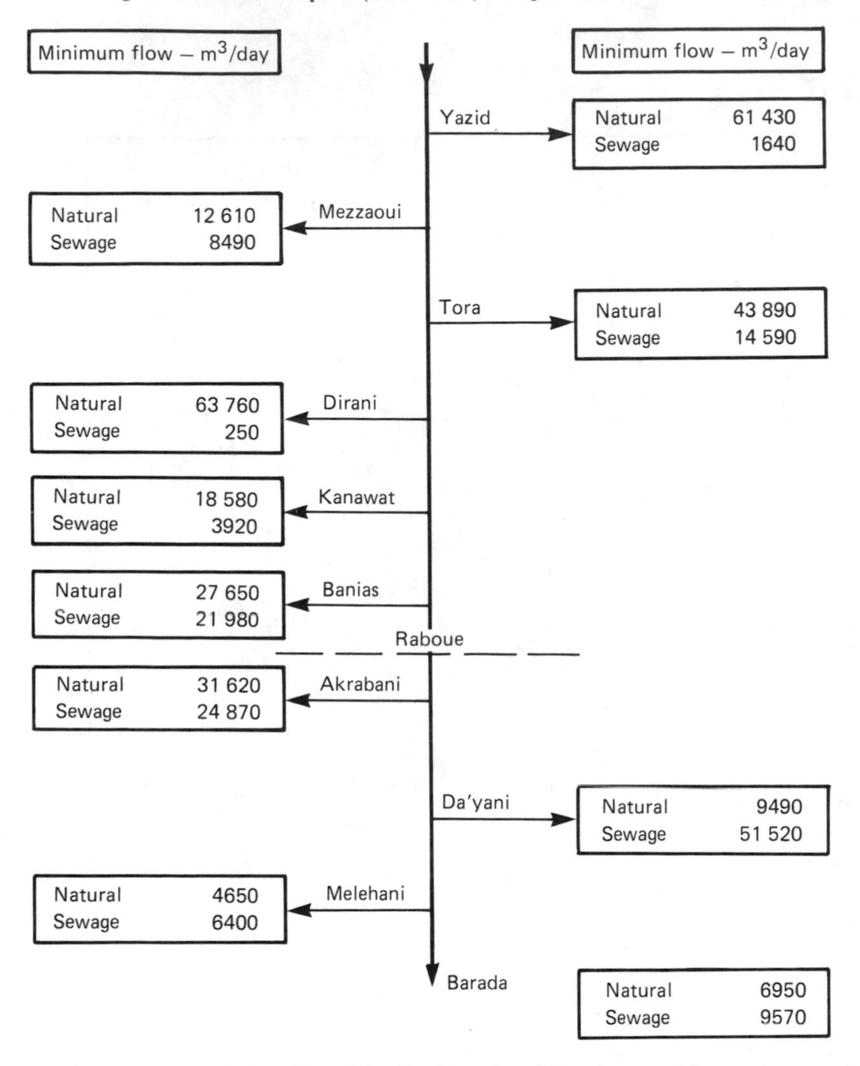

Figure 25.2 Schematic branching of the distributaries of Barada natural flow and sewage flow

Near the narrow valley entrance to Damascus at Raboue, the Barada has six derivative branches which diverge to form a fan flowing through Damascus, finally irrigating about 70 000 ha of the agricultural Ghouta plain before disappearing as minor irrigation canals into the desert or, ultimately, into a depression called Lake Aateibe. *Table 25.1* gives the annual 1 in 5 year (dry) hydrograph of the river Barada at Raboue (*Figure 25.3*), while *Table 25.2* gives the monthly flows in canals and the river Barada at Raboue (1975–76).

- *The Yazid* is the first principal canal to branch from the river Barada north of the village of Hameh; it does so to the left, flows parallel to the Barada until Raboue, then branches to the north side of Damascus through the villages of Kaboun and Harasta.

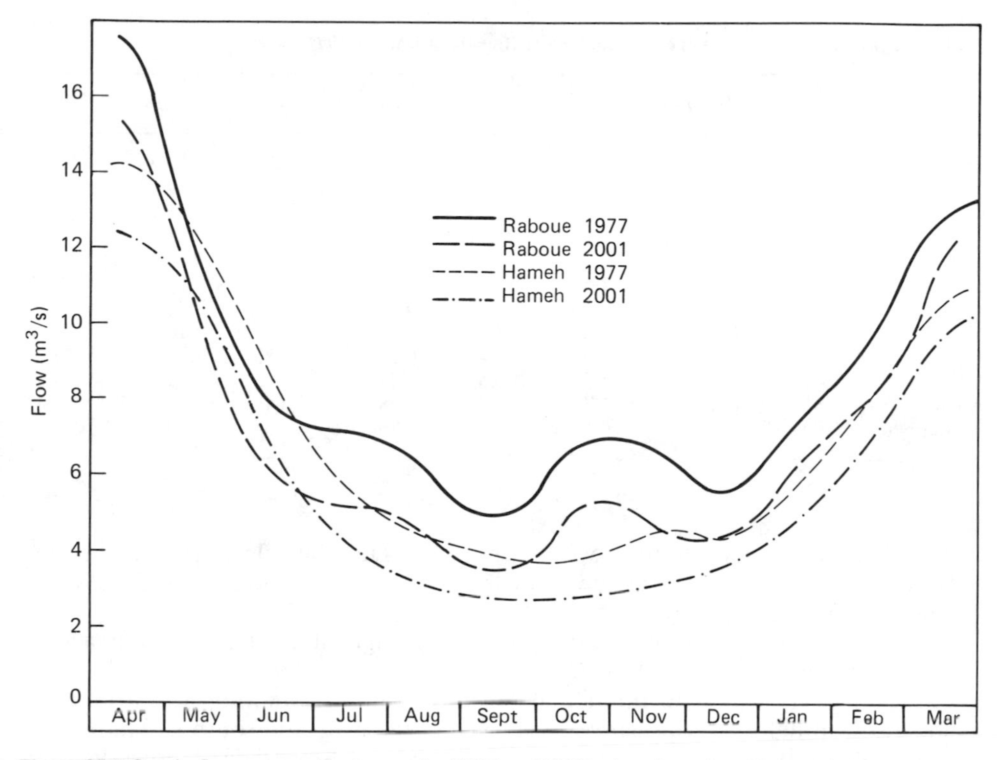

Figure 25.3 One in 5 year annual hydrographs (1977 and 2001); river Barada at Hameh and total flow at Raboue

Table 25.1 Derivation of annual 1 in 5 year (dry) hydrograph of river Barada at Raboue

Month	*1 in 5 year monthly flows at Hameh* (m^3/s)	*Percentage change between Hameh and Raboue*	*1 in 5 year monthly flows at Raboue* (m^3/s)
October	3.89	+78	6.92
November	4.57	+50	6.86
December	4.49	+26	5.66
January	5.91	+26	7.45
February	8.12	+19	9.66
March	10.68	+22	13.03
April	14.22	+22	17.35
May	12.10	−7	11.25
June	8.39	−8	7.72
July	5.60	−27	7.11
August	4.57	+36	6.22
September	4.00	+25	5.00
Mean	7.21	+26	8.69

Table 25.2 Mean monthly flows in canals and River Barada at Raboue 1975–76 (m^3/s)

Month	*Yazid*	*Mezzauoi*	*Dirani*	*Tora*	*Kanawat*	*Banias*	*Barada*	*Total*
October	1.49	0.49	0.74	0.71	0.25	0.35	0.44	4.47
November	1.17	0.21	0.86	0.27	0.24	0.33	0.57	3.65
December	1.14	0.22	0.86	0.94	0.19	0.49	1.16	5.00
January	1.03	0.20	0.67	1.91	0.25	0.62	2.06	6.74
February	1.93	0.33	1.13	2.07	0.28	0.92	2.10	8.76
March	1.46	0.32	1.43	2.18	0.37	1.76	12.11	19.63
April	1.24	0.27	1.83	3.84	0.45	2.02	20.46	30.11
May	1.72	0.25	1.43	3.64	0.55	1.35	10.74	19.68
June	1.63	0.30	1.52	3.43	0.56	0.85	2.37	10.66
July	1.46	0.34	1.53	2.55	0.48	0.83	0.83	8.02
August	1.82	0.32	1.34	1.47	0.38	0.42	0.68	6.43
September	1.80	0.30	0.98	0.88	0.36	0.34	0.65	5.31

- *The Mezzauoi* branches to the right, north of Doumar, flows parallel to the Barada until it reaches Raboue, then branches to the south through the old village of Mezze.
- *The Dirani* is next to branch off, again to the right until it too disappears as irrigation channels in the West Ghouta.
- *The Tora* branches off to the left; it flows through the Kaboun industrial area and Harasta before disappearing as irrigation canals to the south of Harasta.
- *The Kanawat* branches off to the right of the Barada towards El-Keswa district; before reaching that district it splits into many irrigation channels and disappears into the West Ghouta.
- *The Banias* is the last of the major derivative canals to branch off upstream of Damascus; it splits into smaller irrigation canals before disappearing into the East Ghouta.

Other minor branches flowing downstream of Damascus are the Akrabani, the Dayani, the Melehani and the river Barada downstream of Melehani.

Industrial pollution in the Barada river

Table 25.3 lists all the industries considered to be causing serious pollution to the Barada and its distributaries at present (1985 figures). The results of water quality analyses from these water sources are shown in *Tables 25.4* and *25.5*. Estimated concentrations of salinity, chloride, heavy metals and other phytotoxins are shown in *Table 25.6*.

The major industrial discharge concentrations of BOD_5, based on averaged monthly analysis between May 1977 and May 1978, are shown diagrammatically in *Figure 25.4*. At sampling points shown in *Figure 25.5* the percentage of dissolved oxygen saturation is shown in *Figure 25.6a*, the BOD_5 in *Figure 25.6b*, the ammoniacal nitrogen in *Figure 25.6c*, the chloride in *Figure 25.6d* and the conductivity in *Figure 25.6c*.

Table 25.3 Pollution loads and pretreatment recommendations for major polluting industries

Reference number	*Industry*	*Wastewater discharge* (m^3/day)	*Polluting loads* (kg/day)					*Recommended pretreatment requirements*	*Discharge to*
			pH	*BOD_5*	*COD*	*SS*	*Other*		
E. Ghouta tanneries									
38, 39, 40, 41	Tanneries	2000	4–13	–	4000	1000	S high SO_4 high		S.I
98	National tannery number 4	650	–	–	–	–	S 42	Eliminate benzene discharge	S.I
E. Ghouta									
42	Chickens	12	–	500	–	1000		Brush screening, balancing	S.I
43	Abattoir	1000	–	200	5500	–		Blood abstraction plant commissioning imminent	S.I
45	Dairy	407	–	–	–	–		Nil	S.I
46	Soft drinks	50	14	–	–	–		Nil	S.II
50	Fruit and vegetables	201	–	330	650	–			S.II
51	Paint	8	–	–	–	–		In-house discipline	S.II
52	Textile	360	5	–	–	–		Balancing	S.II
54	Pharmaceutical	30	–	–	–	–		Nil	S.II
55	Rubber	90	–	–	–	–		Nil	S.II
56	Matches	400	Batch discharge					Balancing	S.II
58	Soap and oil	143	13	200	2600	360		Brush screening, grease removal acid cracking, neutralization	S.II
Sit Zeinab									
67	Textile	276	–	–	–	–		Nil	S.II
Tarik el Kiswe									
70	Batteries	37	–	–	–	70		Nil	S.II
81	Refrigerators	95	–	–	–	–		Nil	S.II
82	Electroplating	4	12	–	–	8		Provide storage and remove by tanker to dump	T
85	Household utensils	240	Existing settlement					Nil	S.II
87	Weaving	363	Existing settlement					Nil	S.II
88	Ceramics	6	–	–	–	1000		Settlement – 2 h	S.II
90	Cables	Small	2.8	–	–	–		Neutralization	S.II
92	Washing m/c	10	1	Existing settlement				Neutralization prior to settling	S.II

Notes

1. All the above industries are programmed to be connected to the main sewer system by the completion of phase 2 of the Damascus Sewage Master Plan. Pretreatment is therefore recommended to produce effluents of a standard suitable for discharge to sewer.
2. S = Effluent standard suitable for discharge to sewer. S.I. = to be sewered in Stage I. S.II. = to be sewered in Stage II.
3. T = Remove waste to dump.
4. A dash (–) in the table indicates that the concentration of the polluting characteristic is below the limit for acceptance into a sewer.

Table 25.4 Results of water quality analysis of the river Barada and its distributaries

River	*Sample point*	*pH*	*EC* (mmhos/cm)	$CO^{=}_3$	HCO_3^-	Cl^-	Na^+	Ca^{2+}	Mg^+	K^+	*SAR*	*Adj* SAR*
				(mg/l)								
Mezzauoi	9	8.13	450	0	122	11.5	7.6	47.6	7.2	1.2	0.27	0.40
Dirani	10	7.97	410	0	117	11.8	7.4	70.2	8.4	0.7	0.22	0.40
Kanawat	12	7.76	480	0	143	16.0	12.6	65.2	7.5	4.2	0.69	0.70
Akrabani	21	6.90	880	0	238	55.7	64.2	130.0	12.2	15.0	0.84	1.93
Tora	22	7.52	630	0	166	29.9	32.0	66.6	8.3	4.8	0.98	1.86
White Barada	23	7.77	920	0	226	52.1	30.0	121.6	18.4	2.9	0.67	1.54
Barada	23A	7.00	1140	0	254	117.7	69.6	123.6	13.6	10.0	1.58	3.63
Melehani	25	7.13	760	0	203	32.1	34.0	104.6	9.8	8.3	0.41	0.86
Yazid	26	7.91	590	0	165	22.8	22.5	64.2	7.0	4.9	0.71	1.35
Da'yani	28	7.29	850	0	189	60.7	56.6	196.8	13.2	10.5	1.05	2.41
S'beina Canal		8.01	1100	0	287	63.5	49.6	109.2	15.7	16.6	1.17	2.81

* Adj SAR takes into account residual sodium carbonate (Ayres and Westcot, 1976)

Table 25.5 Analysis of samples taken 30 June 1977

River	*Sample point*	*Al*	*B*	*Cd*	*Cr*	*Cu*	*Fe*	*Ni*	*Mn*	*Zn*
		(mg/l)								
Mezzawi	9	0.01	0.04	0.005	0.04	0.013	0.18	0.001	0.004	0.004
Dirani	10	0.08	0.04	0.001	0.07	0.015	0.16	0.010	0.004	0.090
Kanawat	12	0.02	0.21	0.005	0.07	0.021	1.6	0.001	0.031	0.066
Akrabani	21	0.03	0.40	0.002	0.11	0.058	2.9	0.032	0.058	0.185
Tora	22	0.01	0.40	0.001	0.09	0.043	0.65	0.010	0.026	0.150
White Barada	23	0.01	0.04	0.001	0.06	0.021	0.14	0.005	0.004	0.026
Barada	23A	0.02	0.04	0.001	0.34	0.056	0.48	0.013	0.023	0.123
Melehani	25	0.11	0.04	0.003	0.12	0.059	2.8	0.020	0.055	0.180
Yazid	26	0.02	0.58	0.001	0.05	0.037	0.44	0.007	0.035	0.090
Da'yani	28	0.11	0.30	0.005	0.20	0.063	3.9	0.030	0.057	0.184
S'Beina Canal		0.24	0.79	0.003	0.28	0.067	5.3	0.027	0.153	0.240

Table 25.6 River Barada and distributaries – estimated concentrations of salinity, chloride, heavy metals and other phytotoxins in the present 1 in 5 year design flow

River	*EC* (mmhos/cm) at 25°C)	*Chloride* (mg/l)	*Al*	*B*	*Cd*	*Cr*	*Cu*	*Fe*	*Mn*	*Ni*	*Zn*
			(mg/l)								
Tora	845	46.1	0.011	0.653	0.001	0.135	0.065	1.006	0.036	0.016	0.24
Kanawat	529	58.5	0.023	0.265	0.006	0.084	0.024	2.018	0.037	0.001	0.08
Akrabani	1146	80.1	0.039	0.580	0.003	0.152	0.080	4.168	0.056	0.046	0.26
Melehani	891	46.6	0.139	<0.04	0.004	0.148	0.073	3.571	0.067	0.025	0.23
Barada	1484	162.3	0.024	<0.04	0.001	0.474	0.075	0.625	0.025	0.018	0.17
Da'yani	946	68.9	0.129	0.349	0.005	0.222	0.073	4.601	0.066	0.035	0.22

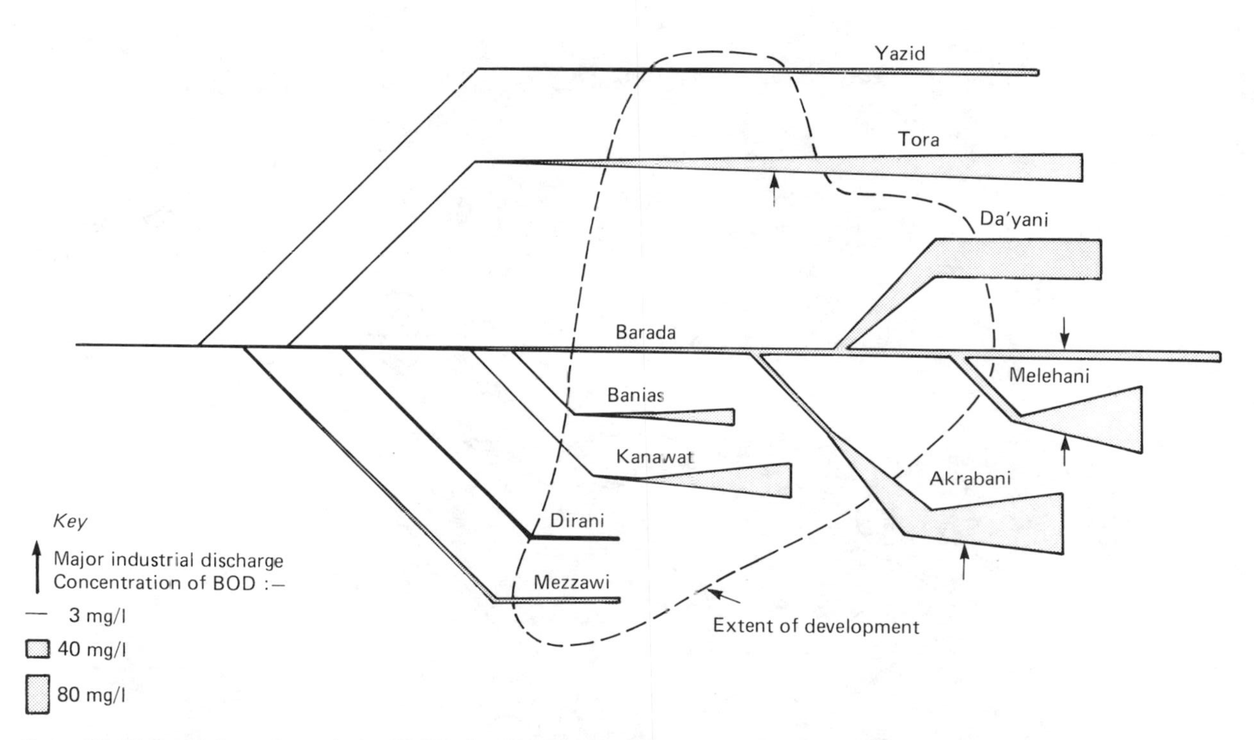

Figure 25.4 Averaged months analysis of BOD_5 from May 1977 to May 1978. Pollution levels are averages of monthly analyses

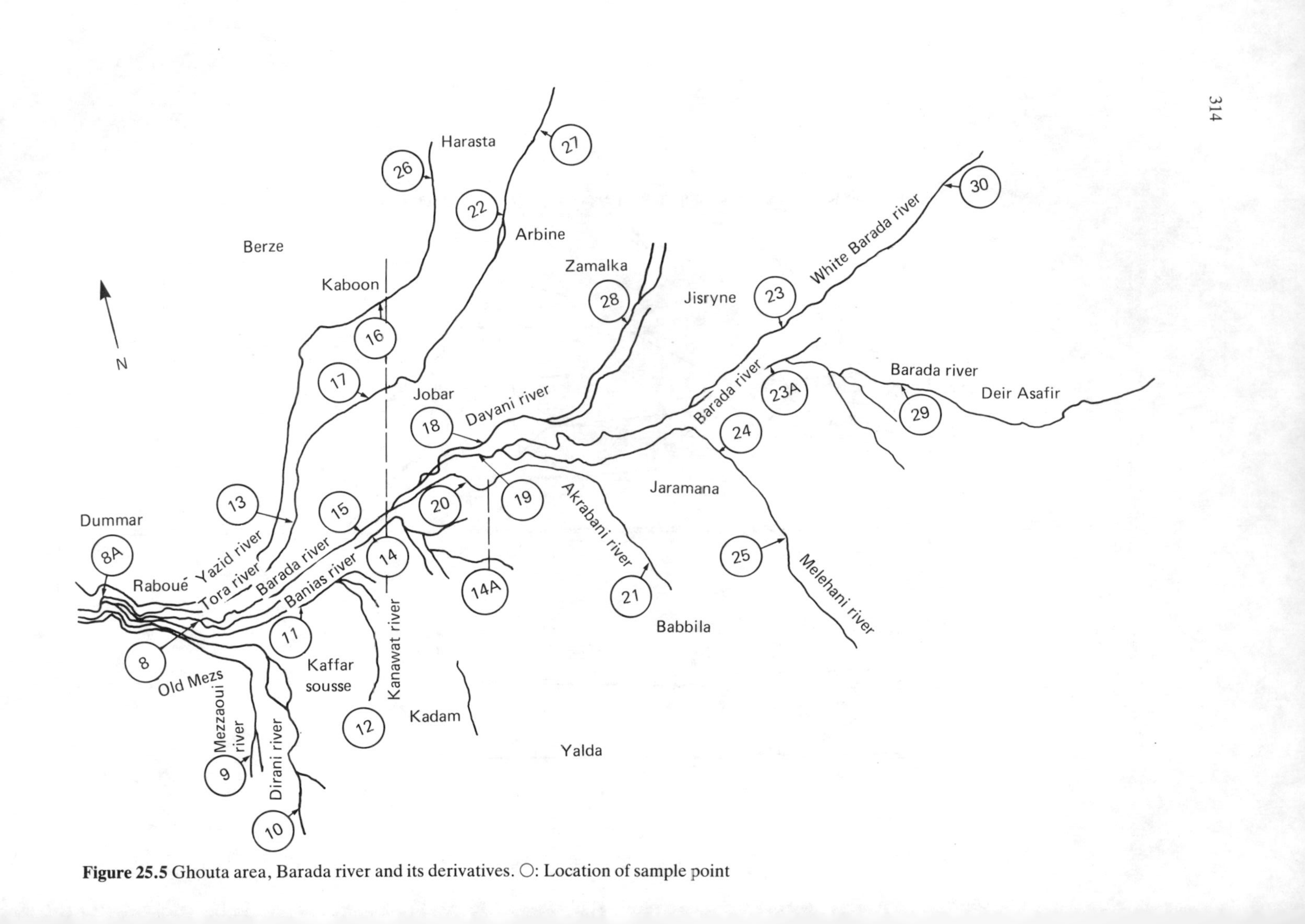

Figure 25.5 Ghouta area, Barada river and its derivatives. ○: Location of sample point

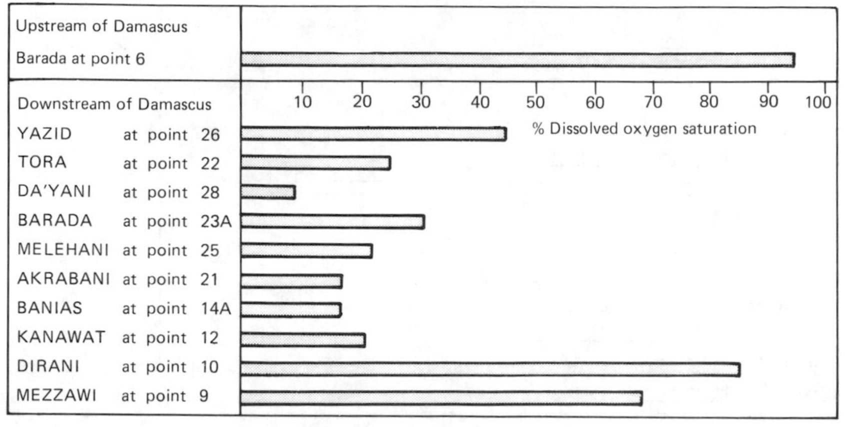

Upstream of Damascus
Barada at point 6
Downstream of Damascus
YAZID at point 26
TORA at point 22
DA'YANI at point 28
BARADA at point 23A
MELEHANI at point 25
AKRABANI at point 21
BANIAS at point 14A
KANAWAT at point 12
DIRANI at point 10
MEZZAWI at point 9
10 20 30 40 50 60 70 80 90 100
% Dissolved oxygen saturation

(*a*)

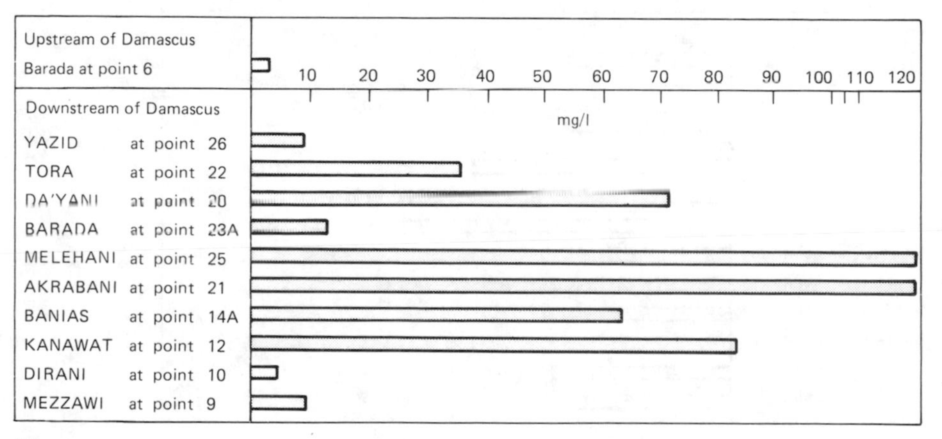

Upstream of Damascus
Barada at point 6
Downstream of Damascus
YAZID at point 26
TORA at point 22
DA'YANI at point 28
BARADA at point 23A
MELEHANI at point 25
AKRABANI at point 21
BANIAS at point 14A
KANAWAT at point 12
DIRANI at point 10
MEZZAWI at point 9
10 20 30 40 50 60 70 80 90 100 110 120
mg/l

(*b*)

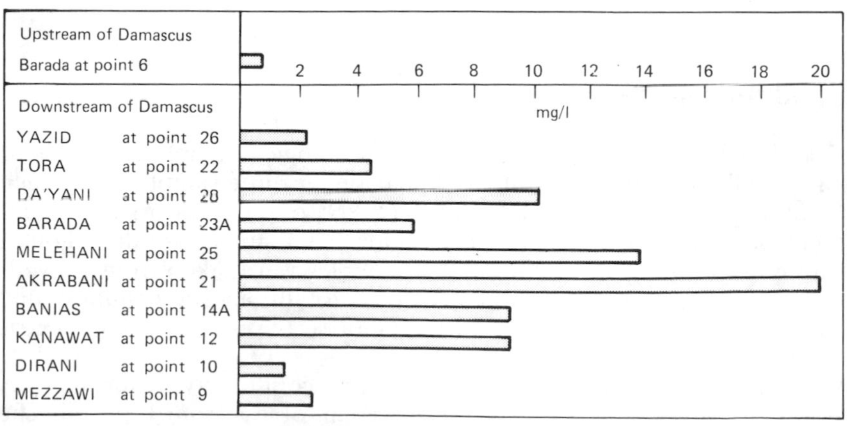

Upstream of Damascus
Barada at point 6
Downstream of Damascus
YAZID at point 26
TORA at point 22
DA'YANI at point 28
BARADA at point 23A
MELEHANI at point 25
AKRABANI at point 21
BANIAS at point 14A
KANAWAT at point 12
DIRANI at point 10
MEZZAWI at point 9
2 4 6 8 10 12 14 16 18 20
mg/l

(*c*)

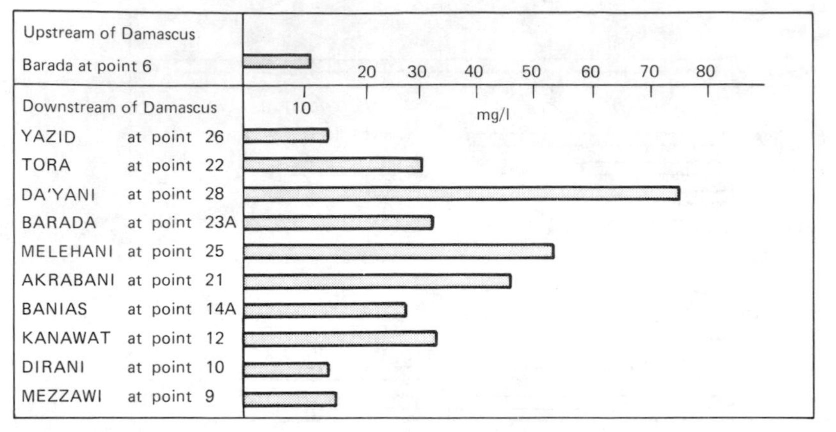

(*d*)

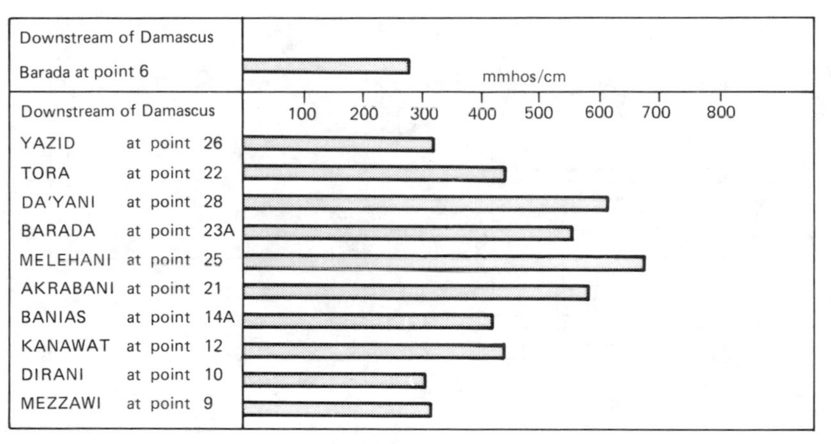

(*e*)

Figure 25.6 Concentrations at sampling points shown in *Figure 25.5: (a)* percentage of dissolved oxygen saturation; (*b*) BOD_5; (*c*) ammoniacal nitrogen; (*d*) chloride; (*e*) conductivity

Domestic pollution in the Barada rivers

Tables 25.7 and 25.8 summarize the results of bacteriological sampling from wells and rivers in the Ghouta, the bacteriological analyses of three samples from each sampling point and the most probable number (MPN) of coliform organisms per 100 ml. As a guide to the potential health hazard for the inhabitants of the Ghouta, especially those using wells for drinking water, samples were taken from 12 wells and from eight river locations. The summary of the total domestic pollution load discharged per day into the river system of the Ghouta, in order of magnitude, is shown in *Table 25.9* as kg/day of BOD_5.

The WHO recommended that individual or small community water supplies from groundwater sources should have a coliform count of less than 10/100 ml. The analyses of Ghouta groundwaters show that most samples exceed this standard and

Table 25.7 Results of bacteriological sampling – wells in the Ghouta

Location	*Most probable number of coliform organisms* (per 100 ml)											
	First sample				*Second sample*				*Third sample*			
	Date	*MPN*	*95% Confidence limit* *Lower*	*Upper*	*Date*	*MPN*	*95% Confidence limit* *Lower*	*Upper*	*Date*	*MPN*	*95% Confidence limit* *Lower*	*Upper*
Wells												
B1 east of Daraya	8.12.77	240	240	infinite	12.3.78	540	180	1400	3.5.78	2400	–	–
B2 north of Sbineh	13.12.77	240	90	850	25.2.78	2400	–	–	3.5.78	2400	–	–
B3 east of Sbineh	8.12.77	240	240	infinite	25.2.78	920	270	2200	3.5.78	1600	390	4200
B4 Akraba Village	8.12.77	10	0	10	25.2.78	20	5	40	10.5.78	10	5	20
B5 north of Mleiha	8.12.77	10	0	10	27.2.78	20	5	40	10.5.78	40	10	90
B6 northwest of Jaramana	13.12.77	60	20	140	27.2.78	10	–	–	10.5.78	10	5	20
B7 south of Jisrine	13.12.77	10	0	10	12.3.78	2400	–	–	13.5.78	20	5	40
B8 north of Sakba	14.12.77	170	60	470	20.2.78	10	5	20	10.5.78	1600	390	4200
B9 northwest of Arbinel	15.12.77	50	20	120	1.3.78	10	5	20	13.5.78	20	5	40
B10 southwest of Harasta	15.12.77	80	30	190	1.3.78	10	–	–	13.5.78	30	10	70

Note: Tests carried out by the Project Team.

Table 25.8 Results of bacteriological sampling – wells and rivers in the Ghouta

Location	*Most probable number of faecal coliform organisms (Escherichia coli/100 ml)* *Date*	*E. coli/100 ml*
A. Wells		
W1	1.8.77	542
W2	1.8.77	542
B. River		
Sample point no. 8	1.8.77	348
Sample point no. 10	1.8.77	348
Sample point no. 21	1.8.77	348
Sample point no. 22	1.8.77	348
Sample point no. 23	1.8.77	542
Sample point no. 23A	1.8.77	542
Sample point no. 25	1.8.77	348
Sample point no. 28	1.8.77	542

Note: Tests carried out by the Central Health Laboratory, Ministry of Health.

Table 25.9 Summary of the total domestic pollution load discharged per day into the river system in order of magnitude

River	*Total BOD_5 load* (kg/day)
Dayani	10 335
Barada	8 136
Akrabani	6 533
Tora	6 408
Banias	2 659
Mezzawi	2 201
Kanawat	851
Yazid	278
Melehani	157

therefore indicate the possibility of disease risk. The sources of well contamination are the foul streams and irrigation ditches. Gastrointestinal and infectious diseases suffered by many of the inhabitants of the Gouta and Damascus area are mostly due to vegetables irrigated with contaminated surface and groundwater in the Ghouta. There are many cases of intestinal worm infections among children and sometimes even among adults.

Removal of most coliform and faecal coliform organisms usually occurs in the first 0.6 m of travel through soil, although total removal is complete only after movements of 90 m through sand and gravel. Irrigation channels in the Ghouta are often less than this distance from wells, and the total vertical travel of infiltration water would only be 12 m, because this is the approximate depth of the water table in the Ghouta groundwater aquifers.

Groundwater replenishment and water balance

The storage coefficient for the Ghouta aquifer is estimated to be about 0.04–0.05 (test at Kharabo Pump Station, Faculty of Agriculture of the University of Damascus). Mean annual rainfall at Damascus is about 220 mm and about 160 mm in the Ghouta area. If it is assumed that the rainfall over the Damascus Basin is about 180 mm per annum, the yearly evaporation losses 1800 mm, the total basin area being 7000 km^2, then it can be estimated that:

- the total surface Barada runoff $\simeq 300 \times 10^6$ m^3/year
- the total yearly replenishment of groundwater $\simeq 200 \times 10^6$ m^3
- the total yearly surface runoff $\simeq 150 \times 10^6$ m^3 in winter.

The annual reported irrigation water requirement met by surface water, in an average year, is 156 Mm^3 and that met by groundwater is 296 Mm^3. Monthly figures are given in *Table 25.10*. Irrigated areas alter throughout the year according to farming practice and there is an increasing tendency to enlarge the irrigated area in the Ghouta by pumping from wells. The actual total irrigated area in the Ghouta is estimated to be about 73 000 ha, of which about 27 000 ha are irrigated in summer and about 30 000 ha in winter. The actual total yearly cropped area irrigated in summer and winter in Syria is about 60 000 ha, according to the official production plan of the Ministry of Agriculture and Agrarian Reform for the year 1983–84. About 80 000 ha is cultivated as rainfed area in the Ghouta.

Table 25.10 Outline monthly water balance for the Ghouta area

Month	*Estimated mean Barada discharge at Raboue (Hameh discharge +30%)*	*Irrigation requirement (evapotranspiration)*		*Excess water*
		River water	*Ground water*	
January	30	1.9	–	28.1
February	38	11.4	–	26.6
March	55	23	6.6	25.4
April	71	28.5	19.6	22.9
May	62	25.5	31.4	5.1
June	42	17	59.4	34.4
July	29	13.5	78.3	62.8
August	24	10.5	61.3	47.8
September	20	9	29.1	18.1
October	20	9.5	9.8	0.7
November	23	5.4	–	17.6
December	23	0.5	–	22.5
Total	436 (+ sewage 60)	156	296	−14 (+46)

Note: All quantities in Mm^3; the excess water is the Raboue discharge less the irrigation requirement.

The greater part of the groundwater resource is derived from the replenishment of the Ghouta aquifer from infiltration of surface water used for irrigation, from surface runoff during flood periods, and from leakage from the earth irrigation canals. The quantity of groundwater that enters the aquifer as underflow from the limestone mountains to the west, and from rainfall infiltration is estimated to supply only about 5% of the groundwater used for irrigation. About 400 Mm^3 per annum is estimated to be actually (1985) abstracted from the groundwater aquifer in the Ghouta by more than 40 000 drilled wells.

The water balance estimates indicate that the Ghouta aquifer is fully developed, at best, and probably overdeveloped. Abstraction of groundwater is equal to or in excess of the perennial yield. A prolonged reduction in the surface-water flows entering the Ghouta area will reduce the perennial yield and result in an increasing decline in groundwater levels. The increasing diversion of drinking water supply from the El-Figeh spring to the increasing population of the Damascus area will greatly affect the irrigation water supply to the Ghouta area. A water balance diagram for the Ghouta aquifer is shown in *Figure 25.7*.

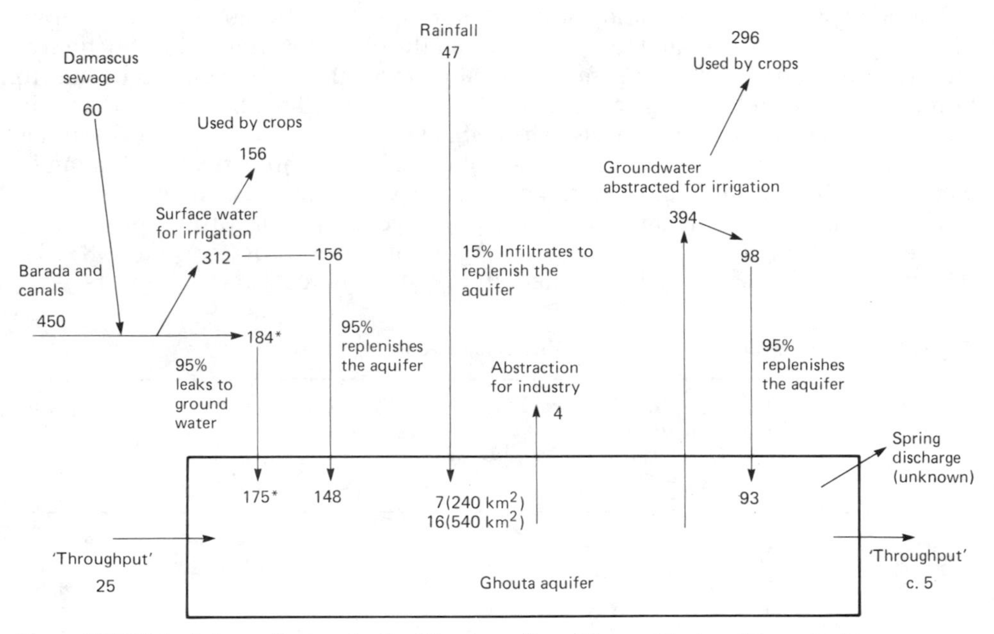

Figure 25.7 Water balance diagram for the Ghouta aquifer. All quantities in millions of cubic metres (Mm^3) per average year. The crop use estimates include direct evapotranspiration from groundwater. Figures with an asterisk assume no surface flow passing beyond Ghouta except spring discharges

River Barada water rights

The traditional right to use water from the canals and waterways in the Ghouta was probably established at the time the canal system was constructed. These water rights have been handed down from owner to owner through the generations. Complications have arisen when properties have been subdivided or amalgamated. Water rights are, in theory, inseparable from land rights but abuse of the rights, sales and exchanges have further complicated water distribution with the passing of time.

The Ghouta irrigation system is a private system managed by the users through management committees for each river distributary. These committees are supervised by an administrative commission which settles differences and fixes rights to water, based on documents of title or on custom. Rearrangement of the whole system is envisaged. It is anticipated that a new study will include recommendations for new offtake structures and proposed redistribution of water according to irrigation requirements rather than traditional water rights. The actual applied water rights in canals and the river Barada, as percentages and average monthly flow at Raboue, are shown on *Tables 25.11* and *25.12*, respectively.

Table 25.11 Mean monthly flows in canals and river Barada expressed as a percentage of total flow at Raboue 1975–76 (%)

Month	*Yazid*	*Mizzauoi*	*Dirani*	*Tora*	*Kanawat*	*Banias*	*Barada*	*Total (%)*
October	33.3	11.0	16.6	15.9	5.6	7.8	9.8	100
November	32.0	5.8	23.6	7.4	6.6	9.0	15.6	100
December	22.8	4.4	17.2	18.8	3.8	9.8	23.2	100
January	15.3	3.0	9.9	28.3	3.7	9.2	30.6	100
February	22.0	3.8	12.9	23.6	3.2	10.5	24.0	100
March	7.4	1.6	7.3	11.1	1.9	9.0	61.7	100
April	4.1	0.9	6.1	12.7	1.5	6.7	68.0	100
May	8.7	1.3	7.3	18.5	2.8	6.9	54.5	100
June	15.3	2.8	14.3	32.2	5.3	8.0	22.1	100
July	18.3	4.2	19.1	31.8	6.0	10.3	10.3	100
August	28.3	5.0	20.8	22.9	5.9	6.5	10.6	100
September	33.9	5.6	18.5	16.6	6.8	6.1	12.2	100

Table 25.12 Annual hydrographs (1 in 5 year dry) of canals and river Barada at Raboue (m^3/s)

Month	*Yazid*	*Mizzauoi*	*Dirani*	*Tora*	*Kanawat*	*Banias*	*Barada*	*Total flow at Raboue*
October	2.304	0.761	1.149	1.100	0.388	0.540	0.678	6.920
November	2.195	0.398	1.619	0.508	0.453	0.617	1.070	6.86
December	1.290	0.249	0.974	1.064	0.215	0.555	1.313	5.66
January	1.140	0.224	0.738	2.108	0.276	0.685	2.280	7.45
February	2.125	0.367	1.246	2.280	0.309	1.014	2.318	9.66
March	0.964	0.208	0.951	1.446	0.248	1.173	8.040	13.03
April	0.711	0.156	1.058	2.203	0.260	1.162	11.798	17.35
May	0.979	0.146	0.821	2.081	0.315	0.776	6.131	11.25
June	1.181	0.216	1.104	2.486	0.409	0.618	1.706	7.72
July	1.301	0.299	1.358	2.261	0.427	0.732	0.732	7.11
August	1.760	0.311	1.294	1.424	0.367	0.404	0.659	6.22
September	1.695	0.280	0.925	0.830	0.340	0.320	0.610	5.00
							Mean	8.69

Irrigated area and main production in Ghouta

The Ghouta soils have been formed on colluvial fans at the foot of the Kassioun mountain, which is directly to the west of Damascus. These soils are generally black, heavy textured clays of varying depths overlying unconsolidated rock. They are highly calcareous and over the centuries have been heavily manured with dung and town refuse. East of the area occupied by colluvial fans, on the fringe of the Ghouta area, reddish-brown soils are found. The reddish brown soils border the Lacustrine Merj soils to the east. Their texture is a heavy clay, which is reported to contain a layer of calcium carbonate accumulation at a depth of 50–80 cm, but gravels do not occur. These soils have a pH value of about 8.

The fact that the Ghouta soils are alkaline in reaction and of heavy texture has an important and favourable influence on their ability to accept water polluted with heavy metals. However, many variables affect the relationship between water quality and crop production such as: soil physical and chemical properties, climatic and plant physiological factors, stage of growth and others. The main Ghouta annual crops, including a wide range of summer and winter vegetables, are commonly grown between the trees of the mixed orchards; 1983–84 production is shown in *Table 25.13* and the annual cropping calendar is given in *Figure 25.8*.

Table 25.13 Annual production of main crops in the Ghouta 1983–84

Crop	*Area* (ha)	*Weight* (tonnes)
Orchard crops (mainly apricots and walnuts)	33 000	
Winter vegetables	10 000	
Summer vegetables	9 000	
Wheat		50 000
Barley		14 000
Cotton		9 000
Sugarbeet		25 000
Tomatoes		110 000
Potatoes		55 000
Garlic (dry)		75 000
Onion (dry)		5 000
Pastoral trefoil and alfalfa		20 000
Pastoral barley and flowering serm		75 000
Maize		5 000
Bitter vetch		2 000
Broad beans (dry)		4 000
Chickpeas		4 000
Haricot (dry)		3 000
Lentils		1 000
Aniseed		1 000
Sumac		1 000

Crop water requirements

Potential evaporation and evapotranspiration (in mm), calculated by the Penman method, for the Damascus area are:

Potential evaporation (E_o) in mm:

Jan	Feb	March	April	May	June	July	Aug	Sept	Oct	Nov	Dec
41	63	126	171	219	260	282	244	170	112	60	40

Total E_o = 1789 mm in Damascus

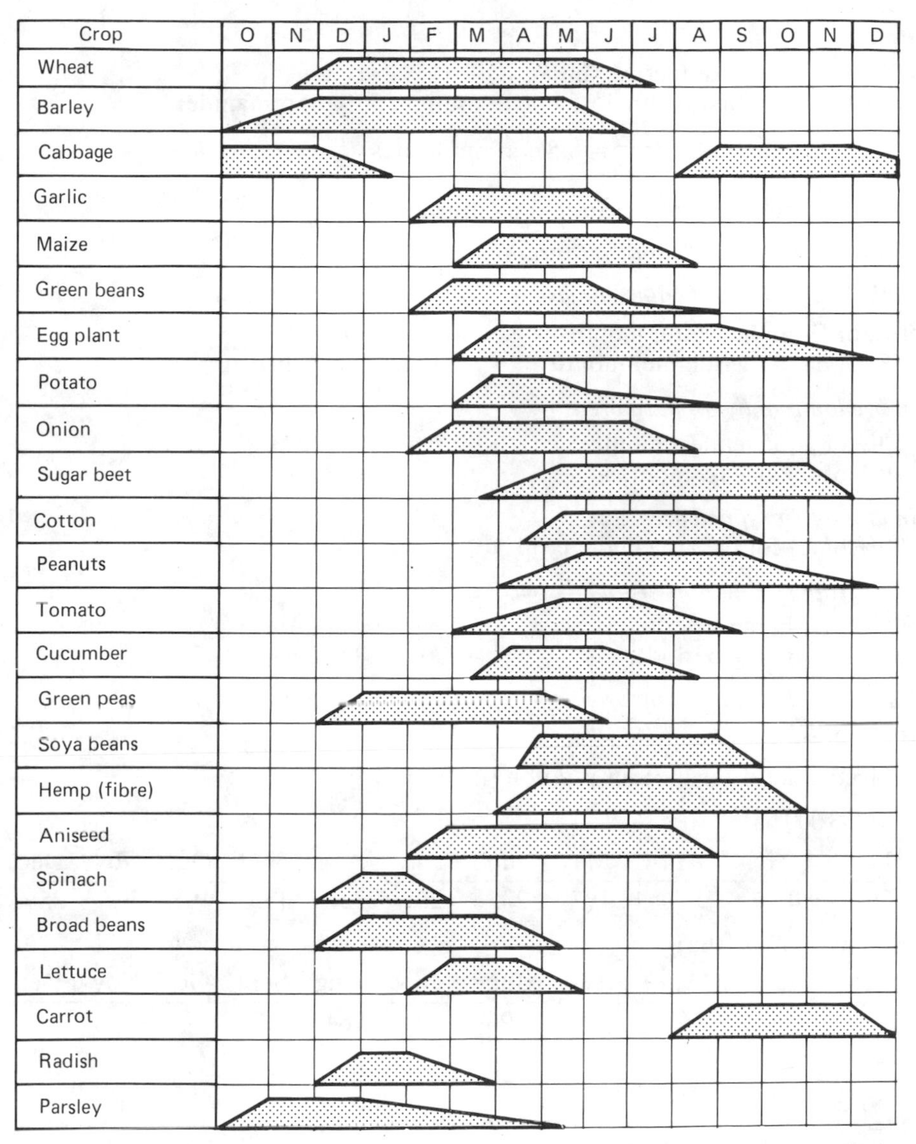

Figure 25.8 Annual cropping calendar for the Ghouta. Based on Kunert and Ali Jabri (1965)

Potential evapotranspiration (E_T) in mm:

Jan	Feb	March	April	May	June	July	Aug	Sept	Oct	Nov	Dec
36	56	114	157	197	235	255	221	153	105	55	35

Total E_T = 1619 mm in Damascus

Mean precipitation (P) in mm:

Jan	Feb	March	April	May	June	July	Aug	Sept	Oct	Nov	Dec
52	31	24	12	9	0	0	0	0	10	27	48

Total P = 213 mm in Damascus

The effective rainfall

July–October: no effective rainfall
November: first 10 mm lost as evaporation, 50% of remainder (i.e.: $\frac{27-10}{2} = 8.5$ mm) infiltrates

December–March: 70% of precipitation infiltrates.

Actual evaporation and evapotranspiration

(i) *In the city, town and village areas:*
50% of E_T in the winter months: *December–May*
25% of E_T in the summer months: *June–November*

(ii) *In predominantly orchard areas:*
70% of E_T in leafless months: *November–March*
100% of E_T in leafy months: *April–October*

(iii) *In annually cropped areas:*
100% of E_T for the area under crop

(iv) *In uncropped and non-cropped areas:*
50% of E_T in rainy period: *November–April*
25% of E_T in dry period: *May–October*

(v) *Lake Aateibe and swamp area:*
100% of E_o

The derived monthly evapotranspiration (in mm) is:

Orchard (33 000 ha)

Jan	Feb	March	April	May	June	July	Aug	Sept	Oct	Nov	Dec
25	39	80	157	197	235	255	221	153	105	39	25

Summer crops (27 000 ha)

Jan	Feb	March	April	May	June	July	Aug	Sept	Oct	Nov	Dec
			26	99	196	255	184	72	77		

Winter crops (30 000 ha)

Jan	Feb	March	April	May	June	July	Aug	Sept	Oct	Nov	Dec
36	56	114	118	49						14	26

Subtracting the effective rainfall

Jan	Feb	March	April	May	June	July	Aug	Sept	Oct	Nov	Dec
26	15	12	6	5	0	0	0	0	9	24	

The following field evapotranspiration requirement is obtained (mm of water/month)

	Jan	Feb	March	April	May	June	July	Aug	Sept	Oct	Nov	Dec
Orchard	–	24	68	151	192	235	255	221	153	105	30	1
Summer crops	–	–	–	20	94	196	255	184	72	17	–	–
Winter crops	10	41	102	112	44	–	–	–	–	–	5	2

Or expressed in Mm^3 for the indicated hectarages (*Table 25.14*)

Table 25.14 Field evapotranspiration requirements in million m^3 (Mm^3)

	Orchard	*Summer crops*	*Winter crops*	*Total*
January	–	–	3.00	3.00
February	7.92	–	12.31	20.23
March	22.44	–	30.62	53.06
April	49.83	5.42	33.78	89.03
May	63.36	25.46	13.21	102.03
June	77.55	53.10	–	130.65
July	84.15	69.08	–	153.23
August	72.93	49.85	–	122.78
September	50.49	19.50	–	69.99
October	34.65	4.60	–	39.25
November	9.90	–	1.50	11.40
December	0.33	–	0.60	0.93
Total	473.55	227.01	95.02	795.58

The mean Barada flows at Hameh (m^3/s):

(Q)/(m^3/s)

Jan	Feb	March	April	May	June	July	Aug	Sept	Oct	Nov	Dec
10	13	17	22	19	13	10	8	7	7	7	8

Or, in Mm^3/month:

27	32	46	57	51	34	37	21	18	19	18	21

Total = 371 Mm^3/year

If irrigation efficiency from surface water is 50% and from groundwater is 75%, then the total estimated yearly crop water requirements for the actual irrigated area (1985) in the Ghouta is about 1000 Mm^3; in other words, surface water from the Barada and distributaries provides 37% whereas groundwater provides 63% of the total irrigation water requirement in the Ghouta area. These figures differ greatly from the official figures mentioned in *Table 25.10* due to extra unreported groundwater pumping from the Ghouta aquifer.

Effects of industrial and domestic waste on irrigation water, soils and crops and public health in the Ghouta area

After this brief review of the hydrological, pollutional and agricultural aspects of the Barada river and Ghouta area in the Damascus plain, the effects of industrial and domestic waste on irrigation water are clearly indicated below:

- The increase in BOD_5 was shown in *Figure 25.4*. Plant growth is impeded by oxygen deficiency at the root zone, caused by high BOD_5 in irrigation water. Empirical evidence suggests that plant growth suffers, and the soils ecosystem may be impaired, at BOD_5 levels of around 300–350 mg/l. Water should be diluted with unpolluted groundwater to avoid damage to plants and soil.

- *Tables 25.4* and *25.5* give the results of water quality analyses of the Barada river and its distributaries. When compared with the recommended maximum concentrations of trace elements in irrigation water, all elements, with the exception of chromium in the Barada and distributaries, are well below the maximum recommended concentrations. Excessive chromium levels occur in the Barada and in the Da'yani rivers, and are no doubt attributable to the metal finishing works and the leather tanneries located in the vicinity of these rivers.
- *Figures 25.6a, b, c, d* and *e* are of particular interest in the study of the variations of percent dissolved oxygen, BOD_5, ammoniacal nitrogen, chloride and, with particular reference to *Table 25.6*, conductivity in the irrigation water along the flow of the Barada and its different distributaries.

  At low concentration, chloride is essential for plant growth. The average chloride level of raw sewage in Damascus is generally below 70 mg/l. Concentrations in excess of 150 mg/l may be found in the Barada and are sufficient to affect sensitive crops like onions, tomatoes and cucumbers.

  Levels of soluble sodium, calcium and magnesium in the irrigation water in the Ghouta are sufficient to produce toxic effects. Nevertheless, the values of sodium absorption ratio (SAR) adjusted to take into account sodium carbonate present in the soil are below internationally accepted limits for potential alkali hazards.
- *Tables 25.7, 25.8* and *25.9* deal with the domestic pollution of irrigation water. The results of bacteriological sampling from the Barada and its distributaries and from wells in the Ghouta are shown in *Tables 25.7* and *25.8* while *Table 25.9* is the summary of the total domestic pollution load discharged, per day, into the Barada system.

The coliform count of most water samples, and in particular from wells used for industrial and drinking purposes, exceeds World Health Organization guidelines. Summer and winter vegetables irrigated with contaminated irrigation water are a good reason for gastrointestinal and infectious diseases, especially dysentery, typhoid fever and some others. Intestinal worms are common among children and even, sometimes, among adults.

Proposed sewerage system and sewage treatment plant for Damascus area

Immediate and urgent work for a wastewater management project to deal with the inherent risks, mainly to health but also to agriculture, to further the development of industry and to protect the environment, are fully recognized and needed. The project will consist mainly of:

- major interceptors and secondary sewers to divert the sewage from the distributary rivers;
- a treatment plant to treat the sewage; and
- a system that maintains the irrigation water rights on the Barada river and its distributaries.

A master plan has been worked out and will be implemented in three phases.

Phase 1 will include the construction of major interceptors and other sewers to collect the wastewater within the present urban area of Damascus City, followed by the construction of a sewage treatment works, with primary and secondary stages of treatment at Ain Terma, for a mean daily flow of 309 400 m^3/day. Recharge facilities to the river will be constructed by way of pumping stations and pumping mains to maintain traditional irrigation flows in the distributaries of the Barada river. The river control weirs on the Barada at the six distributary offtakes upstream of Raboue will be reconstructed.

Phase 2 will include the construction of sewers and pumping stations to convey the wastewater from the outlying suburban areas to the treatment works; then the capacity of the sewage treatment works will be extended to 415 000 m^3/day. The final step of this phase will again be pumping mains to maintain the traditional irrigation flows in the rivers in the vicinity of the outlying areas.

Phase 3 will be to extend the capacity of the sewage treatment works to 485 000 m^3/day and to provide an effluent standard of 10 mg/l BOD_5 and 10 mg/l suspended solids.

Minor sewage treatment works in some specific locations are now being envisaged in order to reduce the number of pumping stations and pumping mains and yet maintain the traditional irrigation flows.

Conclusions

For a long time, municipal and industrial wastewaters have been reused for irrigation in the plain of Damascus, the Ghouta. In 1976, an estimated 60 Mm^3 of domestic wastewater was reused for irrigation and in the replenishment of the groundwater aquifers in the Ghouta area. The water balance of the Ghouta aquifers, including Damascus domestic sewage, the replenishment from surface runoff, from rainfall and from groundwater abstracted for irrigation, is shown in *Figure 25.7*.

In 1985, about 80 Mm^3 of domestic wastewater is being reused annually for irrigation and for groundwater replenishment in the Ghouta area. It is anticipated that this figure will be more than 125 Mm^3 in the year 2001.

The main drawback of this reuse of wastewater from either domestic or industrial areas is the bacteriological count, the concentrations of BOD, the salinity, chloride, heavy metals and other phytotoxins in the irrigation water. This pollution greatly affects the water quality and creates health and environmental hazards. Even the groundwater used by industry or for drinking purposes has a dangerously high level of coliform count which greatly exceeds the guidelines set by the World Health Organization.

The long and historical experience of the Damascus area with this problem has led to the conclusion that an adequate sewerage system and sewage treatment plant should be envisaged before reusing municipal and industrial wastewater for irrigation and for the replenishment of groundwater aquifers, especially in arid or semiarid zones such as the Near East region, in the Gulf States and Arab Peninsula, and in many other Arab countries.

It is true that water rather than soil is the factor limiting expansion of new land under irrigation but health hazards and potential deterioration of the environment, such as soil salinity, sodium alkalinity, etc., should also be considered.

Planned for construction in the Damascus plain under Phase 1 and Phase 2, respectively, are major interceptors and secondary sewers to divert the sewage from the Barada river distributaries, since self-purification is not anticipated. The

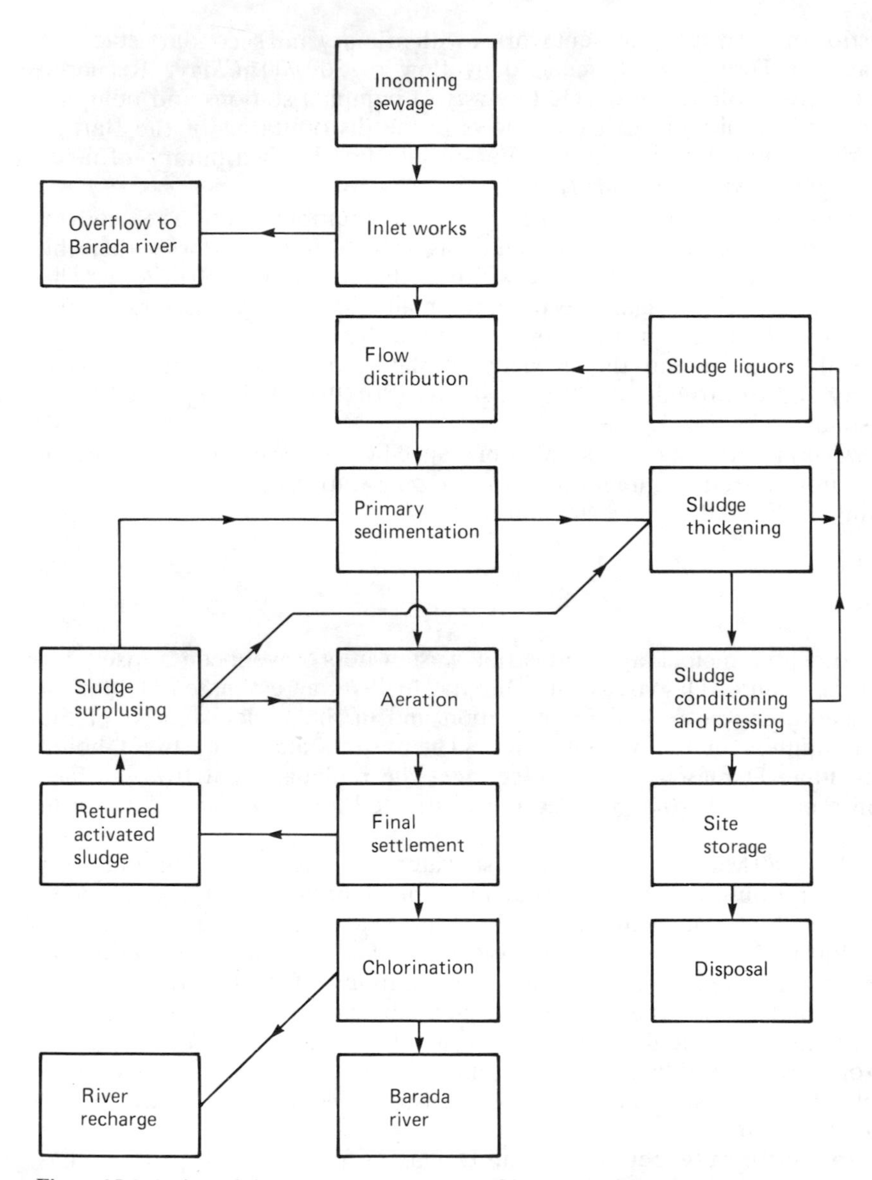

Figure 25.9 Activated sludge treatment process – Ghouta Ain Terma project

flow of water in these distributaries is so low, especially in the summer, that for much of the year they can only be considered as open sewers.

It is planned to construct a sewage treatment works at Ain Terma, just to the east of Damascus City. This plant will be based on the activated sludge treatment process, as shown schematically in *Figure 25.9*. Pumping stations and pumping mains will then be implemented to maintain traditional irrigation flows in the distributaries of the Barada river.

Bibliography

AYERS, R. S. and WESTCOT, D. W. (1976) Water quality for agriculture, *FAO Irrigation and Drainage Paper, No. 29.* Rome: UN Food and Agriculture Organization
CENTRAL BUREAU OF STATISTICS, SYRIAN ARAB REPUBLIC (1984) Statistical Abstract
FAIR, G. M. and GEYER, J. C.(1956) *Water Supply and Waste-Water Disposal.* New York: Wiley
HOWARD HUMPHREYS and SONS (1980) *Damascus Sewage Preliminary Engineering Design,* 4 volumes
IMHOFF, K. and FAIR, G. M. (1949) *Sewage Treatment.* New York: Wiley
KUNERT and ALI JABRA (1965) UNSF Project 101, *Damascus Research Station Report*
MINISTRY OF AGRICULTURE AND AGRARIAN REFORM, SYRIAN ARAB REPUBLIC (1984) *The Agricultural Production Plan for Damascus Area during 1983–1984* (in Arabic)
PHELPS, E. B. (1953) *Stream Sanitation.* New York: Wiley
STEEL, E. W. (1953) *Water Supply and Sewerage.* New York: McGraw-Hill
WALTON, W. C. (1970) *Ground Water Resource Evaluation.* New York: McGraw-Hill

Chapter 26

Study of methodologies regarding the use of treated wastewater for irrigation in Portugal

Maria Helena F. Marecos do Monte*

Introduction

About 50% of Portugal's mainland is a zone of low water resources and this situation is further aggravated by the damaging effects of periodic droughts. To meet present and projected needs, Laboratorio Nacional de Engenharia Civil (LNEC) is undertaking a research project in collaboration with colleagues from other organizations† to obtain experimental data to use as the basis of a draft paper giving national guidelines for the use of treated wastewaters in irrigation.

This chapter describes the methodology followed in the experiments carried out in 1985 in which maize, sorghum and sunflowers were irrigated with wastewater treated by primary sedimentation and high rate trickling filters.

The main object of the study was to determine the effects of wastewater use on the productivity of crops and the possible public health risks, both chemical and microbiological, which might result.

There is, today, a growing awareness throughout the world of the urgent need to maintain, recycle and reuse existing water resources, which are of course limited. During the last 30 years, wastewater reuse technologies have therefore been a matter of great interest to technicians with different specialities, from sanitary and agricultural engineers to biologists, chemical scientists, public health officers and water resources managers.

In Portugal, shortage of water for various purposes, from domestic water supply to energy production and agricultural irrigation, is the hard reality in many regions, particularly in the semiarid regions of Alentejo. Wastewater reuse for agricultural irrigation is, therefore, particularly interesting in regions where water resources are scarce but where the wastewater flow produced by the community is sufficiently large to make reuse economically feasible.

The Alentejo zone falls into this category, since it suffers habitual lack of water and the population is concentrated in a relatively small number of locations. There are several small towns with wastewater treatment plants, discharging treated wastewater directly into receiving water bodies. Use of this treated wastewater for

---

* Chemical Engineer, Sanitary Engineer, Laboratorio Nacional de Engenharia Civil (LNEC), Ministerio da Habitacão e Obras Publicas, Portugal

† Rebelo da Silva Agricultural Chemistry Laboratory (LQARS), Lisbon; Evora, Leeds (UK) and Liverpool (UK) Universities; Evora Regional Health Administration; and Evora Municipality participated in this study.

agricultural purposes would lead to rather profound and extensive changes in the agriculture of Alentejo, where lack of water during a large part of the year is only too evident.

A programme of wastewater reuse would significantly increase the area of land that could be cultivated in a more profitable way. In many cases there could be a change to a more intensive cropping system, with all the advantages this brings, such as better annual income and also the possibility of growing a greater range of crops. For example, an increase in forage crop production would lead to increased cattle production for slaughter or for dairy products. This would not only increase wealth but also might modify the eating habits of the population and improve the diet.

Wastewater reuse for irrigation purposes would also aid the general management of water resources by reducing pollution of water bodies that would otherwise be receiving the wastewater. Furthermore, agriculture benefits not only from the increased availability of water but also from the nutrients that the wastewater contains, such as nitrogen and phosphorus, both of which are the essential elements in most fertilizers.

Objectives of the study

The main objective of the study, at this stage, was to define the most appropriate methodology for the reuse of treated urban wastewater in irrigation (rather than describe results). This necessitated studying such aspects as:

- the required degree of wastewater treatment needed for the types of crops and soils to be irrigated;
- the dilution necessary before reuse of wastewater submitted to only low-level treatment;
- the minimum degree of treatment needed before application to crops usually requiring high quality water.

The minimum safe period between the end of irrigation and the harvesting and use of crops was also investigated.

Therefore, the experimental study comprised sanitary aspects (such as pollution and contamination of soil, irrigated crops and underground water) and assessment of the quality and productivity of crops. The wastewater used in this study was treated in accordance with the processes currently used in Portugal, that is, primary settling tanks, stabilization ponds, trickling filters and activated sludge plants. ‘Controls’ used in the experiments were plots irrigated with drinking water. The same crops were used for studying the effects of the different types of effluents.

In 1984, a preliminary experiment was carried out with effluents from high-rate trickling filters and primary settling tanks to check and modify the methods to be used in subsequent experiments. Currently, an experiment is being carried out with these effluents, the methodology of which is described in detail in this chapter. The study will be completed in 1986, after the inclusion of similar studies using effluents from activated sludge plants and waste stabilization ponds. It will then be possible to assess and compare the effects of various types of wastewaters versus drinking water on the irrigation of cultivated crops. In its simplest form, this will be measured in terms of differences in crop yield. More importantly, it will also consider effects such as possible contamination of the soil and plants by: (i) pathogenic microorganisms transported in the effluents*, and (ii) the accumulation of toxic substances, such as certain metals present in the wastewater from industrial sources. Productivity comparisons on the same crop irrigated with the different wastewaters will be possible. Furthermore, the degree of contamination and sensitivity to pollution by the effluents will be assessed for each crop.

The second major objective of this study is the preparation of a draft Code for the Reuse of Wastewater for Agricultural Purposes in Portugal. This will be submitted to public enquiry.

Experimental study

Characterization of the experimental site

Soil

Soil profiles were studied to assess the suitability for cultivation. The following characteristics were the object of particular attention: depth, colour, moisture content, texture, organic matter content, structure, compaction, porosity, carbonate and general chemical composition.

The physical analyses of the different soil horizons showed it to be of a rough texture varying in depth from loamy sand to sandy loam, therefore easy to till. The values of usable capacity in each horizon are rather low as shown by the value of ⅓ atmosphere (atm) and 15 atmosphere percentages (atm %) indicated in *Table 26.1*.

Table 26.1 Values of usable capacity of the soil

Sample of the horizon	*Values determined*		*Value of usable capacity (%)*
	(⅓ atm %)	*(15 atm %)*	
A1	6.83	1.72	5.11
A2	10.80	2.79	8.01
C	9.82	6.55	3.27

Chemically it is a non-calcareous soil with a pH close to neutrality, a very low organic matter content of less than 1% (by dry weight) and normally low values of assimilable phosphorus and potassium. Micronutrient values were also low.

A value of 10 me/100 ml for the cation exchange capacity of the soil and saturation ratio values between 60 and 80% were obtained. Since good productive soils are characterized by a cationic exchange capacity of about 18–35 me/100 g and a degree of saturation ranging from 60 to 80%, the soil in question would seem to have a moderate productive capacity.

Climate

Characterization of the climate was based on data for the last 30 years, obtained from the climatological station nearest the test site. The characteristics of most concern were those with greatest influence on agricultural production and the establishment of an effective irrigation scheme: air temperature and humidity,

* The term 'effluent' is used for treated wastewater.

Table 26.2 Monthly averages

Month	$T_{average}$ (°C)	*Wind* (km/h)	*HR*		*Insolation* (h)	*R* (mm)
			maximum	*minimum*		
January	9.4	16.4	86	74	155	95
February	9.9	16.0	85	68	161	85
March	11.9	16.4	82	62	196	89
April	13.8	16.6	82	55	245	48
May	16.8	16.9	84	49	297	44
June	20.2	16.5	82	42	329	26
July	23.0	17.5	79	34	386	5
August	23.1	17.7	76	34	361	3
September	21.0	15.6	78	44	267	26
October	17.2	14.9	81	56	233	58
November	12.6	15.9	87	70	162	92
December	9.6	16.4	87	77	154	91

Table 26.3 Potential evapotranspiration (mm) (1931–60)

January	20	July	225
February	35	August	194
March	63	September	117
April	98	October	73
May	148	November	24
June	191	December	14
	Annual 1205		

evapotranspiration rates, precipitation, insolation, cloud cover and wind velocity. *Tables 26.2* and *26.3* contain the main characteristics of the climate at the place where the experiment is being carried out.

Crops tested

The inclusion of several crops in the same experiment will provide a broader view of the effects of irrigating with wastewater. The use of replicate subplots for the different types of wastewater used will increase the number of degrees of freedom and thus the statistical validity of the results obtained. To take account of current regional economic interest in converting some non-irrigated crops into wet crops, the following three crops were selected: maize (grain), sorghum and sunflower.

Field layout

The field layout used was governed by the need to avoid the risk of cross contamination between plots irrigated with different wastewaters. Three types of irrigating water and the crops used were given the following notations:

A_1: effluent from the primary settling tank
A_2: drinking water
A_3: effluent from the final settling tank
c_1: sorghum
c_2: maize
c_3: sunflower

Four replicate plots of each crop were irrigated with each of the three types of irrigation waters (*Figure 26.1*). The actual experimental layout of 36 subplots covering a total area of 1600 m^2 is shown in *Figure 26.2*. Each replicate plot measured 10 m × 3 m.

REP I			REP II			REP III			REP IV		
A_1	A_2	A_3	A_1	A_2	A_3	A_1	A_2	A_3	A_1	A_2	A_3
REP I A_1 C_2	REP I A_2 C_2	REP I A_3 C_2	REP II A_3 C_1	REP II A_2 C_1	REP II A_2 C_2	REP III A_3 C_2	REP III A_1 C_2	REP III A_1 C_2	REP IV A_1 C_1	REP IV A_2 C_1	REP IV A_1 C_1
REP I A_1 C_1	REP I A_2 C_1	REP I A_3 C_1	REP II A_3 C_3	REP II A_2 C_3	REP II A_1 C_3	REP III A_3 C_1	REP III A_1 C_1	REP III A_2 C_1	REP IV A_1 C_3	REP IV A_2 C_3	REP IV A_3 C_3
REP I A_1 C_3	REP I A_2 C_3	REP I A_3 C_3	REP II A_3 C_1	REP II A_2 C_1	REP II A_1 C_1	REP III A_3 C_3	REP III A_1 C_3	REP III A_2 C_3	REP IV A_1 C_2	REP IV A_2 C_2	REP IV A_3 C_2

Figure 26.1 Replicate plots

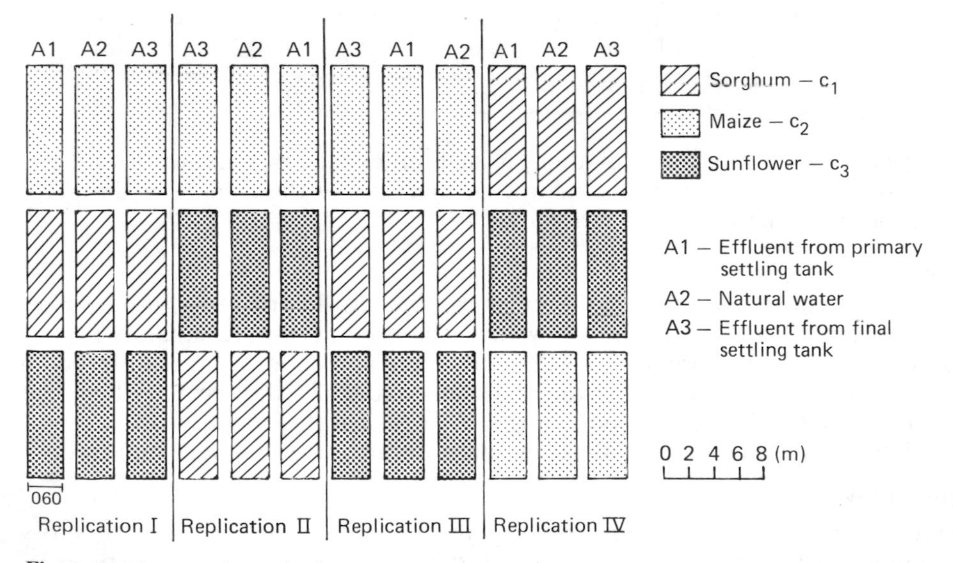

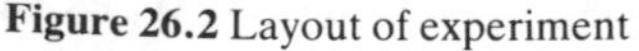

Figure 26.2 Layout of experiment

Setting-up and running the experiment

Analysis of soil before experiment

After ploughing and levelling the soil, a composite soil sample, representative of the overall experimental area, was collected. On the basis of the results of a preliminary analysis of this sample, the types of fertilizer and amounts to be applied to the control subplots, that is, those that are irrigated only with drinking water, were determined. These results also served to indicate whether or not the other subplots irrigated with wastewaters needed additional mineral fertilization to maximize yield. To enable a comparison to be made between the chemical and microbiological characteristics of the soil before and after the experiment, samples were collected from each subplot and analysed for the parameters given below in *Tables 26.11, 26.12* and *26.13*.

Water

From the results of the analyses of various samples from the wastewater treatment plant in Evora (*Table 26.4*) it was concluded that, although the natural water was suitable for the irrigation of any crop, the salinity levels of the effluents of the primary and final settling tanks were such that they should only be used for the irrigation of fairly salt-tolerant crops. Such crops include maize, sorghum and sunflower. Seven out of eight of the drinking water samples tested had salinity values (as measured by electrical conductivity) below 0.75 mmhos/cm and an adjusted SAR lower than 0.6. Only one sample showed values of sodium, chloride and boron sufficiently high to be a potential problem. Thus the water can be taken as almost always being suitable for irrigation of any crop.

Table 26.4 Physicochemical characteristics of the various irrigation waters used

	Natural water	*Effluent from primary settlement tank*	*Effluent from final settlement tank*
Salinity			
Electrical conductivity (mmhos/cm)	0.34–0.52	1.38–2.09	1.19–1.78
Degradation of soil permeability			
electrical conductivity (mmhos/cm)	0.34–0.52	1.38–2.09	1.19–1.78
adjusted SAR	1.19–2.42	3.43–8.65	3.38–8.89
Toxicity of some ions absorbed by roots			
sodium (ppm)	28.00–140.00	68.50–155.0	68.50–160.00
chlorides (ppm)	35.46–163.12	148.93–219.85	163.12–191.48
boron (ppm)	0.15–1.45	0.62–1.34	0.63–1.54
pH	7.20–8.20	4.10–8.20	7.00–7.85

In contrast, the quality of effluents from the primary and final settling tanks had electrical conductivities within the potentially problematic range of 0.75 mmhos/cm and 3.0 mmhos/cm, adjusted SAR between 6.0 and 9.0, sodium absorbed by roots between 68 ppm and 189 ppm, chloride between 142 ppm and 355 ppm and boron between 0.5 ppm and 20 ppm. This leads to classifying them as suitable for the irrigation of fairly salt-tolerant crops. In every case the pH was within the normal limits, that is, between 6.5 and 8.4.

Cultivation techniques

Fertilization It is not possible at the time of writing to give the actual amounts of fertilizer to be applied, since the soil samples whose analyses are required for such calculations were only taken recently. Nevertheless, it can be said that underground and spreading fertilization are carried out in the subplots irrigated with drinking water, whereas in those irrigated with treated effluents only underground fertilization is adopted, since spreading fertilization is directly applied through nutrients carried in the irrigation water. No organic fertilizer (manure) is added, so that the content of the soil can be determined.

Sowing As the crops are watered, and the soil is moderately productive, intermediate sowing densities were selected and proved to be satisfactory on the basis of the results obtained in the preliminary experiment. Nevertheless, these sowing densities may need adjustment as more information is gathered.

The planting densities used were equivalent to:

Maize	67 000 plants/ha
Sorghum	167 000 plants/ha
Sunflower	53 300 plants/ha

Planting details are shown in *Table 26.5*.

Table 26.5 Planting details for the various crops used

	Distance between rows (m)	*Distance between plants in a row* (m)	*Number of plants in each 10 m of row*	*Total number of plants in each subplot*
Maize	0.75	0.20	50	200
Sorghum	0.60	0.10	100	500
Sunflower	0.75	0.25	40	160

Table 26.6 Sowing densities used for each plant species

	Number of seeds in subplot	*Number of seeds in row*	*Total number of seeds to be used*
Maize (grain)	600	150	7 100
Sorghum	1500	300	18 000
Sunflower	480	120	1 440

To ensure that the required planting densities were achieved, the seeds were sown at three times this density. Although this strategy has the drawback of requiring thinning out of the plants, it has the advantage of allowing better control of the number of plants in the subplots. The actual sowing densities used are given in *Table 26.6*.

Irrigation project The irrigation methods most commonly used in intensive farming (because of water use efficiency and economic installation costs) are the sprinkler irrigation method and the furrow method. As the water used in this study is derived from treated sewage, public health aspects are of prime importance. The use of a sprinkler system makes it difficult to guarantee non-contamination of neighbouring plots with water not intended for them. This is particularly true for small-scale experiments, such as those described here. Furthermore, spraying increases the potential for contaminating the aerial parts of the plants with any pathogens present in the wastewater, with a consequent increased risk to man and animals utilizing them. For these reasons a system of closed-furrow irrigation was used. This system is easy to operate, requires little manpower and eliminates many of the problems related to the sprinkler method.

V-shaped irrigation furrows between the plant rows were made by hand with hoes after sowing but before the first watering. For ease of operation, the size of the furrows was constant for all three crops (*Figure 26.3*). Water reaches the plant roots through vertical and horizontal infiltration. To minimize water loss through runoff, the furrows in each subplot were closed off at each end with soil.

From discussions with Professor Zozimo Rego (Instituto Superior de Agronomia Lisboa, ISA) on soil types, the water requirements of the chosen crops and the nutrient loading in the wastewaters (6000 m^3/ha) were indicated as acceptable. To simplify the irrigation schedule used in the experiment, all three crops were watered at a rate equivalent to 5000 m^3/ha. This was more than the 3500 m^3/ha used in the preliminary experiment, which proved inadequate.

The interval between irrigations was identical for all three crops until the beginning of September. Maize was then irrigated twice more. In all, maize was irrigated 16 times, sorghum and sunflower 14 times. Under those experimental conditions it is important that the value of 5000 m^3/ha should not be exceeded because, if the nutrient content in the wastewater is the same as that determined in

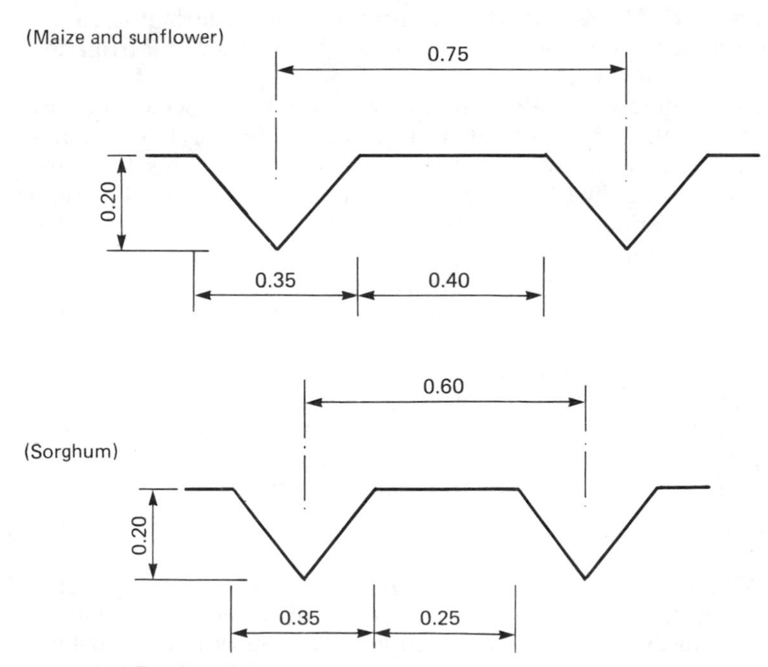

Figure 26.3 Furrow cross-section; the dimensions are in metres

1984, increasing the water volume applied may lead to excess with all the detrimental effects that could result, for example 'lodging' owing to excess nitrogen application.

Monitoring programme

Introduction

The preparation and planning of a suitable monitoring programme for a study like this is very important. It requires as much attention as the setting up of the experimental field plots, since the success of the programme will depend on the correct type of data being collected to allow meaningful interpretation of the wastewater–soil–plant interactions, which are the basis of the experimentation. Monitoring of the study needs to include physicochemical and microbiological parameters of the wastewaters, soil and crops and also measurements of crop productivity.

Physicochemical analyses of wastewaters

There are four types of water to be analysed:

- drinking water (A_2);
- raw wastewater influent to the wastewater treatment plant (A_0);
- effluent from the primary settling tank (A_1);
- effluent from the final settling tank (A_3).

Although only waters from items A_2, A_1 and A_3 were used for irrigation in this study, analysis of the raw wastewater will make it possible also to determine the efficiency of purification carried out by the primary settling tank and of the treatment plant as a whole. This is important as it may have a bearing on future decisions to be made on the use of various types of treated wastewater for irrigation in Portugal.

Only occasional grab samples of the drinking water were taken, since the quality of the water should be relatively constant as it comes from the Evora water treatment plant. Samples of the raw wastewater and of the effluents from the primary (A_1) and final (A_3) settling tanks are composite samples taken at 20 min intervals during the actual periods of irrigation of the experimental plots, and thus are representative of the water applied.

Samples of waters A_0, A_1 and A_3 from each irrigation period are analysed. Water A_2 is only analysed once in a month since, as a drinking water supply, its quality should be constant.

The parameters investigated in the routine analysis of the various waters are given in *Tables 26.7, 26.8* and *26.9*. Some of these parameters were measured in the first set of analyses to determine if their concentrations fell within the usual range of values found in urban wastewaters, or if they exceeded these values, in which case they may represent a risk to the irrigated crops.

Microbiological analysis of wastewaters

The drinking water, A_2, is not analysed microbiologically because it is not expected to contain pathogens. Thus, in the interest of economy the microbiological analyses are only made on raw wastewater and on the effluents of the primary and final settling tanks.

Table 26.7 Parameters analysed in A_2

Organoleptic	Aspect	
	Colour	
	Odour	
Physical	pH	
	pH 25°C	
	Conductivity	
	Turbidity	
Chemical	Alkalinity	
	Alkalinity to $CaCO_3$ saturation equilibrium	
	Total hardness	
	Carbonate hardness	
	Non-carbonate hardness	
	Oxidizability	
	Anions	*Cations*
	HCO_3^-	Ca^{2+}
	$SO_4^=$	Mg^{2+}
	Cl^-	Na^+
		K^+

Table 26.8 Parameters analysed in A_1 and A_3

Parameters analysed					
Routine				*Occasional*	
During irrigation		*After irrigation*			
BOD_5				As	(1)
COD				Hg	(1)
SS				Se	(1)
DS	(3)			CN	(1)
T					
N: NO_3	(2)				
N: NH_4	(2)				
N: org.	(2)				
pH	(3)	Cd	(1)		
Hardness	(3)	Cr	(1)		
Carbonates	(3)	Cu	(1)		
Bicarbonate	(3)	Fe	(1)		
Phosphorus		Ni	(1)		
Orthophosphate		Pb	(1)		
Chloride	(3)	Mn	(1)		
Sulphate	(3)	Zn	(1)		
Conductivity		Mo	(1)		
Calcium	(1)				
Magnesium	(1)				
Sodium	(1)				
Potassium	(1)				
Boron	(1)				

(1) Analyses to be carried out at the end of the irrigation period, on a sample composed of subsamples collected during each irrigation, suitably preserved with HNO_3 at pH = 2, kept in boron-silicate glass flasks with the stopper lined with teflon.
(2) The samples for determining the different nitrogen species will be kept in glass flasks with the stopper lined with teflon. These samples will be acidified to pH = 2 with sulphuric acid and kept at 4°C.
(3) Samples for the determination of the different ionic species cannot be chemically conserved; they have to be kept in plastic bottles at 4°C and analysed as soon as possible.

For microbiological analyses, grab samples were used, collected in sterilized flasks previously supplied by the Laboratory of the Regional Health Administration. These samples were collected every month during the actual irrigation period and kept at a temperature below 4 °C in thermally insulated containers. Transportation to the laboratory occurred within 4 h of collection.

Table 26.9 Parameters analysed in A_0

pH
BOD_5
COD
TSS
FSS
VSS

Table 26.10 Microorganisms analysed

Analytical laboratory	*Microorganism*
Regional Health Administration	Faecal coliforms
Regional Health Administration	Faecal streptococci
Leeds University	*Campylobacter*
Leeds University	*Salmonella*
Leeds University	*Shigella*

Irrigation waters A_0, A_1 and A_3 are analysed for the microorganisms listed in *Table 26.10*. The actual numbers of faecal coliforms and streptococci are determined but *Campylobacter* and *Salmonella* are only analysed on a presence or absence basis.

Physicochemical analyses of soil
For each subplot, samples from the top 15 cm layer and at 50 cm depth in the soil profile were analysed. The soil profile in the experimental area had been previously examined in detail (during the preliminary 1984 experiment) to determine the soil's suitability for the experiments.

All the soil samples collected were composite samples, each one comprising a series of subsamples from an individual subplot. Soil samples for analysis were taken before planting and after harvesting.

The parameters analysed in the surface soil samples and from 50 cm depth are given in *Tables 26.11* and *26.12*. Surface soil samples are analysed after irrigation for the parameters given in *Table 26.13*. These samples are composed of subsamples collected after each irrigation.

Microbiological analyses of soil
The surface layer of all plots is analysed, including those irrigated with natural water. Microbiological analyses of soil (faecal coliforms) were carried out before, during and after completion of the experiment.

Table 26.11 Surface soil parameters analysed

Parameters analysed before sowing and after harvesting	
Physical	Texture
	Structure
	Porosity
	Apparent specific density
Hydraulic	Permeability
	Rate of infiltration
	Field capacity
	Fading coefficient
Chemical	pH
	Cation exchange capacity
	Calcium
	Magnesium
	Potassium
	Sodium
	Carbonates
	Sulphates
	TOC
	Hydrogencarbonates
	Conductivity
	Chlorides
	N: total
	N: organic
	P: total
	K: assimilable
	Boron

Table 26.12 Parameters analysed in soil at 50 cm depth

Parameters analysed before sowing and after irrigation		
Sulphates	N: total	Copper
Carbonates	N: NO_3	Cadmium
Hydrogencarbonates	N: organic	Nickel
Chlorides	P: total	Chromium
Cation exchange capacity	P: P_2O_5	Mercury
Calcium	Conductivity	Lead
Magnesium	Boron	Cobalt
Potassium	Iron	
Sodium	Zinc	

Table 26.13 Surface soil parameters analysed

Iron	Manganese	Lead
Molybdenum	Cadmium	Cobalt
Zinc	Nickel	Mercury
Copper	Chromium	

Physicochemical analyses of crops

As a general rule, it is only necessary to analyse the parts of the plants intended for human and animal consumption, or those with which there may be physical contact by agricultural workers. Nevertheless, as some parts not intended for consumption may specifically accumulate some toxic elements, namely heavy metals, it is advisable to analyse both the parts to be consumed and the non-utilizable parts (for example, leaves of sunflowers).

Samples of crop materials were taken at harvest. Replicate samples of plants were taken at random from rows other than the two peripheral ones, which may be susceptible to so-called 'edge effects'. The chemical parameters determined in the various types of plant material are given in *Table 26.14.*

Table 26.14 Chemical parameters determined in crop material

Part of plant for consumption	*Part of plant not to be consumed*
Boron	
N: total (N: Kjeldahl)	
Potassium	
Iron	
Nickel	
Zinc	Zinc
Chromium	
Copper	
Mercury	
Cadmium	Cadmium
Lead	
Cobalt	

Microbiological analysis of crops

In the type of crops tested in this study, only parts of the aerial section are consumed. The type of irrigation chosen make it likely that only the roots and lower parts of the plants might be expected to come in contact with the wastewater and so become contaminated with pathogenic organisms. However, all parts of the crops should be checked for contamination by the microorganisms referred to in *Table 26.10.* Due to the risks of contamination run by workers who come into contact with the parts of the plants not to be consumed, these parts must also be analysed in addition to the edible parts.

Harvest of experimental crop

After sampling for physicochemical and microbiological analyses the remaining plants in the central rows of each subplot were harvested, those in the peripheral rows being discarded for reasons previously explained. The plants harvested from the subplots were weighed to determine their productivity.

Bibliography

OMS (1982) *La technologie approprié au traitement des eaux usées dans les petites localités rurales.* Lyon

BOTELO DA COSTA, J. (1975) Caracterizacão e constituicão do solo. Fundacão. Lisboa: Calouste Gulbenkian (FCG)

CASIMIRO MENDES E. AND REIS, M. (1980) Contribuicão para o estudo estatistico das series mensais, trimestrais e annuais de quantidade de precipitacão no ano agricola em Portugal Continental. *O clima de Portugal,* Fasc. XXIII. Lisboa: IUMG

FAO (1974) Crop water requirements. *Irrigation and Drainage Paper no. 24.* Rome: Food and Agriculture Organization of the United Nations

INSTITUTO NACIONAL DE ESTATISTICA – ESTATISTICAS AGRICOLAS (1984) Lisboa

PEREIRA, F. A., CONCALVES, E. and DA SILVA, M. E. (1961) *Tabela de composicão dos alimentos Portugueses.* Direccão-Genral de Saude. Lisboa: Instituto Superior de Higiene Dr Ricardo Jorge

MAYER, RUY (n.d.) Estudo agronomico do projecto de rega das campinas de Silves, Portimão e Lagoa. In *Tecnica do Regadio*

SUSPIRO, J. L. (1977) A cultura de soja. Direccão genral de extensão rural, publicacão do LQARS

INIA/ISA (1975) Ministerio da Agricultura e Pescas – factores elementares dos regadios alentejanos, Oeiras Set

Chapter 27

Microbiological testing of soils irrigated with effluents from tertiary lagooning and secondary treatment

M. Cadillon and L. Tremea*

Introduction

Whenever wastewater is used for agricultural purposes, it is the sanitation problem that remains most worrying because of the risks of contamination of:

- soil;
- crops and consumers;
- agricultural workers on the land irrigated with such effluents.

This chapter summarizes the results of a microbiological study of soils irrigated with effluents:

- from tertiary lagooning, used to irrigate orchards and some market (truck) gardening crops (on the island of Porquerolles, France), with irrigation both by row and by local irrigation;
- from secondary treatment, used to irrigate a forest by sprinkler or local irrigation (Cogolin, France).

The purpose of these experiments was to try to answer the following questions:

- What is the behaviour of the various bacteria of faecal origin supplied to the soil by the effluents? Do they survive? If so, under what conditions?
- Is there an accumulation of bacteria of faecal origin in the soil over successive irrigations?
- How are the bacteria distributed in the soil profile compared to a control soil?
- What is the influence of cultivation methods such as type of irrigation and ploughing on the survival of these bacteria?

After explaining the experimental set-up and the analytical conditions, a summary of the results on the microbiological evolution of the soil will be given from p. 348 onwards.

---

* Societé du Canal de Provence, et d'Amenagement de la Region Provence Institut Mediterranéen de l'Eau, BP 100, Le Thoionet, 13603, Aix-en-Provence, Cedex, France

Experimental framework

Experiment on Porquerolles

The island of Porquerolles has a network of sewers to collect the wastewater and convey it to:

- a treatment plant, activated sludge with average loading; and
- a natural tertiary lagoon system with the characteristics shown in *Table 27.1*.

Table 27.1 Characteristics of three lagoons used in the Porquerolles experiment

	Area (m^2)	*Depth* (m)	*Average volume* (m^3)
Lagoon 1	4000	1 ± 0.10	4000
Lagoon 2	2000	0.30	1500
Lagoon 3	4000	0.30	1300

Note: Minimum residence time in the lagoons = 30 days.

The treated effluents are used to irrigate approximately 6 ha of orchards on the farm of the Porquerolles botanical conservatory. All plots studied are on this farm and they are:

- either irrigated with water from the tertiary treatment lagoons;
- irrigated either by row or by 'BRL drip-irrigation system'; or irrigated with groundwater: control plot;
- homogeneous as to soil type: not very altered, on colluviums–alluviums from micaschists, deep, sandy (0.60 cm), on clayey sand (60–120 cm).

The water from the lagoon contains the following counts of indicator organisms:

- 10^3–10^4 total coliforms/100 ml,
- 10^2 faecal coliforms/100 ml,
- 10^2 faecal streptococci/100 ml

Groundwater has very low concentrations of faecal bacteria.

Experiment at Cogolin

The Cogolin plant treats the wastewater from 7000 inhabitants in winter and 10 000 in summer using the activated sludge extended aeration process. Effluent used for irrigation is drawn off at the outlet of the secondary sedimentation tank and distributed either:

- by sprinkler system over 0.5 ha;
- by local irrigation over 2 ha,

at a dosage of 1 or 2 ETP, depending on the plot.

The forest has cork oak, white oak, pine and about 15 species of non-indigenous plants. The soil is very light altered, shallow (0–40 cm) to superficial, originating from the weathering of micaschists. The sandy-silty texture at the surface becomes clayey-silty 40 cm down, consequently promoting hypodermic runoff.

The effluent had the following bacteriological features:

10^6–10^7 total coliforms/100 ml,
10^5 faecal coliforms/100 ml,
10^4–10^5 faecal streptococci/100 ml.

Pathogenic organisms researched were pathogenic staphylococci and salmonellae.

Table 27.2 Comparison of the experimental conditions at Porquerolles and Cogolin sites

Characteristic	*Porquerolles*	*Cogolin*
Origin of the effluent before irrigation	Tertiary lagooning	Secondary treatment
Bacterial content		
total coliforms	10^3–10^4/100 ml	10^6–10^7/100 ml
faecal coliforms	10^2/100 ml	10^5/100 ml
faecal streptococci	10^2/100 ml	10^4–10^5/100 ml
staphylococci	Not tested for	Present
salmonellae	Not tested for	Present
Soil	Deep	Shallow: 0–20 cm sandy-silty 20–40 cm sandy-clayey
Vegetation	Citrus and peach orchards, artichokes	Cork oak, pine forest, deciduous species introduced
Type of irrigation	By row, drip irrigation	Drip irrigation, sprinkling
Irrigation dosage	Dependent on evapotranspiration	1 ETP and 2 ETP plots
Cultivation methods	Ploughing	Clearing of undergrowth
Control	Irrigated with groundwater	Not irrigated
Samples taken at depth	0–20 cm, 40–60 cm, 80–100 cm	0–20 cm, 20–40 cm
Length of study	1 year	3 years

Analytical conditions

The various bacteria under investigation were cultivated on specific media under precise incubation conditions. *Tables 27.3* and *27.4* show data for water and soil analyses, respectively. The seeding of these various media was carried out as follows.

Table 27.3 Researched bacteria, culture media and incubation conditions for the water samples

Researched bacteria	*Culture media*	*Incubation conditions*	
Aerobic mesophilic heterotrophic microflora	Trypcase – Soja–Agar (Biomerieux)		
at 37°C for 24 h		37 °C	24 h
at 20°C for 72 h		20 °C	72 h
Total coliforms	Tergitol TTC (Sartorius)	37 °C	24 h
Faecal coliforms	Endo (Sartorius)	44 °C	24 h
Faecal streptococci	D-coccosel (Biomerieux)	37 °C	24 h

Table 27.4 Researched bacteria, culture media and incubation conditions for the soil samples

Researched bacteria	*Culture media*	*Incubation conditions*	
Total coliforms	Presumptive test:		
	bactolactose broth (Difco)	37 °C	48 h
	Confirmative test:		
	brilliant green bile broth (Biomerieux)	37 °C	48 h
Faecal coliforms	Presumptive test:		
	bactolactose broth (Difco)	37 °C	48 h
	Confirmative test:		
	brilliant green bile broth (Biomerieux)	44 °C	48 h
	peptone water without indole (Pasteur Institute)	44 °C	48 h
Faecal streptococci	Presumptive test:		
	azid glucose broth (Biomerieux)	37 °C	48 h
	Confirmative test:		
	ethyl violet azid broth (Biomerieux)	37 °C	48 h
Pathogenic staphylococci	Champan agar (Biomerieux)	37 °C	24 h
Salmonellae	Enriched culture medium		
	Selenite broth (Biomerieux)	37 °C	18 h
	Isolating medium		
	Hektoen-agar (Biomerieux)	37 °C	24–28 h

For water samples

For the water samples, dispersion and filtration methods were used as a function of the microorganism under investigation and their presumed concentrations; 0.2 ml of raw samples and successive dilutions down to 10^{-6} were spread on specific agar plates, to enumerate the aerobic heterotrophic microflora (Porquerolles) and the indicator organisms of the secondary effluent (Cogolin).

For the other water samples in both experiments, indicator microorganisms were determined by filtration of 1 ml, 10 ml, 50 ml and 100 ml of raw sample. The same was true for the pathogenic staphylococci. Salmonella spp. were recovered in the Cogolin waters by filtration of 2 l of raw sample. The enrichment phase was made by inoculation of Selenite broth, then incubated for 18 h at 37 °C. Several loopfuls of each enrichment broth were streaked onto two or three plates of Hektoen-agar. The plates were incubated overnight at 37 °C. The presumptive salmonellae colonies were purified on the same selective media. After purification, the isolates were plated on a conventional nutritive agar before identification by selective biochemical tests (API system).

If the various biochemical and nutritional tests indicate the Salmonella type, the presence of Salmonellae is presumed. Serological tests alone can permit identification of these germs.

For soil samples

For the soil samples, a first dilution of 10^{-1} is done by weighing 10 g of wet soil and dispersing it by magnetic agitation for 30 min in 90 ml of normal saline solution at pH 7.5. Plating of the successive dilutions 10^{-2}, 10^{-3} and 10^{-5} was then carried out at a rate of 0.2 ml on the nutritive agar for the aerobic heterotrophic microflora.

The most probable number method was adopted for indicator organisms, at a rate of five tubes per dilution. The investigation for pathogenic staphylococci was done by spreading of the successive dilutions 10^{-1}, 10^{-2}, 10^{-3}, 10^{-4} and 10^{-5} at a rate of 0.2 ml on the specific medium. Salmonellae were removed by inoculating 10 ml of Selenite broth/10 ml of the 10^{-1} dilution. After incubation in this enriched medium, the method used was the same as for the water samples.

In order to express better the qualitative and quantitative evolution of the indicator organisms researched in the soil as a function of irrigation water, the results are expressed as a function of a unit of volume of wet soil. This unit of volume was set at 20 ml or 1 cm^2 of soil 20 cm high (the thickness of the layer under study). Transformation of the initial data is achieved by applying the following multiplier:

$$\frac{100 - \%\ \text{soil moisture}}{100} \times 20\,\text{ml} \times \text{soil density}$$

$$\text{or } (100 - \text{moisture content}) \times \text{density} \times 0.2$$

Microbiological evolution of the soil

The soil microbiological evolution shows particular characteristics, depending on whether the effluent comes from a lagoon or a secondary treatment plant. However, some points are common to both.

For an effluent from lagooning

The microbiological evolution of the soils under cultivation on Porquerolles shows the following:

- There is no significant difference between the densities of the indicator bacteria in soil irrigated with wastewater and soil irrigated with groundwater. This is a highly favourable element for the use of this method.
- The total coliform concentrations are greater by 2 log units than those of the faecal coliforms and faecal streptococci, which are of equal concentration. In some cases, the faecal streptococci have densities slightly greater than those of the faecal coliforms. The nature of the species involved may explain such behaviour.
- The evolution of these densities as a function of the weather is apparent, especially for the faecal coliforms. At the end of the irrigation season, the faecal coliform concentration increases for all the profiles studied, both for the controls and for the soil irrigated with wastewater. A slow gradual migration of the faecal coliforms seems to occur toward the deep layers.
- The evolution as a function of the soil depth shows that:

- although low, the organic material contents have an influence on the densities of aerobic heterotrophic microflora. The upper layers, which are slightly richer in organic material, have more numerous heterotrophic bacteria (2 log units) than the poorer, deeper layers;
- the total coliforms have densities that increase with soil depth, except in August. At this time, the upper and lower layers have basically the same concentrations, whereas the intermediate level has the highest concentration, no matter what

type of irrigation is used and consequently no matter what the moisture content is in the various layers studied;

- the indicative role of faecal contamination is better observed for the faecal coliforms, which are subject to great fluctuations. This is especially noticeable for the surface layer, which is more directly influenced by environmental conditions.

- The survival of the indicator bacteria, and particularly of the faecal coliforms, depends on a number of interrelated factors. Those that seem to be most important in this study are:

- temperature: in the summer, the high temperatures recorded promote the survival of these bacteria since they approach their optimal growth. The temperatures allow these bacteria to withstand conditions of very low moisture;
- soil moisture: this is a basic component for bacterial life. However, the evolution of the profiles located near the irrigated areas, but not directly receiving water, is identical to that of the irrigated profiles. Moreover, these bacteria can regenerate when the initial moisture of the soil is low and they are more numerous than at the beginning of the irrigation season;
- soil aeration: ploughing contributes to a great degree to the aeration of the soil, even if it involves only the surface levels. It causes a considerable increase in the ratio of faecal coliforms to faecal streptococci. The coliforms seem to find more favourable conditions for their survival;
- organic material in the soil: it appears that during ploughing, the better aeration conditions bring about an increase in the biodegradation reactions of the organic material.

- There does not seem to be any significant difference between the various types of irrigation with wastewater. Only the evolution of the faecal coliform to faecal streptococci ratio shows higher values when drip irrigation is used compared with row irrigation. Between two nozzles, the values obtained are intermediate.
- The comparison between the bacterial densities caused by irrigation water and those encountered in the soil does not allow the conclusion that there is a significant change because of the irrigation water. Indeed, the densities of faecal coliforms and faecal streptococci brought by the wastewater for the whole irrigation season are, respectively, 10^2 and $10^3/cm^2$ of soil. These same values are found on average in the soil for the irrigation season but with peaks of 10^4–10^5 during some sampling campaigns (number of bacteria in 20 ml of wet soil). But they are also the densities found in the soil irrigated with groundwater, whose faecal contamination is very low or nil.

Indicator bacteria are thus not introduced only by irrigation water. Other factors must play a role in this, such as the use of fertilizers with very high bacterial contamination.

In a forest soil irrigated with effluent having undergone only secondary biological treatment

The microbiological evolution suggests that:

- No pathogenic bacteria (staphylococci, salmonellae) were detectable with the analytical method used, even during periods of intensive irrigation. This may indicate the short survival of these bacteria outside their original faecal habitat.

- During the irrigation season, the effluent produces significant increases in bacterial concentration compared to the non-irrigated control. These increases are, on average:

  from 3 to 4 log units for the total coliforms,
  from 2 to 3 log units for the faecal coliforms,
  from 1 to 2 log units for the faecal streptococci.

  The results show an increasing order of influence, as follows:

  the sprinkled plot where it is primarily the top layer that is influenced by the irrigation,
  the microirrigated plot, where the dose is approximately 1 ETP,
  the microirrigated plot with a dose of 2 ETP.

- Irrigation also influences the soil between two nozzles, not directly irrigated by wastewater, because the movement of the water in the soil can contaminate neighbouring areas not irrigated.
- The highest bacterial count occurs during the hot period and the lowest bacterial count in April. Concentrations decrease gradually during the winter; the levels recorded in winter and at the beginning of spring are sometimes at the limit of analytical detection and are not significantly different from those of the control. There is no increase in bacterial count from one year to the next.
- The comparison between the concentrations brought by the irrigation effluent and the bacterial densities counted in the soil shows a good correlation.

Points in common at Porquerolles and Cogolin

The indicator organisms studied may be characterized by their more or less strict faecal origin and, consequently, by their greater or lesser specificity as indicators of such contamination. Faecal coliforms are the most representative of faecal contamination, both qualitatively and quantitatively (*Table 27.5*).

Faecal streptococci are much less specific. Moreover, they are present at relatively stable contents over time. They show a better adaptation to conditions encountered outside their usual faecal habitat. This is probably due to the presence of streptococci species that are more widespread in the natural environment (particularly on the plant level).

The total coliforms are the least indicative of faecal contamination. Their number in the soil is influenced by external factors such as temperature. They have the highest densities during the highest summer temperatures. In winter, their concentrations are not negligible.

Conclusions

These studies dealing with the microbiological analysis of soil irrigated with effluents from tertiary lagooning and secondary treatment answer centain questions concerning the sanitation risks incurred by such irrigation practices:

- Effluents from a finishing lagoon present sanitary guarantees and are similar in quality to classical irrigation water. There is no significant difference between the densities of the indicator bacteria in soil irrigated with wastewater and those

Table 27.5 Comparison of major findings of the study at the two sites

	Porquerolles	*Cogolin*
Increase compared to control		
total coliforms	Insignificant	3–4 log units
faecal coliforms	Insignificant	2–3 log units
faecal streptococci	Insignificant	1–2 log units
Pathogenic germs in soil		Absent
Differences traceable to type of irrigation	Insignificant	Aspersion < local irrigation 1 ETP < local irrigation 2 ETP
High concentration periods	Hot period	Hot period
Low concentration periods	Period of no irrigation	Spring
Comparison with the bacterial concentration in irrigation water	Insignificant	Good correlation
Accumulation from one year to the next	None	None
In-depth migration	Very slow migration of faecal coliforms	Faecal coliforms: leaching at depth by rain

in soil irrigated with groundwater. This is, consequently, a very favourable element for the use of this method in a Mediterranean climate.

- No pathogenic bacteria (staphylococci and salmonellae) were removed in the soil, even during intensive irrigation. This may reveal their short survival outside their original faecal habitat and in a given pedological context.
- There is no accumulation of indicator bacteria in the soil from one year to the next. Initial conditions are re-established over the winter.
- The excess water due to irrigation or rainfall can cause a very gradual and low penetration of these bacteria down to the deeper soil layers.
- The type of irrigation has no direct influence on soil contamination. It does, however, play an indirect role in the case of secondary effluents, where the quantity of water applied influences the total content.
- It would be worthwhile to extend these experiments:

– to continue the study of strict faecal bacteria such as faecal coliforms and faecal streptococci;
– to continue investigation of pathogenic bacteria (such as salmonellae, vibrio and pseudomonas) and other bacteria indicative of former faecal contamination (such as clostridium) in the effluent, the soil and the crops; and
– to study the structure of the various bacterial communities living on or in the various substrata so as to determine exactly the interactions that may exist between them, especially antagonism phenomena between the indigenous telluric bacterial populations and the aquatic bacteria introduced.

Chapter 28

Treatment and use of sewage effluent for irrigation in Western Asia*

C. Ertuna**

Introduction

Population growth, urban expansion and rapid development of industry and agriculture have started being considerably constrained by water shortages, in particular in the member states of the ECWA† region with limited water resources (*Figure 28.1*). Scarcity of supplies has given cause for concern in formulation of national development plans, and decision-makers are being increasingly involved in devising ways to optimize the use of available supplies as well as augmenting available water resources by non-conventional means.

The limited nature of water supplies in many countries of Western Asia, located in one of the world's most extensive arid zones, has led to growing interest in rational use and conservation of this resource, essential to human needs and economic growth. Augmenting conventional water supplies by non-conventional water development techniques has already been applied in Western Asia and is gaining wider support. Present technology has made it possible, with a few exceptions, to treat wastewater to a quality level equal to that of high-quality natural water resources. Agricultural uses of treated sewage effluents include: irrigated agriculture; watering of gardens, open spaces, trees and forests and grasslands; fishponds and other aquaculture; and watershed development.

The potential of water reuse in meeting various demands was recognized in the ECWA region quite some time ago. As early as the 1950s in Qatar treated sewage effluents were used for irrigation of municipal areas. The reuse of wastewater after treatment was then put into application in Kuwait in a restricted manner, to be followed by the United Arab Emirates and Saudi Arabia. Currently, this non-conventional method of water supply is in practice on a limited scale in Egypt but related research work is continuing. In Iraq, two sewage treatment plants near Baghdad treat municipal effluents before they are discharged to the Tigris for reuse downstream. Bahrain has plans to use treated wastewater in irrigated agriculture. Jordan presently treats municipal effluents from Amman and discharges them to the King Talal Dam for reuse. In the region, some conventions and techniques have

---

* This chapter is based on the ECWA report 'Wastewater Reuse and Its Applications in the ECWA Region, E/CECWA/NR/84/2'.

** Natural Resources, Science and Technology Division, Economic Commission for Western Asia (ECWA).

† The ECWA region consists of Lebanon, Syria, Jordan, Iraq, Kuwait, Bahrain, Qatar, Saudi Arabia, The United Arab Emirates, Oman, The People's Democratic Republic of Yemen, the Yemen Arab Republic and Egypt (*see Figure 28.1*).

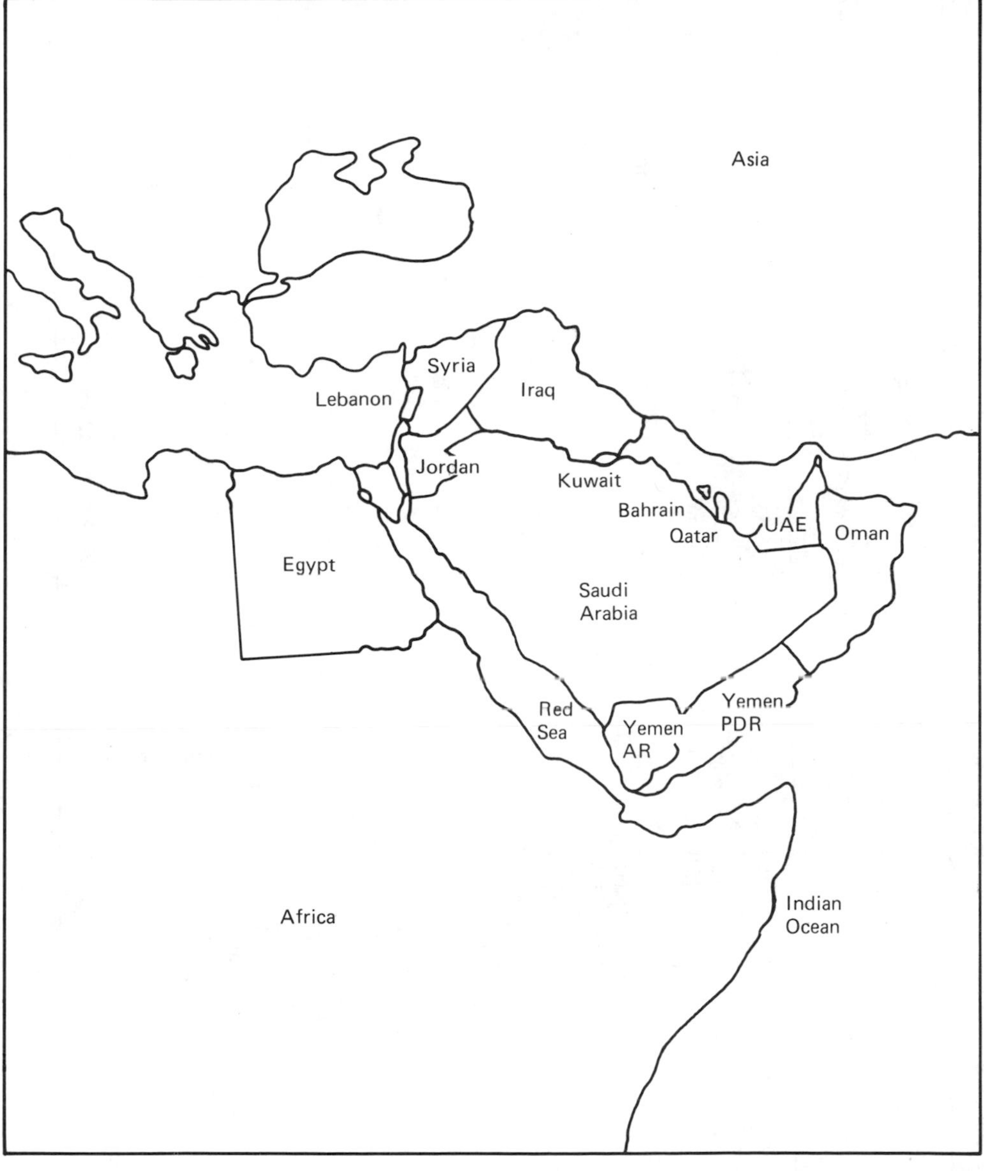

Figure 28.1 Location map of the ECWA region

been developed and experience gained, and some cost figures are also available. Although the health aspects of water reuse have not been completely defined yet, experience has shown that confining the use to agricultural and industrial purposes provides a fairly safe supply.

Treatment and use of sewage effluent in irrigation

There must be a central water supply system and a sewage collection network in cities and urban centres in order to have municipal sewage in adequate amounts for

Table 28.1 Typical physical and chemical characteristics of domestic wastewaters

Constituent	*Concentrations*, mg/l *(except as noted)*							
	*USA**			*Western Asia*				
	Range		*Average*	*Baghdad*	*Doha*	*Dubai*	*Bahrain*	*Riyadh*
	Low†	*High†*						
BOD_5	82 (100)	297 (300)	181 (200)	450–550	200	380	138	360
COD	204 (250)	871 (1000)	417 (500)	800–900	350	1130		620
TSS	64 (100)	406 (250)	192 (200)		150	370	94	400
NH_3–N	13.8 (12)	35 (50)	20 (25)	25–30	22	39	20	35
NO_3–N	0.09 (0)	1.52 (0)	0.61 (0)					0+
Org–N	7.0 (8)	15.3 (35)	13 (15)				10.6	
Ortho-P	3.0 (4)	13 (15)	6.8 (7)					
Total-P	5.96 (6)	132 (20)	9.4 (10)			24	7.3	10
Alkalinity	91	426	211				277	350
Oil and grease (units)	35 (50)	104 (150)	61 (100)				45.5	100
pH	7.0	7.6	7.2	7.0–7.2	7.5	7.0		7.5
Arsenic	0.002	0.020‡	0.007					
Barium	0.041	0.700‡	0.235					
Cadmium	0.002	0.02	0.008					
Chromium	0.015	0.560	0.167					
Copper	0.058	0.217	0.117				0.10	
Fluoride	0.6	1.36	0.864					
Iron	1.24	3.54	2.25				0.86	
Lead	0.017	0.206	0.148‡				0.10	
Manganese	0.044	0.195	0.117					
Mercury	0.66‡	0.001‡	0.112					
Selenium	0.001	0.014	0.006					
Silver	0.004	0.044	0.022					
Zinc	0.125	1.195	0.419				0.18	
TOC	63 (100)	171 (300)	102 (200)					
Hardness	84	672	362				1358	
Color P-C units	39	67	55					
Turbidity, FTU	78	152	2.16					
TDS	554 (250)	(850)	577 (500)	400–450	2500	2420	5235	1300

* US information is adapted from: US Department of the Interior (1979).
† Values in parentheses adopted from Metcalf and Eddy, Inc. (1972).
‡ From limited data.

treatment and reuse. The processes required for treatment of wastewater will largely depend on the constituents present in the sewage, the quality requirements of the treated effluent, availability of treatment facilities and manpower, and choice between capital and operating cost combinations.

Typical domestic wastewater characteristics of some urban centres in Western Asia and the range in the United States are presented in *Table 28.1*. Water constitutes about 99% of the volume of sewage and the composition of the solids varies according to the dietary habits of the population. The organic matter in raw sewage may be as much as 60–70% of the total solid matter and is the most active portion of the sewage. This organic matter is normally composed of proteins, carbohydrates, fats and oils, surfactants and agricultural trace compounds.

The most critical element in wastewater reuse is that of health. Effluent must be treated and then used in such a way that it will not be a danger to human beings or to the environment. The major constituents of sewage effluent that need to be considered in selection of treatment processes for a reuse scheme are:

- biological: pathogenic bacteria and viruses; parasite eggs; worms and helminths
- chemical: nitrates and phosphates; salts; toxic chemicals (including heavy metals)

In Western Asia, particularly in the Gulf area where wastewater reuse has been in practice for a considerable period of time, the quality requirements have been established, generally as 10 mg/l BOD_5 and 10 mg/l total suspended solids (TSS), mainly for conservation of public health and environment. Standards in Saudi Arabia were established based on the guidelines laid down by the convention of Moslem leaders who met in 1979. Accordingly, in order to meet the requirements for 'clear, odourless, colourless and tasteless water', the drinking water criteria of WHO were used as the basis for the process design of new projects.

The oil-rich countries of the region do not experience energy problems, obtaining of spares is not difficult, and land area required for treatment plants is readily available. Treatment methods selected for the plants reflect the requirements of the climatic and topographic conditions of the locality. *Table 28.2* presents the characteristics of major operational and planned municipal sewage treatment plants in the ECWA region. Generally, sedimentation and activated sludge methods are preferred for secondary treatment and rapid gravity filters followed by chlorination or ozonation for tertiary treatment. The flatness of land requires pumping in both collection and distribution systems.

Treated sewage effluent applications in irrigation in Western Asia

Most of the ECWA Member States, in particular those located in the arid region, have very limited natural water resources. Growth of population, rising standards of living and increasing demands of the fast developing economies have prompted the governments, especially in the Gulf area, to turn to the reuse of treated sewage effluent mainly for irrigation, thus freeing fresh water supplies for drinking and for other uses requiring high quality standards. Despite current limited application of wastewater reuse, plans have been formulated and projects activated for significant development of this non-conventional source of supply.

The main projects described below indicate the extent to which the countries in Western Asia are considering treated sewage effluents as a complementary source of water supply for irrigation. The characteristics of the sewage treatment plants are presented in *Table 28.2*.

Bahrain

When the project is completed, involving transport of treated effluent by pipeline and pumping stations to the new agricultural area to the west of Isa Town, the wastewater treated up to the tertiary stage will irrigate at least 500 ha – between a third and a half of the agricultural area on Bahrain island. There are also plans to increase the treatment capacity of the Tubli plant to 163 000 m^3/day by the year 2000 (Ukayli and Husain, 1983).

Jordan

Presently, there is indirect reuse of wastewater in Jordan, mainly for irrigation. Effluent from the sewered areas of Amman is treated and then discharged into a wadi where it is mixed with flow from stormwater during the wet season before entering the King Talal Reservoir. After mixing with the waters of the reservoir and detention, it is used for irrigation. The annual flow from the treatment works into the reservoir is about 18 million cubic metres per year (Mm^3/year). The live storage capacity of the reservoir is 48 Mm^3.

Jordan has plans to set up 36 sewage treatment plants serving 65% of the population. It is expected to treat 60 Mm^3/year of effluents up to the quality levels with which it will be possible to irrigate a total of 30 000 ha.

For further information on wastewater reuse in Jordan, *see* Chapter 21 where the present and projected wastewater reuse activities in the country are given in considerable detail.

Kuwait

The use of effluent in agricultural and forestry schemes has been in practice in Kuwait for many years. Initially, substantial amounts of septic tank contents were collected by tankers and used on government-controlled enclosed afforestation areas where there is no public access. The Agricultural Development Programme started experimental studies on the use of clarified sewage effluent for irrigated agriculture in the 1960s, and the Food and Agriculture Organization of the United Nations provided Kuwait with assistance. In 1977, an irrigation scheme with a projected total area of 920 ha was put into operation utilizing the 24 000 m^3/day available to the project, and irrigating alfalfa, winter forage crops, barley, and a small amount of different vegetables (United Nations Food and Agriculture Organization, 1978).

The present strategy of Kuwait is to utilize treated sewage effluent in development of irrigated agriculture by intensive cultivation and environmental forestry. The plans aim at developing 2700 ha of intensive agriculture and 9000 ha of forestry.

Treated sewage, after tertiary treatment, is at present used at the experimental farm and the irrigation project in irrigating a total of 900 ha of which 75% are fodder crops and the rest field crops, fruit trees and some vegetables, utilizing 45 000–55 000 m^3/day of treated effluent.

Table 28.2 Characteristics of major operational and planned municipal sewage treatment plants in Western Asia

Location	*Capacity* (m^3/day)	*Primary and secondary treatment*	*Tertiary treatment and disinfection*	*Effluent quality* *BOD* (mg/l)	*SS* (mg/l)	*Status*
Bahrain						
Tubli (present plant)	65 000	Mechanical, extended aeration activated sludge system	Dual media filters, chlorination	10	10	Operating
Iraq						
Rustomiyah (Baghdad)	90 000	Mechanical, sedimentation, activated sludge, settling, chlorination	Not applied	40	60	Operating
Kuwait						
Four plants	320 000 (total)	Sedimentation, activated sludge	Rapid gravity sand filters, chlorination	10	10	Operating
Qatar						
Al Naijah	65 000	Sedimentation, activated sludge	Rapid gravity sand filters, chlorination	<5	<5	Operating
Saudi Arabia						
Riyadh	200 000	Sedimentation, high rate trickling filters, aerated lagoons, chlorination	Not applied. Rapid gravity sand filters planned	20	40	Operating, will be upgraded
Jubail	115 000	Sedimentation; secondary (not finalized)	Not finalized	15	15	Planned
Jeddah	50 000 (each unit)	Coagulation, lime softening, sedimentation, recarbonation	Dual media filters, reverse osmosis, chlorination	1.5	0	Designed
Mecca	50 000 (total)	Coagulation, lime softening, sedimentation, ozonation, recarbonation	Dual media filters, chlorination	1	7	Designed
United Arab Emirates (UAE)						
Abu Dhabi	200 000 (eventual)	Sedimentation, activated sludge	Rapid gravity sand filters, chlorination and ozonation	Varies	Varies	Operating
Dubai	130 000–200 000	Mechanical, settlement, activated sludge, aeration	Rapid gravity sand filters, chlorination and ozonation	10	10	Design complete
Al Ain	30 000–50 000	Mechanical, extended aeration	Rapid gravity sand filters, chlorination	10	10	Operating
Other Governorates	320 000	Mechanical and activated sludge	Not applicable	40	40	Under construction

Sludge from the treatment plants, after digestion, thickening and drying, is used as soil conditioner and fertilizer supplied to government and private agricultural farms.

More information is available in Chapters 23 and 24.

Qatar

Wastewater reuse has been in practice in Qatar since the 1950s and effluent is mainly used for watering of roadsides, parks and municipal gardens. Due to the limited fresh water sources and increasing cost of desalination, use of treated effluents for agricultural purposes has gained increasing acceptance over the years (Balfour-Halcrow, 1981; United Nations Food and Agriculture Organization, 1980).

The treatment works in Al Naijah, constructed to serve a population of 180 000, makes 65 000 m^3/day of treated effluent available for reuse, mainly in agriculture. According to existing plans, part of this will be pumped by a pipeline over 40 km to an area where it is proposed to develop gradually a total of 7000 ha by the year 2000 under various crops. The remainder will be used for irrigating trees and landscaping Doha.

Saudi Arabia

In order to cope with the shortage of water, steps have been taken to develop new sources of water supply as well as to utilize treated effluent in irrigation and for other limited purposes. In particular, the following projects are of special interest.

Riyadh

Riyadh Wastewater Treatment Plant, with a capacity of 200 000 m^3/day at present, treats over 120 000 m^3-day of wastewater, serving the capital of Saudi Arabia with an estimated population of over 1 000 000 (Al-Dhowalia *et al.*, 1984). The General Petroleum and Minerals Organization (Petromin) utilizes 20 000 m^3/day of Riyadh's treated wastewater. Petromin's water reclamation plant produces three grades of water: utility water for hose stations and firefighting; process water for crude oil desalination and cooling tower use; and boiler feed water. The balance of the treated effluent from Riyadh plant is available for agricultural irrigation at Dirab and Dariyah.

Presently at Dirab and Dariyah, 57 000 m^3/day are used to irrigate 2000 and 850 ha, respectively. At Dirab the main crops are wheat, fodder crops and some vegetables and, at Dariyah, date palms, fruit trees, fodder and vegetables. Very soon, an additional 80 000 m^3/day will be available for irrigating 1200 ha at Ammariyah, once the Riyadh Wastewater Treatment Plant starts to operate at full capacity.

Jeddah and Mecca

The historical water shortage in Jeddah (Singley *et al.*, 1981), situated on the Red Sea, will be largely alleviated upon completion of an advanced wastewater purification plant designed to meet the drinking water criteria of the World Health Organization. Although the treated effluent will be used for non-potable purposes, this project is the first of its kind in the ECWA region to treat wastewater up to the standard of drinking water supplies. To service an anticipated population of 2.25 million by the year 2000, a number of treatment plants will be constructed, each with an initial capacity of 50 000 m^3/day extendable to 100 000 m^3/day.

The Holy City of Mecca (Singley *et al.*, 1981), which has also suffered from a traditional lack of adequate water supplies, will implement a project similar to that in Jeddah. Treated effluent in both cities will be suitable for municipal, industrial, and agricultural purposes, thus relieving the demand on the potable water supplies.

Jubail Industrial City
In the new industrial city of Jubail there are extensive plans for a green landscaped city where demand for non-potable water will be considerable (John Taylor and Sons, n.d.). Plans indicate that all domestic effluent and some industrial wastewater will be treated to be used in municipal irrigation and for other purposes. The city is planned to have a population of 400 000 by the year 1992 and the treatment plant will have a capacity of 115 000 m^3/day.

Madinah
The present wastewater flow from the Madinah treatment plant is about 27 000 m^3/day and is expected to reach 100 000 m^3/day by the year 1995 to irrigate about 1750 ha. By the year 2000, 141 000 m^3/day of treated effluent are expected to irrigate 3100 ha.

Detailed information on present and projected uses of treated wastewater for agriculture in Saudi Arabia is presented in Chapter 22.

United Arab Emirates

Abu Dhabi
An extensive scheme for using treated effluent to irrigate municipal areas and for landscaping has existed in Abu Dhabi since 1976 (John Taylor and Sons, n.d.; Banks, 1980). The system was designed eventually to provide 70 million m^3/year of treated effluent from a population of 665 000. The treated wastewater distribution system is clearly marked to avoid cross-connection with the potable water network. Municipal areas readily accessible to the public are not watered by treated effluent and, where applied, the treated sewage irrigation systems operate under low pressure to avoid the spray irrigation being carried by the wind into populated areas.

Al Ain
The population of this growing city was over 150 000 in 1984 and is expected to reach 250 000 by the year 2000 (Deane *et al.*, 1983). The city has a sewage treatment plant designed to produce treated effluent suitable for selected irrigation purposes. Irrigation of crops that are likely to be eaten raw or partially cooked is not allowed. Surplus activated sludge is aerobically treated and dried on open beds. Sludge cake is then taken into the nearby composting plant for production of compost to provide humus for plants at roadsides and in limited locations.

Dubai
The existing sewage treatment plant in Dubai (GWE Consulting Engineers, 1982) is not adequately equipped for production of treated effluent of good quality, and the product is at present used only to water some parks by means of water hoses. Therefore, a new sewage treatment plant has been designed and construction work is underway. According to the design the plant will produce increasing quantities of treated effluents, reaching its ultimate capacity of 200 000 m^3/day by the year 2005.

There are also plans for a sludge treatment plant producing digested, conditioned and dried sludge suitable without restriction for agricultural and market garden uses.

Other water reuse activities in Western Asia

A sewage treatment plant with a capacity to serve a population of 150 000 is planned to be constructed in Sana'a, the capital city of the Yemen Arab Republic. The treated sewage effluent will be suitable for irrigation and/or aquifer recharge purposes.

In Iraq, treated effluent directed to the Tigris from treatment plants upstream along the river and the Diyala River goes through additional natural flow purification processes before it is abstracted for irrigation downstream.

Damascus, the capital city of Syria, has also considered utilizing treated sewage effluent for irrigation; however, no plan of action has so far been developed.

Public health, cost and policy considerations in wastewater reuse

The most critical elements in treatment and use of sewage effluent in irrigation are public health and the cost of treatment. Wastewater must be treated and used in such a way that it will not be hazardous to human beings or to the environment. Pipes transporting treated effluent and potable water must be clearly differentiated to avoid drinking by human beings and animals. Bodily contact by agricultural workers with the treated wastewater and exposure to aerosol drift must be minimized to reduce short-term and long-term effects of any remaining contaminants.

In order to protect public health, the effluent should either be properly treated before irrigation application, or its use should normally be restricted only to certain crops so that improperly treated wastewater will not come into contact with any part of a plant used for human food or animal feed. In the ECWA region, wastewater reuse in irrigation has been in practice, though on a limited scale, for a number of decades. However, so far no epidemics or serious infections have been reported due to the limited use of the treated effluent and the safety precautions undertaken.

Although the technology for wastewater reclamation is available, economic considerations limit its use to special locations or particular purposes. As demands for existing resources in water-scarce areas become greater, use of treated effluent gains more potential in certain uses and releases natural sources of water for potable supplies. Wastewater reuse becomes particularly attractive when it is planned in conjunction with environmental protection. Pollution control measures are becoming stricter all over the world and such regulations are also being set up in Western Asia. Therefore, an increasing number of sewage treatment plants are being established where the effluent is treated to certain levels that will prevent pollution and will allow some direct use or, after additional treatment, almost any use. Where treatment plants are planned for environmental protection, the marginal cost of advanced treatment and the distribution system is only about a quarter of the total cost of the scheme. The component cost estimates for the treatment plant under construction in Dubai are presented in *Table 28.3*; the combined cost of the tertiary treatment and effluent pipeline makes up only 26% of the total.

Table 28.3 Cost breakdown for the 200 000 m^3/day wastewater treatment plant planned for Dubai municipality

Components of the scheme	*Component cost, 1981 prices*	
	Dirham (million)	*US dollars* (million)
Power supply	6.7	1.84
Access road	1.5	0.41
Drinking water supply	1.3	0.36
Telephone	0.5	0.14
Influent pipeline	215.0	58.90
Effluent pipeline	170.0	46.58
Pumping stations	40.0	10.96
Inlet works	16.9	4.63
Primary settlement	22.5	6.16
First stage biological treatment	60.1	16.47
Second stage biological treatment	79.2	21.70
Coagulation	4.6	1.26
Tertiary treatment	50.1	13.73
Sludge treatment	45.3	12.41
Administration and service buildings	25.1	6.88
Power supply (internal)	4.3	1.18
Pipework	56.4	15.45
Ancillaries	30.9	8.47
Siteworks	24.6	6.74
Total cost (1US$ = 3.65 Dh)	855.0	234.25

Source: GWE Consulting Engineers (1982)

Table 28.4 Comparison of unit costs of treated municipal sewage effluents and desalinated water in Western Asia

Location	*Base year*	*Treated effluent unit cost range* (US$/m^3)	*Desalinated water unit cost range* (US$/m^3)	*Remarks*
Bahrain	1984	0.28		Without a reverse osmosis unit
	1984	0.84		With a reverse osmosis unit
Egypt	1980	0.05–2.90		Primary treatment cost of industrial wastewaters
Kuwait	1985	0.33	2.50 (1984)	
Qatar	1981	0.24	1.14–1.16	Energy supplied at zero cost
	1981		1.45–1.64	Energy supplied at market cost
Saudi Arabia	1980		0.04	Desalination of brackish water by reverse osmosis
	1979–82		0.71–1.23	Desalination of seawater by multistage flash system
United Arab Emirates (UAE)	1982	0.30–0.41	1.00–1.45	

However, in the ECWA region it appears that even if the whole project cost is charged to treated wastewater, still the unit cost is quite reasonable. The available unit cost estimates for treated wastewater in the region, together with the cost of desalination, the other prominent non-conventional source, are presented in *Table 28.4.* From this information it can be concluded that, in the early 1980s, the cost of complete treatment of municipal wastewater for reuse was not above US$0.40/m^3 in Western Asia. This is a very favourable unit cost compared with that of desalination of seawater, though the latter produces a drinking water level product. Desalination of brackish water by reverse osmosis-type membrane processes produces fresh supplies at comparable or cheaper prices than wastewater treatment, but an unlimited supply of brackish water and easy access for brine disposal are necessary. However, as complementary alternatives, the cost comparison of desalinated water and treated sewage effluent gives a fair indication of the potential for wastewater reuse in Western Asia.

However, the economic analysis of reuse projects is not always the determining factor. In many arid lands, such as in the Gulf countries, there is a desire to develop and maintain public green areas for amenity purposes. This type of utilization, as a policy, is often adopted as the sole use when beautifying the environment creates another burden on the limited water resources of a country.

In order to initiate any development in the wastewater reuse field, policies must be established ahead of other activities. After the assessment of the need for this practice, the policies and options can be laid out according to the requirements and characteristics of the country. In planning for reuse, it is essential to consider the sources of the effluent and the purposes for which it will be used after treatment. Priority in policies must be given to public health and environmental considerations due to the potential hazards associated with this practice.

In Kuwait, the policy related to wastewater reuse allows its use for the irrigation of forestry, fodder crops and certain horticultural products. In Qatar, some municipal areas have been irrigated by treated sewage effluent. In Saudi Arabia and the United Arab Emirates, uses of treated wastewater are restricted, according to the present policies.

The policy for wastewater reuse must be backed up by adequate legislation, clearly establishing the legislative measures governing the collection, treatment and distribution systems and defining the responsibilities and scope of various government agencies. Furthermore, the institutional setup for wastewater should be adequate to provide a framework for the allocation, use, quality and health and safety aspects and there should be continuous coordination and cooperation between the concerned agencies. Personnel at different levels must undergo thorough training to make wastewater reuse projects a success by maintaining safe and economic operations. Research is usually needed on developing improved management techniques and institutional arrangements for increased efficiency in the use of treated effluents for various purposes. Other research work may be required on ways to reduce the cost of treatment processes and to find answers to possible virological hazards.

Conclusions and recommendations

Wastewater reuse has been in practice in the ECWA region for a considerable period of time, however, it has been limited in application and only recently have plans been formulated for large-scale development of this non-conventional source of supply. In part, lack of knowledge of the long-term effects of use of treated effluent for various purposes, and also the availability of other sources of water, have prevented wider application. However, with the development of new technology and the rising cost of water developed by conventional methods or by desalination, there is a new outlook in the field of wastewater reuse.

The most important considerations in wastewater reuse are the cost of treatment and health. Treated effluent will not be used to replace more conventional or other non-conventional sources, unless it is proved that the cost of treated effluent is less. Water reuse, whether for agriculture or other purposes, still entails certain risks to human beings and to other living species and must be practised under careful control.

A policy for wastewater reuse must be clearly established prior to any development activity in this field. An obvious basis for such a policy is to limit the use of treated effluent for purposes outside potable uses, which would avoid or minimize the risk of human contact. The necessary legislation must be enacted to promote the control related activities and to create an institutional framework, and this latter should be adequate to provide for the allocation, use, quality, and health and safety aspects of effluent reuse and there should be continuous coordination and cooperation between the concerned agencies. Thorough training of personnel is essential to make wastewater reuse a successful practice by maintaining safe and economic operations. Training is recommended at different levels. Research work on wastewater reuse in the ECWA region should be directed towards finding ways to reduce cost and to reduce hazards, largely virological.

The following have been recommended for wastewater reuse applications in Western Asia in the comprehensive ECWA study entitled 'Wastewater Reuse and its Applications in the ECWA Region', on which this chapter is based:

- Use of treated effluent for potable and domestic purposes is not recommended. When the wastewater is to be used for other purposes, where human contact is unavoidable, then it should be treated close to the established standards required for drinking water.
- In the ECWA region, use of treated wastewater is recommended for the following purposes, provided that the minimum standards required by each particular user are met:

- Agricultural uses: irrigation of forests, fodder crops and others that will be not be eaten raw or will not contaminate kitchen utensils. Fishponds and aquaculture may also be considered as possible areas of use.
- Municipal uses: watering of roadsides and divides, municipal parks, gardens, golf courses, etc.; firefighting and streetcleaning may be considered provided that the minimum standards are strictly met.
- Industrial uses: largely for cooling and boiler feed and in some industries as process waters; ore separation is another major use. Treated wastewater reuse is not recommended in food industries since a mix-up between the treated wastewater and fresh water supply systems may cause a major disaster.
- Other uses: streamflow augmentation; groundwater recharge, and prevention of saltwater intrusion into aquifers are other possibilities. Use of treated sewage effluent in recreational ponds is not recommended in the ECWA region.

- In the early 1980s, the unit cost of treated municipal effluent in the region was established as US$0.40/m^3. In planning for augmentation of the present water resources for industrial, agricultural and non-domestic municipal purposes, if the cost of the additional supplies is estimated to exceed the above figure (after updating for inflation), then wastewater reuse should be seriously considered as a possible alternative, especially if there is a population centre nearby with over 100 000 people and with an existing wastewater collection system.
- The survey of new treatment plants that will enable reuse of municipal wastewater has indicated that 10 mg/l BOD_5 and 10 mg/l suspended solids are established as minimum standards in the Gulf countries. These criteria are recommended for application in treatment plants elsewhere in the region, as well as where there are plans for the use of treated effluents. In planning for the use of treated wastewater for various purposes, the common minimum requirements should be met in the principal treatment plant, and the standards required for each particular use must be met at the separate plants. When deciding whether to use a centralized plant or a number of individual plants and in the selection of the required train of processes, local characteristics will be the most important factors.
- If a government takes the decision to use treated wastewater as a source of supply, a policy regarding the use of treated effluent must be established. A policy once established may be revised later, according to new circumstances, but at the beginning the policy must spell out the uses, the conditions governing the uses and users, and it must establish the required standards to be met. A certain recommended base for such a policy in Western Asia is to limit the use of treated wastewater for purposes other than domestic uses. The necessary legislation and institutional framework for wastewater reuse must also be established.
- It is recommended that the following precautions and safeguards be taken in wastewater reuse, particularly in the ECWA region.

- Wherever treated wastewater is used, the standards and strict regulations should be followed from the beginning and, if necessary, these standards should be revised after long-term monitoring of the activities. Continuous and safe operations should be secured and any end-product not meeting the standards should not be released for use.
- The pipes and other system components carrying fresh water supplies and the treated sewage effluent should be clearly differentiated for domestic purposes.
- Municipal workers and agricultural workers who have to use treated wastewater as part of their work should be well-informed about the particulars of this practice and required to minimize their contact with the treated effluent. They should not be allowed to work barefoot and sanitary facilities for washing and eating should be provided. They should also be subjected to periodic medical check-ups in order to detect any long-term effects of exposure and, in protection of their health, personal hygiene should be promoted.
- Industry should be encouraged or, if necessary, pressured to recycle part of its wastewater and to treat the rest to acceptable standards so as to minimize pollution. Industrial wastewaters should not be mixed with municipal sewage since they have distinct characteristics and mixing them would make the job of treatment very costly. However, relatively clean cooling-water may be treated

for reuse in irrigation. In the ECWA region such waters are estimated to amount to 4 billion m^3/year.

- Return flows as drainage water to streams should be strictly monitored because the length of the natural purification process may not be adequate before reabstraction downstream, causing accumulation of salts and undesirable substances. If necessary, the drainage water should be given adequate treatment before allowing its return to streams. Purification of industrial wastes by natural processes in streams is not recommended.
- In streamflow augmentation, the treated effluent must meet the established criteria so as not to endanger the health of human and animal life downstream.
- In groundwater recharge and prevention of saline water intrusion to an aquifer, treated wastewater should be close to drinking water standards. The period of natural purification during percolation through the soil layers may not be long enough and should not be counted upon without adequate research work. If the groundwater becomes contaminated it is a mammoth, if not impossible, task to purify it again, incurring tremendous cost. This application of wastewater is very important for the Gulf countries in the region. When treated wastewater is used for watering gardens or irrigating crops and not directly used for recharge there will still be some percolation. Therefore, the quality of the treated wastewater should be very good to start with (that is 10 mg/l BOD_5, 10 mg/l SS), and contamination with insecticides and pesticides should not pose a problem. Furthermore, continuous percolation of irrigation water may cause a steady rise in groundwater levels, leading to accumulation of salts and other undesirable factors if the groundwater level is not controlled.
- The necessary precautions to be taken on the reuse of treated effluents must be presented to the public through the information media.

In general, one must conclude that wastewater reuse has a promising future in Western Asia, and that ECWA Member States should start to adopt this practice as a serious alternative to conventional sources of water, if they have not already done so. Agricultural use of treated sewage effluent stands out as the largest potential user since it involves large quantities covering vast areas. Under strict controls, as mentioned above, wastewater reuse in irrigation will certainly continue gaining importance, as the limited water resources of the region become unable to meet the requirements of all the competing demands.

References

AL-DHOWALIA *et al.* (1984) Reuse of treated wastewater for agricultural irrigation at Riyadh. *Proceedings of the Water Reuse Symposium III, San Diego*

BALFOUR–HALCROW (1981) 'Reuse of sewage effluents' and 'Potable water sector'. Master Water Resources and Agricultural Development Plan, Qatar, Ministry of Industry and Agriculture

BANKS, P. A. (1980) Effluent utilization projects in the Middle East. *Seminar on Health Aspects of Treated Sewage Reuse, Algeria, June*

DEANE, A. N. *et al.* (1983) Sewage treatment and effluent reuse in arid regions: a case study, Al-Ain, United Arab Emirates. *Arab Water Technology Conference, Dubai, March*

GWE CONSULTING ENGINEERS (1982) New Dubai sewage treatment plant feasibility study. *Dubai Municipality, May*

METCALF and EDDY, INC. (1972) *Wastewater Engineering*. New York: McGraw-Hill

SINGLEY, J. E., BRODEUR, T. P., DOUGHERTY, C. W. and BEAUDET, B. A. (1981) Wastewater reclamation at Jeddah and Mecca, Saudi Arabia. *Proceedings of the Water Reuse Symposium II,* Washington DC: AWWA Research Foundation

TAYLOR, JOHN and SONS, CONSULTING CIVIL ENGINEERS (nd) *Reuse of Sewage Effluent*. London: Internal Publication

UKAYLI, M. A. and HUSAIN, T. (1983) Feasibility study of water resources alternatives in Saudi Arabia. *Arab Water Technology Conference, Dubai*

UNITED NATIONS FOOD AND AGRICULTURE ORGANIZATION (1980) Feasibility study of the agricultural use of treated sewage effluent water in the State of Qatar. *Technical Note 12*

UNITED NATIONS FOOD AND AGRICULTURE ORGANIZATION (1978) *Soil and Water Resources and Agricultural Potential in Kuwait*. Cairo

US DEPARTMENT OF THE INTERIOR (1979) *Water Reuse and Recycling*. Washington, DC

Chapter 29

Discussion

A questioner from Cyprus indicated that Dr A. S. Abdel-Ghaffar had stated that in an experiment, the pH of the soil dropped from 7.7 to 6 after 3 years. Previously, he had said that heavy metals were not available after irrigation for 45–50 years with sewage effluent. Could he comment on this?
Dr Abdel-Ghaffar explained that the information on the El Gabal-El Asfar farm is based on experience since 1911 but in the case of the pH change from 7.7 to 6, that is a recently started experiment in the Abou Rawash area. It is expected that the pH drop will be temporary because the microorganisms are highly activated by the addition of sewage and produce a lot of acid but in time this will stabilize and the pH should rise again.
A questioner from Egypt suggested that Dr Papadopoulos should reconsider his thinking on trickle or drip irrigation because, unless there is abundant leaching, salinity will tend to be a problem. Perhaps the reason why it had been found that irrigating with sewage effluent had reduced soil salinity, compared with irrigating with good quality water, was because of the more generous application of the effluent than the more expensive or limited river water or groundwater.
Dr J. Papadopoulos confirmed that trickle irrigation, under Cyprus conditions, is the best type of irrigation to fight against salinity. This is true because it can keep the salinity below the dripper at low levels, at least at the level of the irrigation water. During winter it is expected that some leaching of salinity which is accumulating on the periphery of the wetting zone will occur. However, at least during the growing season, trickle irrigation will keep the soil and soil water at least at the concentration of salinity in the irrigation water used. In the first experiment in Egypt the results were similar to those in Cyprus where an improvement in soil salinity occurred with effluent irrigation, rather than deterioration. Nevertheless these processes, particularly in respect of some components of salinity such as high concentrations of sodium, are very dynamic, so the results up to now are not yet conclusive.
Dr A. Arar, FAO, pointed out that trickle irrigation is the most suitable irrigation method for saline water. The United Arab Emirates are using drip irrigation in forestry with water up to 10 000 ppm salinity. The soil moisture stress in the soil can be controlled more conveniently with trickle irrigation than with sprinkler or surface irrigation because there is no long period between irrigations when the soil moisture decreases and tension increases. However, the other problems, especially clogging of drippers with sewage effluent due to algal growth and bacterial slime may be more of a problem than salinity.

Mr A. Bouzaidi, Tunisia, asked Dr Abdel-Ghaffar if high calcium soils act as an antidote to high sodium in irrigation water.

Dr Abdel-Ghaffar drew attention to the fact that new types of organic matter, new nutrients and other additives were being introduced to the soil and, consequently, it is very hard to reach a balance. Concerning calcium soils, the north coast of Egypt is very rich in calcium carbonates, sometimes at a level of about 96%, and for such soils sewage effluent can be used without problems. It is a very coarse-textured soil, rich in carbonates, and will neutralize any acids in the sewage or produced by microorganisms. There is no fear of it going alkaline because of high SAR in the effluent.

Dr H. A. C. Montgomery asked Dr J. Papadopoulos to comment on the fact that *Table 19.2* in his chapter shows that the sewage effluent was actually less saline than the fresh water used. How did this relate to his mention of problems of salinity? He also asked Dr Shende if he had experienced in India iron and manganese deficiencies when secondary-treated effluents were used as a source of nutrient for crops.

Dr J. Papadopoulos replied that it was true that the electrical conductivity of the effluent was 2.4, compared with 3 in the case of the local water used for irrigation. However, this was the average salinity of the effluent. It ranged from in excess of 2 to, this year, a conductivity in excess of 3. The results presented are the average conductivities of the previous year. This year, the conductivity is slightly higher, so the average for the whole irrigating season is going to be around 3.

Dr G. B. Shende agreed that iron and manganese are necessary nutrients for plants, but as there is quite an abundant supply of these elements in the soil in India it is not normally necessary to apply them.

R. O. Cobham, Oxford, United Kingdom, questioned Dr Shende about the use of treated effluent for irrigating forest areas. He drew attention to *Table 16.14*, in which are shown some quite impressive gross returns from the irrigation of eucalyptus. He asked if there was a control plot as part of that experiment and, if so, were there different yields and returns between the effluent and the sweet water? He also asked Dr Shende to comment on the choice of eucalyptus as a species, which is non-native to India and, therefore, must have been used with the expectation or intention of widening the range of species under this sort of trial?

Dr G. B. Shende indicated that the data given in his chapter on the irrigation of eucalyptus was from other organizations, it being a sort of review. It seems they did not use any control in their experiment but, in India, forest trees are one of the best alternatives for wastewater irrigation since there is no health hazard involved. The choice of eucalyptus was because it is a fast-growing tree and, particularly under Indian conditions, within 4 or 5 years it grows to the extent that it can be cut and the wood used for either fuel or timber purposes. There are some disadvantages with eucalyptus, particularly the effect of depleting the groundwater, but there are controversies over that. Other species of forest trees are also being popularized for afforestation and wastewater utilization.

Dr P. Economides, Cyprus, asked Dr S. Al-Salim in which form alfalfa and barley irrigated by treated effluents is given to animals? Is the health of the animals fed on such fodder monitored or checked and are the meat and milk of these animals inspected by veterinarians before being passed for human consumption?

Dr S. Al-Salim, Jordan, replied that in Jordan they were not using wastewater for direct irrigation. Sewage effluent, after treatment in waste stabilization ponds, travels about 40 km in a wadi until it reaches King Talal Dam where it is mixed with stormwater and only after a long detention time is it used for unrestricted irrigation. The capacity of the King Talal reservoir is about 56 Mm^3 and the quantity of water from the waste stabilization ponds is about 16–18 Mm^3/year. Since the quality of the water from the dam is the same as potable water, there is no need for monitoring of the type suggested.

Dr K. Al Salem, Kuwait, stated that alfalfa is irrigated by sprinkler irrigation in Kuwait. Health is monitored and checked by cooperation between the authority for agriculture and the Ministry of Public Health. They also check the animals, the meat and the milk. There is no production of meat products from animals raised in Kuwait at the present time, they only provide dairy products. All livestock for meat are imported from other countries under the supervision of the Ministry of Public Health and the Municipality, which is in charge of the slaughterhouse, and meat products are checked daily. Livestock raised in Kuwait are all checked because of the requirements of the Ministry of Public Health. There is a doctor on every farm who has to be present on these farms to check daily and report to the Ministry of Public Health. This is coordinated with the local authority in charge of inspection of all livestock in Kuwait.

Mr Z. Ali Khan, Pakistan, addressed Mr Al Shatti from Kuwait. He had indicated that the treated effluent is being used, apart from experimental purposes, in the private sector also. Could he say whether the private sector is being charged for the supply of this treated effluent and, if so, what is the relationship between the cost of production of the effluent and the charge for supplies made to the private sector?

Mr J. M. K. Al-Shatti, Kuwait, pointed out that no treated effluent is supplied to the private sector, as a private individual user. The company concerned with use of effluent in agriculture in Kuwait is not considered to be in the private sector but is in partnership with the government. After the policy decision was taken to utilize the treated effluent for the production of crops, either vegetables or animal fodder, it was decided to give support to the public, who originally were farmers or interested in farming, to initiate farming in the country. So, besides the effluent, technical support, advice, financial support, land and sometimes even seeds, were provided. The effluent is supplied to the United Agriculture Company at the present time. In the past, before tertiary treatment was introduced, they were charged 10 fils (which is equivalent to 5 cents in Cyprus money) for every 4500 litres. Of course, this includes the cost of transporting the effluent from the treatment plant, including treatment. In fact, allowing for sewage collection and treatment as well as transport to the irrigated land, the total cost is about 1 Cyprus pound for every 4500 litres.

Mr Y. Stylianou, Cyprus, questioned Mr Cobham on whether any uprooting of the trees had been observed in the desert areas when the drip method of irrigation had been used? If such a thing had occurred, had it been necessary to support the trees to a greater extent than with the older method of furrow irrigation?

Mr R. O. Cobham said he was aware of the problem but had not seen any widescale devastation or windblow more than might occasionally be experienced in the United Kingdom. There appeared to be no distinct difference between furrow and drip irrigation.

Mr Y. Stylianou, Cyprus, went on to suggest that the drip method of irrigation should be practised when applying sewage effluent to eliminate the danger of contaminating the aquifer lying below. This suggestion resulted from his experience in Cyprus.

Dr M. Abu-Zeid, Egypt, raised the issue of water resources planning. He said it was clear, listening to several countries' experiences in wastewater reuse, that the economic value of water varies from one country to another; even within the same country it varies with time. Some of the techniques used are quite expensive and yet provide minimum control of health or environmental hazards. Others are a little cheaper but have more hazards. The future will, no doubt, bring increased, or expansion in, wastewater reuse and there is a need for guidelines covering economic evaluation or feasibility studies of wastewater reuse. The environmental hazards or health impacts require evaluation and quantification before the decision can be taken on which approach is most cost-effective. In Alexandria, for example, a decision has to be taken on whether to discharge a large volume of sewage to the sea or to reuse the effluent in irrigation. Economic factors must be taken into account but little data on the economics of systems are available. Only Chapter 24 (on Kuwait) in this book gives any details of costs and normally the question will be very hard to answer.

Dr S. Al-Salim reviewed the decision-making process in Jordan. Originally, when the decision was taken to instal sewers and dispense with septic tanks, etc., there were two factors. First, were the cesspools affecting the aquifers or not? If a city overlies a groundwater aquifer, and on-site waste disposal is affecting the quality of the water, the groundwater must be protected. There is no need to study how much it will cost in an arid, water-short region. The problem is how best to protect the groundwater and the first objective was to construct sewers. After the sewers were built, the wastewater collected had to be treated again to protect the groundwater which supplies the city. The investment to build sewers and treatment plants had already been made so why not take advantage of the effluent? It is better to use the effluent and in Jordan research on the potential for reuse in irrigation is proceeding.

Mr J. M. K. Al-Shatti suggested that the decision on reuse depends on the particular situation, how great is the need, what is the effluent to be used for and whether or not there is another alternative. In the case of Kuwait there was no other alternative. The effluent had either to be used or dumped into the sea, with consequent sea or beach pollution. There is no other additional source of water to be utilized for agriculture and it costs too much to use desalinated water for agricultural purposes. Groundwater is limited and not up to the salinity standards for agriculture; it has a higher salinity than the treated effluent. The other factor is environmental protection; the sea, the land and public health must be protected, so treatment of wastewater is necessary. There are additional costs to achieve the standard required for the purpose. Each country is obligated to invest in treatment, unless the use of septic tanks is to continue with adverse effects on groundwater. If the groundwater aquifer is polluted, there will be additional costs for treatment before supply as potable water.

A speaker from Saudi Arabia agreed there should be guidelines for the economics of treatment but these should be flexible. Every country has its own situation and although criteria from WHO or other sources should be studied it is not necessary to apply them completely. They should be altered to suit the local situation. For example, desalination is very costly, but when it comes to water the social value is something which cannot be measured. Whether the price is high or not, water has to be provided.

Dr A. Arar referred to the cost of treated sewage effluent in Kuwait which had been stated to be about 1 Cypriot pound per 4500 litres, or about 45 cents per m^3. How does this cost compare with the cost of desalinated water in Kuwait at the moment?

Mr Al-Shatti replied that the cost of desalinated water in Kuwait is now less than 2 Cyprus pounds, exactly 3½ dinars Kuwait per 4500 litres. The government subsidizes the cost of water and it is sold to the public at much less than it costs. Similarly, the effluent is supplied to the agricultural company at a cost which is subsidized by the government. Unsubsidized, the cost would be the same as that of desalinated water.
Mr Z. Ali Khan addressed a question to Dr Kalthem of Saudi Arabia. Having learned of the very ambitious programmes of utilization of effluent in the cities of Riyadh and Medinah, what are the plans for Mecca, Jeddah and Dammam? Is it intended to wait for the results of the projects which are in hand or have definite plans been drawn up for immediate implementation in these cities also?
Dr M. S. Kalthem, Saudi Arabia, replied that the cities mentioned, Riyadh, Qassim and Medinah, are only examples of the national programme and there is no time to go over all the projects being undertaken. Chapter 28 from ECWA mentions activities in Mecca, Jeddah, Jubail and other cities in Saudi Arabia.
Mr Y. Stylianou suggested that plans to reuse sewage effluent should create maximum influence on the appearance of the landscape. In this respect oleander is one of the plants to be considered; this plant grows wild in Cyprus. Another local plant is *fidolaka* and, of course, there is the olive tree, which produces fruit for the region because it does not look ugly with the dust of the desert.
Dr S. Al-Salim mentioned that *Nerium oleander* is widely used in Jordan and in addition, *Tamarix. Ziziphus* and other trees are planted in parks.
Mr Y. Stylianou, addressed Ms Tremea, who appeared to have compared in the second experiment the full evapotranspiration (ETP) amount of water as one treatment and two times ETP as another. If the intention is to get rid of water, 10 times ETP might be applied. In the Mediterranean region, aquifers are considered to be very precious and by applying more than evapotranspiration the risk of groundwater contamination increases. More than the ETP amount might be required to control salinity, for example, but not twice as much. Colleagues in the Cyprus group could give advice on the specific extra amounts required to correct salinity, if there is a salinity hazard.
Ms L. Tremea, France, replied that the purpose of the experiment in Cogolin was to try to find the best irrigation rate for the shallow soil and three dosages were tested: 0.5 ETP, 1 ETP and 2 ETP. With 2 ETP little percolating runoff occurred but this year, after 3 years irrigation with 2 ETP, runoff resulted. Now the 2 ETP plot has been converted to a 1 ETP plot. Although the intention is to find the best dosage the aquifer must be protected. Analyses of the quality of percolating runoff from the plots have been carried out because the soil is not very homogeneous or level. A high level of pollution has only been observed after the first rain, which is the most important in the Mediterranean region. However, with the direct factor at this time the aquifer is not highly polluted.
M. D. Fougeirol, France, drew attention to Dr Al-Rifai making specific mention of the bacterial pollution of rivers and aquifer used for irrigation of vegetables in the Ghouta, creating health hazards for the population. Then a plan for constructing a secondary activated sludge treatment plant had been mentioned. In view of the fact that secondary treatment will not remove bacterial pollution and health hazards will still exist when irrigating vegetables with secondary treated effluent, was the possibility of tertiary treatment, by either technological or natural processes, considered?

Dr Al-Rifai, Syria, replied that tertiary treatment was being considered and that no bacterial hazard will be allowed to take place. The problem of bacteriology will be considered in the treatment plants. Replying to a question on alternatives, Dr Al-Rifai indicated that all possible schemes had been considered for Damascus, which now has over three million population. After looking into alternative systems and taking account of the bacterial problem and others, the chosen system was considered the best. Details concerning the choice are all documented.

Mr A. Bouzaidi, Tunisia, asked Ms Marecos do Monte the reasons on which the selection of crops was based.

Ms M. H. Marecos do Monte, Portugal, replied that salinity was not the only factor in the decision. The selected crops are resistant to moderately saline waters but these crops are also interesting from the point of view of the economy of the local population. They are now farmed as dry crops but can also be farmed as water crops. In addition, they minimize public health risks in that the harvested part of the crop does not come into direct contact with wastewater. Of course, this depends on the irrigation method but even if spray irrigation is used the risk would be minimum because maize and sunflower are harvested after a certain period of drying, so pathogens would certainly die.

Mr R. O. Cobham asked Mr Ertuna if he could elaborate on the rather important statement he had made concerning the ECWA region reuse recommendation reviewed in Chapter 28. Listed on p. 363 under municipal use is 'watering of roadsides and divides, municipal parks, gardens, golf courses, firefighting and street cleaning, may be considered provided the minimum standards are strictly met'. Mr Cobham explained that, in his experience, this would appear to be a significant departure from some of the rather stringent attitudes taken in a number of Middle Eastern countries up to now. Why does there appear to be this relaxation, because reuse for a golf course, for instance, could present very severe health hazards, with a lot of human contact, even if the irrigation takes place during non-user periods? Could this point be explained and the minimum standards referred to?

Mr C. Ertuna stated that, as he had mentioned in his presentation, in some places in the region drinking water criteria are being met so, if this is the case, the sewage effluent could be used for irrigating golf courses.

Mr Cobham then suggested that this should be called a minimum standard, rather than a maximum.

Mr C. Ertuna defended his statement by suggesting that minimum standards were required for each use. Perhaps the last line could be modified, provided that the minimum standards are strictly met for each use.

Another questioner drew attention to Mr Ertuna's tables showing the cost for treatment of 1 m^3 in which only the capital cost is taken into consideration, especially for Abu Dhabi. He pointed out that the capital cost is not enough to judge the total cost of treatment; operational costs, including the cost of running tertiary treatment, must be included. Secondly, he asked Mr Ertuna to explain why the cost of desalination to treat 1 m^3 in Kuwait is higher than in other countries in the same area? Concerning the minimum standard, the questioner knew of only Saudi Arabia where they are using the WHO potable water standard for treated effluent. Other countries in the Gulf area are using the 10:10 effluent standard.

Mr C. Ertuna, pointed out that his chapter was based on another study produced earlier by ECWA and which now needs updating. However, even in that study, the economic analysis included operational costs as well as capital costs. He referred to

Table 28.4 which gave, for various project lives, the unit cost of different capital amortization rates, 6%, 8%, 10% and 12% and for project lives of 20, 30 and 40 years. The amortization plus 2.75% of the total cost is taken as operation and maintenance costs, and these are actually incorporated in the final figure. For desalination, 1981 costs in Qatar were over $1.50/m^3$. In some of the Gulf countries the fuels used for desalination plants are either subsidized or they are the 'waste' gas from refineries. If the energy cost is taken at market price, a high cost is arrived at for desalination. Economic accounting makes the difference in estimating the cost of desalination.

Mr J. Al-Shatti, Kuwait, also responded to the question by pointing out that most of the information available in his paper had not been updated to the time of presentation but he had done his best to revise the information presented. He suggested that international organizations like ECWA should not collect information from scientific magazines and present it in scientific seminars. It would be preferable for ECWA to make direct contact with the countries to collect more reliable information.

Mr Ertuna followed this up by saying that the United Nations had very limited travel funds, which did not allow agencies such as ECWA to go to each country and spend extended periods collecting information. Frequently, questionnaires are devised and distributed. Sometimes responses are received and sometimes they are not. He hoped that more responses would be received in future so that reliable information from direct sources could be presented, rather than having to resort to secondary sources, as has often been the case in the past.

A telegram from the *United Agricultural Production Company, Kuwait* read:

> I am pleased to convey to you and to the participants my warmest greetings on the occasion of this important seminar. As we are utilizing treated sewage effluent for irrigation in our project measuring 17 000 hectares to produce fodder crops and vegetables the outcome of this seminar is vital to producers like ourselves. We hope the seminar will lead to good results. Wishing you and the seminar every success.

List of participants

Country delegates

ALGERIA	K. Quanfouf, Engineer, Hydro-Agricole, Sous Directeur de l'Agriculture et de Peches a la Wilaya de Bilda
BAHRAIN	J. H. Ahmed, Director of Projects, Ministry of Commercc and Agriculture
CYPRUS	C. Serghiou, Director, Agricultural Research Institute, (Head of Delegation), Nicosia
	I. Papadopoulos Agricultural Research Officer A', Agricultural Research Institute, Nicosia
	P. Marcou, Senior Agricultural Officer, Department of Agriculture, Nicosia
	G. Grivas Senior Agricultural Officer, Department of Agriculture, Nicosia
	P. Kalimeras, Senior Agricultural Officer, Department of Agriculture, Nicosia
	P. Economides, Senior Veterinary Officer, Department of Veterinary Services, Nicosia
	C. Andreou, Senior Water Engineer, Water Development Department, Nicosia

D. Kypris,
Senior Hydrogeologist,
Water Development Department,
Nicosia

S. Zomenis,
Senior Geological Officer,
Geological Survey Department,
Nicosia

A. N. Charalambous,
Senior Health Inspector,
Ministry of Health,
Nicosia

C. Akelidou,
Senior Analyst,
State General Laboratory,
Nicosia

DJIBOUTI

D. Mohamoud,
Adjoint au Chef du Service de l'Agriculture,
Ministry of Agriculture

EGYPT

Helmy M. Imbrahim,
Under Secretary,
Ministry of Irrigation

Nabil Sief El-Yazal,
Professor, Soil and Water Research Institute,
Agricultural Research Centre,
Cairo

JORDAN

A. Subbi Yousef,
Irrigation Specialist,
Ministry of Agriculture,
PO Box 226,
Amman

KUWAIT

Sh. A. El-Hadary, MD,
Head, Environmental Health Unit,
Ministry of Public Health,
PO Box 12129, Kuwait

Jasem M. K. Al-Shatti,
Director of Irrigation and Project Department,
Sanitary Engineering,
Ministry of Public Works

Khalil I. Al-Salem,
Head of Plant Production,
Kuwaiti Agriculture Affairs and Fish Resources
Authority

A. Al-Nakib,
Superintendent of Agricultural Wealth,
Kuwaiti Agriculture Affairs and Fish Resources
Authority

MOROCCO

M. Lamtiri-Laarif-Mohamed,
Engineer and Chief of Division,
Aménagement Hydro-Agricole,

	Minstère de l' Agriculture Rabat A. Azizi, Sanitary Engineer, Ministère of Public Health, Rabat
PAKISTAN	Javed A. Aziz, Professor of Public Health Engineering, Engineering University, Lahore M. Zakir Ali Khan, Managing Director, Karachi Water and Sewerage Board
QATAR	M. M. Youssef Murad, Chemist (Doma Sewage Treatment Works)
SAUDI ARABIA	A. T. Shammas, Expert in Reuse of Treated Wastewater Adnan Gur, FAO Wastewater Expert, Ministry of Agriculture and Water
SYRIA	Abdalla Kendakji, Deputy Planning Director/Civil Engineer, General Company for Irrigation, Construction
TUNISIA	A. Bouzaidi, Chef du Project d'Utilisation des Eaux Usées Traitées en Agriculture
YEMEN ARAB REPUBLIC	Eng. Ali Mohamed Alzumeir, Irrigation Department, Sana'a Ministry of Agriculture and Fisheries
YEMEN, PEOPLES' DEMOCRATIC REPULIC	N. M. Abdul-Whab, Agricultural Engineering Supervisor, Agricultural Soils and Water Analysis

Observers

M. Abu-Zeid, Chairman, Water Research Centre, Egypt
C. Chimonides, Ex. FAO Staff Member, Irrigation Consultant, Cyprus
A. Hamdy, Director of Research, IAM, Bari, Italy
Petros Loizides, Ex. FAO Staff Member, Cyprus
J. Margeta, Professor, PAP/RAC-SPLIT, Yugoslavia
R. Oliver, EMAYA, Spain, 3, Oan Aragall – (07066) Palma de Mallorca
R. A. S. Robertson, UNHCR, Cyprus
A. P. Rocroi, Irrifrance, Cyprus (Unofficial)
H. Salaymen, Agri-Engineer, PLO, PO Box, 12789, Damascus, Syria
R. Tortajada, Ministry of Health, Direcion General de Salud Publica Paseo Del Prado, 20, 28014 Madrid

Agricultural Research Institute, Cyprus

G. Eliades,
Agricultural Research Officer A′
Chr. Metochis,
Agricultural Research Officer A′

Department of Agriculture, Cyprus

A. Gregoras,
Agricultural Superintendent 2nd Grade
C. Hadjiparaskevas,
Agricultural Officer
Chr. Koudounas,
Agricultural Officer, Class I
L. Markides,
Agricultural Officer A′
Chr. Motites,
Agricultural Officer, Class I
Y. Nicolaou,
Chief Agricultural Superintendent
G. Oroundiotis,
Agricultural Officer A′
A. Phocaides,
Agricultural Superintendent 1st Grade
Chr. Photiou,
Agricultural Officer
A. Savvides,
Agricultural Officer A′
N. Tsappis,
Agricultural Officer A′
Chr. Tsiattalos,
Agricultural Officer A′
S. Zemenides
Agricultural Officer, Class I

Nature Conservation Service, Cyprus

A. Antoniou,
Biologist
A. Pissarides,
Agricultural Officer, Class I

Water Development Department, Cyprus

Chr. Artemis,
Senior Water Engineer
Chr. Kambanellas,
Executive Engineer I
E. Kambourides,
Executive Engineer I

Chr. Marcoullis,
Senior Water Engineer
M. Michaelidou,
Executive Engineer I
K. Savvides,
Chemical Engineer
N. Stylianou,
Senior Water Engineer
S. Theodosiou,
Mechanical Engineer

Geological Survey Department, Cyprus

S. Afrodises,
Senior Geological Officer
D. Ploethner,
Expert

Ministry of Health, Cyprus

A. Cleanthous,
Health Inspector

State Laboratory, Cyprus

E. Ioannou,
Analyst, Class I
G. Kathidiotis,
Analyst, Class I

Department of Town Planning and Housing, Cyprus

P. L. Christodoulou,
Sanitary Engineer
G. Koullapis,
Sanitary Engineer
A. C. Kourouzides,
Public Health Engineer, Grade II
E. Markides,
Public Health Engineer
M. Pantis,
Public Health Engineer

Public Works Department, Cyprus

P. Papasozomenos,
Executive Engineer

Foreign Experts in Cyprus

H. Luken,
Soils Expert,
Director of the Soils Division in the Federal Institute for Geosciences and Mineral Resources,
Hanover, West Germany

Lecturers

A. S. Abdel-Ghaffar	University of Alexandria, Faculty of Agriculture, Alexandria, Egypt
S. Al-Salem	Director, Treatment Plants Department, Water Authority, PO Box 150793, Amman, Jordan
Jasim Al-Shatti	Director of Irrigation and Project, Department of Sanitary Engineering, Ministry of Public Works, Kuwait
M. N. Al-Rifai	Professor of Irrigation and Sanitary Engineering and Dean, Engineering College, University of Damascus, Syria
A. Arar	Senior Officer (FAO Secretariat) Water Resources, Development and Management Service, Land and Water Development Division, FAO, Rome
H. Bouwer	Director, US Water Conservation Laboratory, USDA, 4331 East Broadway Road, Phoenix, AZ 85040, USA
R. O. Cobham	Senior Partner, Cobham Resource Consultants, 19 Paradise Street, Oxford OX1 1LF, UK
N. Sanchez Duron	Director General, Department of International Relations, Ministry of Agriculture and Water Resources, Mexico, DF
C. Ertuna	Senior Economic Affairs Officer, Natural Resources Science and Technology Division, ECWA, Baghdad, Iraq
Dominique Fougeirol	Ingenier Civil Des Mines De Paris, BURGEAP Ingenieurs-Conseils, 70 Rue Mademoiselle, 75015 Paris, France
P. J. Hillman	WHO (CEHA), Amman, Jordan
A. M. Jamaan	Hydrogeologist, Water Resources Development Department, Ministry of Agriculture and Water, Riyadh, Saudi Arabia
M. S. Kalthem	Director, Water Research Institute, PO Box 50080, Riyadh, Saudi Arabia

A. Kandiah	Technical Officer (FAO Secretariat) Water Resources, Development and Management Service, Land and Water Development Division, FAO, Rome
Rolf Kayser	Institut für Stadtbauwesen, Abteilung Siedlungswasserwirtschaft der TU Braunschweig, Pockelsstrasse 4, D-3300 Braunschweig, Federal Republic of Germany
V. D. Krentos	Agricultural Research and Development Consultants, 55 Dighenis Akritas Avenue, Nicosia 135, Cyprus
Duncan D. Mara	Department of Civil Engineering, University of Leeds, Leeds LS2 9JT, UK
M. H. F. Marecos do Monte	Sanitary Engineer, Laboratorio Nacional de Engenharia Civil Ministerio de Habitacão e Obras Publicas, Lisbon, Portugal
T. Morishita	Associate Professor, Institute of Applied Biochemistry, University of Tsukuba, Ibaraki-ken, 305 Japan
H. A. C. Montgomery	Consulting Water Scientist, 2 Ashacre Lane, Worthing, West Sussex, BN13 2D, UK
J. Papadopoulos	Agricultural Research Officer, Agricultural Research Institute, Ministry of Agriculture, Nicosia, Cyprus
M. B. Pescod	Head of Department and Tyne and Wear Professor of Environmental Control Engineering, University of Newcastle upon Tyne, Department of Civil Engineering, Cassie Building, Claremont Road, Newcastle upon Tyne, NE1 7RU, UK
G. B. Shende	Assistant Director and Head Waste Water, Agriculture Division, National Environmental Engineering Research Institute, Nehru Marg, Nagpur-440 020, India
Y. Stylianou	Agricultural Research Officer, Agricultural Research Institute, Ministry of Agriculture, Nicosia, Cyprus
M. Talhouni	Secretary General, Water Authority, Amman, Jordan
Lidia Tremea	Microbiologist, Mediterranean Water Institute, Canal de Provence, BP 100, Le Thoionet, 13603, Aix-en-Provence, Cedex, France
George B. Willson	Agricultural Engineer, Soil-Microbial Systems Laboratory, Agricultural Environmental Quality Institute, USDA, Agricultural Research Service, Beltsville, Maryland 20705, USA

Local organizing committee

C. Andreou (Transport)
G. Grivas (Registration)
A. Ioannou (Transport, Registration)
P. Kalimeras (Registration)
D. Kypris (Registration)
Chr. Loucaides, Chairman (Coordination)
Y. Stylianou (Transport, Accommodation)

FAO Secretariat

A. Arar, Senior Regional Officer, AGL
M. Farrell, Secretary, AGL
H. M. Horning, Director, AGL
A. Kandiah, Technical Officer, AGL
R. Mahmoud, Disbursing Officer, RNEA
H. Tonkin, Meetings Officer and Editor, AGL

Index